GENDER
PARADOX

DISCRIMINATION AND DISPARITIES
IN THE POSTMODERN ERA

GENDER PARADOX
DISCRIMINATION AND DISPARITIES IN THE POSTMODERN ERA

ZACHARY ELLIOTT

This book was printed using Lulu.com, a self-publishing company for authors. They offer high quality printing at low costs, allowing anyone to print, publish, and sell their own book in a few simple steps. Thanks Lulu!

The cover design on the book is a second edition, and the content inside remains unchanged from the original release (first edition).

www.zacharyaelliott.com
www.theparadoxinstitute.com

© 2020 Zachary A. Elliott. All rights reserved.
ISBN 978-1-79486-870-0

The Gender Paradox is crafted for a specific type of person: it is made for those who have a curious mind never content with the commonplace or the mundane but whose intellectual curiosity and depth of thought extend into the farthest reaches of human knowledge and understanding; to those who are looking for the answers to complex questions which are hidden from view by our illiberal gatekeepers in the media, academia, and politics; and to those who see our society descending into the depths of chaos and confusion. Finally, and perhaps most importantly, this book is crafted for those whose innate uniqueness has been suppressed, marginalized, and alienated by bankrupt ideologies.

The Gender Paradox is a book which champions individual liberty and provides a scientific case for the innate desires, interests, and values of every human being which must be protected from the evils of illiberal forces. The book does this not through arguments from the transient sources of society or culture, but through the more primordial sources of biology which bestow upon each individual innate and priceless qualities. *We as a society mess with these qualities at our peril.*

Special thanks goes to clinical psychologist and professor of psychology Dr. Jordan B. Peterson, whose exceptional lectures on personality psychology, philosophy, meta-narratives, and individual differences inspired a large portion of this book. *The Gender Paradox* could not have been put together without his incredible depth of knowledge and insight into the complex topics of sex and gender.

CHAPTERS

The Gender Paradox ix
1. *Sex as a Social Construct* 1
   - The Social Constructionist Perspective 3
   - Language Shapes Reality 7
   - The Argument from Indefinite Variation 10
   - Sex is Subjective 17
2. *Philosophy of Gender* 23
   - The Blank Slate 28
   - Gender as a Social Construct 32
   - Constructing Gender 'Essentialness' 38
3. *The Postmodern Tyranny* 55
   - The Theory of Rejecting Theories 56
   - Paradoxes of Postmodernism 70
   - The Truth about Hierarchies 72
4. *Discrimination and Disparities* 77
   - Gender Inequality in Liberal Democracies 80
   - For Every 100 Girls… 87
   - The Lie of Equity 90
5. *Nature as Formwork* 97
   - Human Universals 100
   - Dangers of Blank Slate Ideology 102
6. *Exploring Gender Differences* 107
   - Sex and Gender as Dials 111
   - Origins of Sex and Gender Differences 118
   - Three Unique Theories 121
   - The Biopsychosocial Model 125

7 *Sex as a Biological Mechanism* 129
- Sex Determination............131
- Effects of Sex Hormones 134

8 *Conception to Birth*............143
- Development of the Zygote............144
- Prenatal Androgen Exposure............146
- Psychological Effects of Androgens............152
- Hormones and Socialization............158

9 *Development of Gender Identity* 165
- Gender Identity as a Blank Slate 171
- Gender Identity as a Biological Reality 184

10 *Gender Stereotype Pathology*............199
- Origins of Gender Stereotypes............200
- The Feminine and Masculine Archetypes 206
- When Stereotypes Become Pathological............214

11 *The Activational Stage*............225
- Three Endocrine Events............226
- Puberty Initiates Morphological Changes 229
- Activational Effects on the Brain 235
- Identifying Sex Through the Brain 250

12 *Personality and the Big Five*............259
- What is Personality?............262
- How Do We Measure Personality?............266
- Gender Differences in the Big Five 272
- Biopsychosocial Model of Personality 286

13 *Interests and Inequality* 299
- The Things-People Dimension............303
- Women in STEM............310
- Shattering the Glass Ceiling............314
- Sex Differences in School and Work............317

14	*Solving the Gender Paradox* *323*	
	• The Solution: A Biopsychosocial Model 329	
	• Universal Sex Differences 336	
	• Ideology and the Denial of Science 343	
15	*Critiquing Social Constructionism* *347*	
	• Biological Sex is a Social Construct 347	
	• Gender is a Social Construct 353	
	• Gender Stereotypes are Pathological 357	
	• Socialization at Puberty Makes Boys and Girls Different 360	
	• Gender Differences in Personality Do Not Exist 361	
	• Gender Differences in Occupational Preferences Do Not Exist 363	
	• Gender Gaps are Indicative of Injustice 364	
16	*The Gaslighting Ideology* *367*	
	• J.K. Rowling and the Denial of Sex 370	
	• Jordan Peterson and Compelled Speech 385	
	• Medicalization of Gender Non-Conformance 400	
17	*Beyond Gender* *415*	
	• Paradoxes of Social Constructionism 418	
	• Beyond the Blank Slate 421	
	• Beyond Ideology 429	
18	*Bibliography* *439*	
19	*Glossary* *453*	
20	*Index* *469*	

AN INTRODUCTION TO...
THE GENDER PARADOX

The Penrose Triangle is an impossibility in its purest form. And yet, such an impossibility is simply an optical illusion. Shift your perspective slightly, and you will find that the sides of the Penrose Triangle don't actually connect. You can easily build and model this paradox in real life, and with careful consideration of perspective, an illusion of impossibility can be created. However, there is nothing impossible about the Penrose Triangle in its physical form. There is no actual magic, sorcery, or the most advanced form of witchcraft which can make it true, nor is there any mathematical or philosophical way of solving it.

Trying to solve it through any type of mathematical means is impossible. Rather, the solution to the triangle is simple: it's all about getting the right perspective. A shift up or a shift down. Left or right. With just a slight adjustment in one's view, the optical illusion of the Penrose Triangle disintegrates into the realm of the possible. Like its other paradoxical cousins such as the Penrose Stairs, such a seemingly impossible object has a simple explanation. Ironically, a similar paradox exists inside the social constructionist theory of gender.

Social constructionist theory states that all differences between boys and girls, men and women, are the result of socialization. Boys and girls become different, not through genetics, hormones, or chromosomes, but through behavioral reinforcement and punishment from parents, teachers, peers, and the media. This socialization continues into adulthood, causing men and women to choose different career paths. In other words, society is what defines the similarities and differences between the sexes, not biology. Any differences that do exist are linguistic constructions of society which solidify, maintain, and perpetuate binary and heteronormative structures of what it means to be a man or a woman. Therefore, any gender inequality must be the result of socialization forces which interact with stereotypical gender norms that discriminate and suppress behavior deemed *inappropriate*.

Utilizing this theory of social constructionism throughout the 1970s, 80s, and 90s, social scientists studied the development of gender equality in hundreds of nations around the world. The hypothesis was that if men and women differ because of society, then the *more* gender-equal a country becomes, the *less* gender differences there will be.

The level of gender equality across 160 nations was measured using a United Nations metric known as the Gender Inequality Index (GII), which rank orders its member nations through an analysis of four factors: reproductive health, proportion of parliamentary seats occupied by females, the proportion of adult females and males aged 25 years and older with a secondary education, and the labor force participation rate among females and males aged 15 and older.[1] The lower the GII value is in a given country, the more gender-equal the society. A country with a low GII value is therefore a country which has heavily pursued egalitarian policies which encourage and facilitate women's participation in the economy and the culture.

In 2008, Norway was ranked number one for the GII (lowest score), meaning they were the most gender-equal country out of 160 measured. Other EU Member States were ranked among the top, including Switzerland, Denmark, Sweden, and Finland. Outside of Europe, however, countries such as Yemen, Ethiopia, South Sudan, India, Iraq, Iran, Libya, Egypt, and Afghanistan had some of the worst GII rankings with the highest scores.

---

[1] Gender Inequality Index (GII). *United Nations Development Programme, Human Development Reports.*

Yet throughout the many decades of GII scores, something strange was happening. As countries became more egalitarian, adopting policies which encouraged and facilitated women's participation in the economy and society, the differences between men and women didn't minimize, they didn't go away, and they didn't disappear. Paradoxically, the countries with the best GII scores, such as Norway, Finland, and Sweden, had the *largest* gender differences between men and women. From personality, occupational preferences, and interests, the differences between men and women did not just stay the same; *they grew and maximized over the course of four decades.* And social scientists were shocked. Such a prediction that gender-equal societies would cause gender differences to minimize was proven not just incorrect, but precisely *opposite* of the data. Why were the most gender-equal countries seeing the largest differences in men's and women's occupational choices, interests, and personalities?

Dismayed at their seemingly impossible findings, social scientists dubbed this phenomenon the *Gender Equality Paradox,* which is to say, *the more gender-equal a society becomes, the more gender differences there will be.*[2] Such a phenomenon seems counter-intuitive to common sense. It seems outside of any reality, truth, or explanation. In a word, it just seems impossible! And yet, such a phenomenon is not impossible. In fact, it's only impossible if you look at it from the *single* perspective of socialization. However, if we adjust our perspective just slightly, thereby integrating a variety of causes, then the paradox fades away as easily as the optical illusion in the triangle. So, if we can solve the paradox with a change in perspective, what perspectives can we take?

Social constructionism is one possible perspective, but can it solve the paradox? In college, I took a class called Sociology of Gender, where we studied how gender is socially constructed through behavioral reinforcement and punishment. Taught by Dr. Heather McLaughlin, a professor in sociology and an expert in gendered institutions, the class focused on the influences of socialization for the origins of gender

---

[2] Schmitt, D., et al. (2008). Sex differences in Big Five personality traits across 55 cultures. *Journal of Personality and Social Psychology, 94(1),* 179.

differences, claiming that gender extends to not just our behavior but our societal institutions. Gender, Dr. McLaughlin taught us, is embedded in the construction of unequal divisions between men and women, the proliferation of images and symbols of the feminine and masculine, the regulation of interpersonal relationships and interactions, the conceptualization of individual identity, and the use of gendered organizational policies.[3] Such gendered institutions seem to construct sex and gender differences as 'innate' and perpetuate discrimination, injustice, and oppression.

And yet, while such an explanation into the origins of sex and gender differences has truth to it, can it account for *all* the differences between men and women? And can it account for the existence of the Gender Equality Paradox? If we take the perspective of Dr. McLaughlin, that gender differences are merely the result of sociocultural forces and gendered institutions, it seems as though we would arrive at the same exact perspective from which we began (namely, that gender differences should minimize as countries become more gender-equal), and the paradox would remain unsolved. While the forces of socialization certainly exist and *can* be powerful, such forces cannot create this paradox on their own. Gender-equal countries such as Norway have the *largest* gender differences, and these differences have only grown larger throughout the 1980s, 90s, and 2000s despite egalitarian policies which consistently push for the elimination of all gender gaps. Therefore, if men and women only differ due to society, then the paradox seems to become something unsolvable, an impossibility in its purest form.

However, if men and women differ due to a *mix* of biological and societal factors, then perhaps biology has some role to play in the existence of the paradox. Maybe men's and women's innate interests, rather than minimizing, actually *maximize* once societal forces are diminished. Such an explanation seems to solve the paradox instantly: *if you eliminate sociocultural forces which cause gender differences, then you allow for any*

---

[3] McLaughlin, H. (2018). Gendered institutions. *Sociology of Gender, Oklahoma State University.*

biological variation to express itself. That may explain why gender differences in gender-unequal countries such as India and Iran are minimal; perhaps people are not given the freedom to express their innate selves. After all, when people *are* given the freedom to choose their own unique path in life, innate differences between individuals may triumphantly ascend from the depths. But such an explanation seems counter-intuitive and even overly traditional. Do men and women really differ, in part, from biology? Isn't such a view outdated?

Around the same time I was taking Sociology of Gender, an internal memo was circulating through Google. James Damore, a Google software engineer, was responding to a suggestion by his upper management that opinions on their diversity seminar be expressed. Using contemporary scientific research from the fields of biology, psychology, neuroendocrinology, Damore explained that one reason for women's underrepresentation in Google's coding sector may be due to average differences between men and women in *interests*.[4] Because of this, he was fired.

To claim that some differences between men and women may be due to differing interests was blasphemy to social constructionist dogma. It didn't matter that Damore wrote he wanted to increase women's representation in coding at Google, and it didn't matter that he claimed diversity and inclusion should be valued. Both sentiments I agree with! What mattered is that he dared to challenge the social constructionist idea that all differences between men and women are the result of socialization and not innate interests. He was not saying women lack the intelligence, the ability, or the drive to excel in coding. Rather, the claim was that women, on average, tend to choose coding less than men. Such a statement was unacceptable.

And yet, ironically, such a statement solves the Gender Equality Paradox instantly. If average differences in interest exist between men and women, then the paradox can be easily solved. After all, perhaps the differences between men and women are the result of complex factors

---

[4] Damore, J. (2017). Google's Ideological Echo Chamber. *Damore-Google Manifesto*.

which all interact together to form an individual. Perhaps it is not just society which influences our decisions. Maybe biology has some role to play.

And so from this paradox we are left with a question: *does* biology affect, not just sex, but *gender* and its behavioral expressions? This is why I wrote this book, to answer such a question. Despite being an architecture student who loves to design the built environment, I've been fascinated with understanding human psychology, and more specifically, how differences between individuals manifest themselves. Perhaps such a fascination may help me understand the perspectives of others so that I can design better buildings, or perhaps sharing such perspectives can help others understand their own interests and behaviors. For years I have continually followed the developments in our society: that sex and gender are social constructs, that reality and truth are subjective, and that there is nothing outside of language structures.

Why has a society like ours, a civilization which once championed reason, logic, and scientific inquiry, fallen so far away from our fundamental axioms? Such a question is complex and multifaceted. Such a question requires an understanding of philosophy, biology, and psychology. And such a question is perhaps one of the most important questions our contemporary society can possibly ask. If reason, logic, and scientific inquiry are abandoned, then what are we left with?

My goal for this book is to study the causes of the Gender Equality Paradox (*why do gender differences grow larger as a country becomes more gender-equal?*). Throughout its pages I will provide a framework for understanding the origins of sex and gender, and as I do, we will explore the composition of sex and gender differences; how these differences develop throughout an individual's life; how differences between men and women create disparities across society; how gender stereotypes can turn pathological; and how social constructionist theory has become the dominant ideology in our contemporary societies.

I will explore how there are more differences *within* groups of people than there are *between* them; how males and females are much more alike than different; how statistically significant differences on the extremes result in degrees of inequality; and how individuals should be judged as individuals, not as members of a group.

By exploring the complexity of sex and gender, I will provide one of the most holistic arguments for the origins of sex and gender differences, a framework which utilizes many variables from a variety of fields while integrating biological, psychological, and sociocultural theories. Such an interactionist approach is known as the *biopsychosocial model*.

While we explore how sex and gender differences originate from biopsychosocial factors, we will also explore the pathologies inherent in bankrupt ideologies which relegate boys and girls, men and women, to regressive and illiberal stereotypes. Thus, I hope this book can help others understand the dangers of erasing sex, the pitfalls of viewing all gender differences as social constructions, the pathology of judging individuals through gender stereotypes, and the evils of treating children as malleable blank slates.

I want us to understand why differences between men and women still exist in the most gender-equal societies; I want us to explore how understanding sex and gender differences and similarities is important to our civilization's physical, psychological, and sexual health; and lastly, and perhaps most importantly, I want us to understand the complexity of sex and gender research so that we can fight inaccurate and oppressive stereotypes which constrain the free expression of individual liberty. Ultimately, I want to explore this sex and gender research not because I wish to constrain individual behavior, but because I wish to champion the rights of all human beings and fight for those who are striving to express their innate interests and desires.

With courage and curiosity, I dare you to venture into the complex frontiers of our society's scientific discoveries. There may be surprises you're not ready for, there may be evidence you haven't seen, and best of all, there may be arguments which challenge your initial assumptions, dismantle your existing judgments, and connect once dark gaps in your knowledge with a light and clarity only evidence and wisdom can provide. Perhaps, at the end of this terrifying and exhilarating adventure, we can finally solve the impossibility of *The Gender Paradox*.

To argue with a man who has renounced the use and authority of reason, and whose philosophy consists in holding humanity in contempt, is like administering medicine to the dead, or endeavoring to convert an atheist by scripture.

- Thomas Paine

SEX AS A SOCIAL CONSTRUCT

Solving the Gender Paradox requires that we develop a model for sex and gender which integrates a variety of theories and research. By doing so we can gain a more holistic understanding of the complexities of sex and gender, and with this depth of knowledge, we can reveal the properties behind the optical illusion of the Gender Equality Paradox. If gender differences in behavior, interests, and personality grow larger in the most gender-equal countries, as I discussed in the intro, then how can such a phenomenon be explained? Out of all the theories available to us, one theory holds the most provocative explanation for the origins of sex and gender differences. This theory we call *social constructionism*--the ultimate representative of our culture's contemporary spirit.

Social constructionism states that differences in behavior, personality, and interests among individuals are the result of learned behaviors. Behavioral reinforcement and punishment, combined with societal expectations, produce the uniqueness of the individual, and with it, the differences between men and women. For the social constructionist who studies sex and gender, boys and girls are not born different but are *made* different. Socialization, cognitive social learning, and cultural practices shape boys and girls into exhibiting different behaviors, interests, and personalities. Such a theory can be considered the spirit of our age, as the differences between individuals are not viewed as the sum total of biology, psychology, and society, but rather the byproducts of more transient forces of society and culture from which behavior is molded and formed.

Because social constructionists emphasize the influence of sociological factors on an individual over biology and psychology, the categories of both sex and gender are subjected to intensive scrutiny: "Are the categories of sex and gender legitimate?" they ask, "Or are such categories in need of restructuring and redefining?" Perhaps our culture has too narrow a view of sex and gender, and perhaps such definitions of boy and girl, man and woman, and male and female are overly simplistic and highly constraining. Perhaps these binary categories do not fully explain the diversity of human experience. If the binary categories of sex and gender are proven to be scientifically inaccurate and ill-defined, then perhaps we can discover effective methods for their reformation.

For the social constructionist who believes that the binary categories of sex and gender are constructed through societal conditioning, there exists an enticing theory: if socialization alone can account for the differences between men and women, then perhaps the properties of sex and gender can be dismantled. Furthermore, if we can deconstruct the categories of sex and gender and show that these categories are societally-defined, then we can also escape the paradox. For if we can show that gender differences in the most gender-equal nations are the result of treating boys and girls differently, we can also develop ways of fixing these gender gaps in behavior, personality, and interests. After all, if boys and girls are not born different but *made* different, then socializing them the same should produce less differences. If we can understand how we socialize our boys and girls into gender-typical roles, then perhaps we can eliminate the gender gaps between men and women.

To the social constructionist, such a solution to the Gender Equality Paradox seems promising. Maybe the differences between men and women *are* socially constructed, and maybe, if we deconstruct the binary category of sex into its societally-defined elements, then perhaps we can show that the Gender Equality Paradox is the result of a society which socializes boys and girls into stereotypical gender roles. Maybe it is this perspective, the social constructionist one, which can most effectively dismantle the optical illusion of sex and gender by showing that male and female, man and woman, and boy and girl are socialized (rather than innate) categories.

THE SOCIAL CONSTRUCTIONIST PERSPECTIVE

We will begin our exploration of the Gender Equality Paradox through the lens of *social constructionism*. I do this not because I agree with the theory in all aspects, but because social constructionism represents the mainstream view of our culture--a culture which views gender, and *even biological sex*, as social constructs.

Ironically, it is in the category of *biological sex*, not gender, where the arguments for social constructionism begin. While it has been accepted for decades that aspects of gender are socially constructed, social constructionists of the 1990s began arguing that biological sex is also a societally-defined binary category through which differences between men and women are constructed and maintained. There is no essence of *male* or *female*, the strong social constructionists argue; rather, these categories are socially constructed labels.

According to the social construction theory of sex, no objective criteria exists to reliably sort human beings into the two categories of male and female due to the high variability of traits and the subjectivity of the dividing line. From the formation of the genitalia, levels of sex hormones, development of the gonads (testes and ovaries), and chromosomal variation, there is an indefinite amount of variability across populations. Because of this, sex is theorized to exist on a spectrum, not a male-female binary switch.[5]

Here the social constructionists understand a critical aspect of biology: that on every level of analysis, variability of individuals exists. However, the very presence of categories and labels seems to eliminate the concept of variability. By their very nature categories seem constraining, discriminatory, and even oppressive. Categories between groups were the status quo not long ago, justifying the institutional and systemic discrimination and oppression of peoples and cultures. Yet beyond this, categories seem inherently biologically determinate. It isn't difficult to see the ultimate conclusion of the biological essentialism witnessed in the 20th century, as groups of people were arrested, imprisoned, sterilized, and even

---

[5] Brusman, L. (2019). Sex isn't binary, and we should stop acting like it is. *Massive Science*.

exterminated by their governments due to traits which were viewed as fixed, innate, and unchanging. Because of the history and context of categories, it makes sense why such a push against them exists on the mainstream in the current culture of our Western societies.

Dismantling the binary category of sex is a major thoroughfare to the elimination of categories as a whole. If innate categories can be eliminated, then perhaps injustice, discrimination, and oppression can be overcome. Medical technology has increasingly elucidated the variability of biological sex *characteristics*, rendering them near-infinite in their possibilities and ultimately meaningless of categorical description according to gender theorists. This variability is pushed to the philosophical extreme by mainstream social constructionists who reject the concept of biological sex, as I will show.

But first, some clarity on definitions. What exactly is sex? Is it a set of characteristics which exist on a spectrum? Is the dividing line just subjective, as social constructionists claim? Or is there more to it? For the most part, it's actually rather simple to define biological sex, which is differentiated into two unique reproductive capacities based in two unique gametes:

> **Female:** *of or denoting the sex that can bear offspring or produce eggs, distinguished biologically by the production of gametes (ova) which can be fertilized by male gametes.*
>
> **Male:** *of or denoting the sex that produces small, typically motile gametes, especially spermatozoa, with which a female may be fertilized or inseminated to produce offspring.*

Despite the fact that sex is defined through those who produce ova (females) and those who produce spermatozoa (males), and that all of the animal kingdom can be differentiated through female gametes and male gametes (even in plants), this has not stopped social constructionists and gender theorists from arguing that biological sex does not exist.

Take, for example, a quote from a professor of Transgender Studies at the University of Toronto, Nicholas Matte, who said, "It is *not correct* that there is such a thing as biological sex."[6]

Another example comes from a 2019 video in the pop-culture magazine, Teen Vogue:

> "The binary is bullshit," one woman said, "Sex typically refers to your biological traits: your gonads, your genitalia, your internal sex characteristics, your hormone production, your hormone response, and secondary sex characteristics. Gender is about your identity, your expression, and it's often based on ideas about sex."
>
> Another person added, "It's important we really break down what we are talking about when we talk about sex and gender and if there is something called 'biological sex' and what that means."
>
> One woman added, "This idea that the body is either male or female is totally wrong, and I am living proof of that."
>
> Another said, "We know that intersex people exist and breakdown this binary. We all have characteristics that are typically male and typically female, and it is really about political choices, social choices, and ideological choices, that we really assign meaning to different parts of our body."[7]

The video received over 100,000 views on YouTube, and yet, to the dismay of the creators, received only 700 likes to its 17,000 dislikes. Perhaps the dismantling of the male-female category is just not ready to be accepted.

The idea that sex is something fluid, transient, and malleable has also made its way into elementary schools. In the UK, this has quickly entered the mainstream, where the social constructionist theory of sex and gender is taught to ten-year-olds in the context of sex education. For example, one school had two sex educators come in to teach the young students that sex

---

[6] Scheermeijer, J. (2016). No Such Thing as Biological Sex. *YouTube*.
[7] (2019). 5 Misconceptions about Sex and Gender. *Teen Vogue*.

and gender are not just social constructs, but always transient and fluid based on an individual's internal feelings:

> "Gender is how you feel on the inside about whether you're a boy or girl, a man or a woman, if you're non-binary, feel like neither, or both. People can also be fluid, feel more like female or more like male based on a different day or time...it's really individual."[8]

The educators then asked the kids a question: "Everyone born with a vulva is a girl, true or false?"

Each of the kids shot their paper plate signs into the air, with some answering true and some answering false. According to the sex educators, the kids who answered *true* were wrong: "Everyone is not exactly sure," she said, "and that makes sense. *But our genitals don't determine our gender. Some people born with vulvas can be boys.*" Having already failed basic social constructionism, the kids put their paper signs back down in defeat. We've progressed so fast that boys can now have vulvas and girls can now have penises. If you're confused, then perhaps you need to be re-educated.

If you think these views are outside of the mainstream, then you are not up-to-date on the recent advancements in gender theory. The idea that politics, society, ideology, and an individual's feelings are the determining factors of sex is the core of social constructionism and is the key to understanding the philosophy. If biological sex exists on a spectrum like the gender theorists say, then the male-female category is useless; or, in the words of Dr. Heather McLaughlin, professor of sociology: "For strong social constructionists, this means that the categories themselves just don't work."[9] For the social constructionist, sex is not something which fits into a binary category, but rather something which is "*an ever-changing psychosocial construct created in the symbolic social worlds that allow people*

---

[8] Questioning LGBT/CSE Education, (2019). Sexualizing Children with Sex Education. *Twitter.*

[9] McLaughlin, H. (2018). What is sex? *Sociology of Gender, Oklahoma State University.*

to interact with each other. Each individual gives meaning to sexual and self-concepts through the complex interaction of external discourse and social relations with the existing power structure."[10]

Strong social constructionists do not view the category of male and female as a biological reality, but rather a system which reinforces an existing power structure between individuals. Because of this, the very existence of categories is in direct opposition to the fundamentals of social constructionism. It is true that in the past categories have been used to constrain, discriminate, and oppress, and it is this oppressive nature of the category that social constructionists wish to dismantle. While such a sentiment is understandable, things quickly go off-the-rails when this philosophy is taken to its extreme. And it is here where we arrive at a serious impasse with the structure of reality.

The fundamentals of social constructionism state that humans are born as blank slates, that culture and society construct biological sex, that actions can only be understood in the context of sociocultural forces, that words can be used as tools to shape reality, and that humans can be molded in preferential ways. In other words, anyone can be whatever they want to be. Here an individual's subjective perception reigns supreme, and the fight to restructure the category of biological sex through the use of language begins.

LANGUAGE SHAPES REALITY

Take for example an activist named Cass Clemmer. Cass is a female individual who does not identify with the category of female. Instead, she identifies as transgender. As a newborn, she was examined by a doctor who, deciding based on subjective criteria of sex, *assigned* her to the female category. Now as an adult, she chooses to assign herself a new identity. In fact, Cass asks people to use *they/them* pronouns when referring to her. For this discussion, however, we'll refer to Cass using her assigned female pronouns of *she/her*. Cass is a period activist who works to dismantle what

---

[10] Bullough, V. (2003). The Contributions of John Money: A Personal View. *The Journal of Sex Research, 40(3),* 232.

she views as a categorical error: *that women are the only ones who can get periods.*

In 2017, Cass posted a photo of herself sitting on a bench with menstrual blood soaking her pants. A sign she held read, "Periods are not just for women. #BleedingWhileTrans." Responding to why she posted the photo of herself bleeding, Cass said: "As an activist in the menstrual health space, I've noticed a concerning lack of space and recognition of people who menstruate and do not identify as women, whether that's because they're trans, non-binary, and/or intersex."

She also crafted a coloring book for children featuring Toni the Tampon, a "gender-neutral character meant to illustrate that people of all genders get periods."[11]

As Cass correctly points out, the idea that only women can get periods is a natural byproduct of the constraining and prejudicial nature of categories. To claim that only women can get periods denies the fact that some people who do not identify as women also get periods, which is to say, an indefinite number of potential identities can menstruate. Such a statement is true, but only *linguistically*. An individual can identify as whatever they prefer, but such an identification does not change one's biological ability (or inability) to menstruate.

Once again, the idea of category dismantling utilizes language to shape perception. If we assign the label "cat" to the animal known as a *dog*, and then say that "this cat loves to play fetch," this doesn't mean that the dog is suddenly a cat. While it's true we can call a dog by something else, such as "cat," it does not mean the dog *is* a cat. We can call a table a chair and a chair a table, and yet it does not change their different functions or their material makeup. Yet we can surely see how a fundamental confusion may arise when talking to a friend about your cat who is, in reality, actually a dog. Just because there are an indefinite number of identities or ways to call something does not mean categories do not exist or are in constant flux. At the same time, categories do not preclude the existence of identities, which

---

[11] McNamara, B. (2017). Period Activist Cass Clemmer Responds To Hate After Posting Period Photo. *Teen Vogue*.

can exist in the mind of the subject. A person can identify however they like, but the biological category still exists.

The same principle applies when saying men can get periods too. Despite what you identify as, there exists a fundamental reality: an adult who has functioning XX chromosomes, a functioning reproductive system of fallopian tubes, ovaries, and a uterus, will experience menstruation, the monthly discharge of blood and tissue from the inner lining of the uterus through the vagina. People who do not have such internal structures will not experience menstruation. To state such an obvious fact, such as *women are the only ones who get periods*, is a form of category assignment: some people will fit into the category, while others will not. And this assignment is seen as alienating, marginalizing, and oppressive to those who do not identify with the category of *woman* but still experience menstruation.

Championing non-conforming identities who exist outside the binary category of male and female is at the core of Cass's motives:

> "I hope people start thinking about menstruation as more of a multidimensional issue, and consider the different ways other marginalized identities intersect with the experience of getting your period. Specifically around trans-identities, I hope that people start remembering to shift their language to be more inclusive, fight for less gendered advertisements/product design, and ensure that menstruators of all genders have access to period products in safe bathrooms."

You can see how the very idea of category assignment onto biological sex is being challenged here. Menstruation is no longer a univariate category belonging to women. Instead, it's a multidimensional issue which comprises a variety of intersecting identities, from people who identify as men or even transwomen who identify as women. In the culture, companies are beginning to catch on to this way of thinking. For example, the menstrual hygiene product brand *Always* took the female symbol off their packaging to "be inclusive of transgender and non-binary customers."[12]

Much of the decision was influenced by users on Twitter such as Jocelyn, who writes, "I understand that you guys love girl positivity but please understand that there are trans-men that get periods, and if you could please do something about the ♀ symbol on your pad packaging, I'd be happy. I'd hate to have any trans-males feel dysphoric."[13]

The company responded by issuing a statement saying, "We're committed to diversity and inclusion and are on a continual journey to understand the needs of all of our consumers," adding that the female symbol will be taken off their products. Similar efforts from other companies and governments have launched Trans and non-binary issues into the mainstream as the culture begins to dismantle the supposed subjective and arbitrary category of male and female in place of increasingly fragmented marginalized identities.[14]

THE ARGUMENT FROM INDEFINITE VARIATION

Here social constructionists champion an important truth: that the categories of biological sex are much more complicated than they appear. We'll call this the *argument from indefinite variation*. For instance, not all females menstruate, and not all males produce viable sperm. Hormone levels vary within and between the sexes. Chromosomes sometimes constitute themselves in unusual formations, and genitals may not always be fully or normally developed. All of these statements are true, but it is in the analysis where the flaws of social constructionism appear.

Take intersex individuals, for example, who are born with ambiguous genitalia. An intersex male might have an unusually small penis or missing/deformed testes, while an intersex female might have an unusually large clitoris or ambiguous ovaries. These abnormalities are

---

[12] Wolfe, E., Krupa, M. (2019). *Always* Is Taking The Female Symbol Off Its Packaging To Be Inclusive Of Transgender And Non-binary Customers. *CNN*.

[13] Twitter User Jocelyn. (2019). Female Symbol on Packaging Is Not Inclusive. *Twitter*.

[14] Dragicevic, N. (2018). Canada's Gender Identity Rights Bill C-16 Explained, *CBC*.

developmental conditions which result from unusual chromosome configurations, hormone insensitivity, and gene expression failure. Some intersex individuals have chromosomal abnormalities such as XXY in Klinefelter's syndrome, where an extra X creates hormonal and development issues in what would be a typical male. Another condition, called Congenital Adrenal Hyperplasia (CAH), renders an XX individual with completely ambiguous female genitalia, such as an enlarged clitoris and a near-completely fused labia. Reconstruction surgery is often an option for these girls to reform the vulva correctly. Others object to these surgeries as examples of category enforcement, forcing someone to fit into the binary category of male or female. After studying the conditions of intersex individuals, the social constructionist argues that because intersex individuals have unusual genitalia or chromosomes, they constitute new sexes.

These variations of sexual characteristics constitute what social constructionists deem a failure of the biopsychosocial model (a model which, among other things, states that humans can be objectively and reliably sorted into the discrete categories of male and female). While it is true a certain amount of sexual variation exists across human beings, this does not eliminate or dismantle the category of biological sex. While there are disorders of sexual development, such as intersex, this does not mean that intersex individuals have a third gamete, or a third reproductive capacity; they simply fail to develop properly functioning male or female gametes (often times). If disorders of sexual development (DSDs) actually developed into new gametes outside of the male-female category, then it would be correct to say that sex is not binary. But such a third gamete has never existed and will never exist in sexual reproduction. Reproductive capacities are defined through sexual dimorphism (they are split into two). To say that intersex people are not male or female is to 1) deny the biological mechanisms which differentiate males and females in the womb and 2) relegate intersex individuals to the category of 'Other.'

Thus, when it comes to intersex people, their sexual characteristics often fail to develop into normally functioning male or female reproductive structures. Because of this, they are often developmentally sterile. Though social constructionists often conflate intersex individuals (who have abnormal sexual development) with trans individuals (who have abnormal gender identities), intersex individuals are usually not transgender and trans individuals are almost always born with healthy male or female gametes. As expected, chromosomal abnormalities and ambiguous genitalia for intersex occur very infrequently, at around 1 in 2000 births, a miniscule *0.05 percent.*[15] However, although these individuals have ambiguous genitalia, most of them identify as male or female as they grow older even in the absence of genital surgery.[16] This is not because society socializes them as males or females, but because intersex people *are* male or female. This shows that while their genitals might not be fully functioning or morphologically normal, their gender identity (their sense of self) almost always aligns with their birth sex, which is always either male or female. For transgender individuals, however, gender identity is a very different issue.

Therefore, elimination of a category such as biological sex must require significant evidence of chromosomal, hormonal, and gamete variation to be warranted. 0.05 percent is not significant. Other estimates, which take into account more than just the intersex population but all types of DSDs, range from 1 to 2 percent of newborns having unusual chromosomal or genital structures. Even if these sexual development variations grew in large numbers, it would still not mean they constitute new sexes. A new gamete type would have to be introduced for such a change in biological sex to be reached. Despite the 98 to 99 percent of the population whose chromosomes, gametes, and hormones are aligned with a typical male or female (including most trans individuals), social constructionists still use the exceedingly rare disorders of sexual development to argue the male and female category is unreliable and defunct.

---

[15] (2018). How common is intersex? *Intersex Society of North America.*
[16] Berenbaum, S., Bailey, J. (2003). Effects on gender identity of prenatal androgens and genital appearance: Evidence from girls with congenital adrenal hyperplasia. *Journal of Clinical Endocrinology and Metabolism, 88,* 1102-1106.

When confronted with this fact, the social constructionist uses hormone levels as another variable to argue for the dismantling of the biological sex binary. While it is certainly true hormone levels are variable *within* and *between* males and females, this does not negate their averages or effects. Typical levels of testosterone in a human male at the age of 18 range from 300 to 1200 nanograms per deciliter, while females tend to have 20 to 75 nanograms per deciliter. This means the typical young adult male has about fifteen times more testosterone than his female counterpart.[17]

The effects of hormones on the development of the fetus, the neural structure, and the physical attributes of a male or female body cannot be overstated. Prenatal sex hormones differentiate male and female brain structure, bone structure, sexual desire, sexual arousal, mood, cognitive function, blood pressure regulation, motor coordination, sensitivity to pain and stress, ovulation in females, sperm production in males, memory retention, spatial-affinity, interest in people versus things, and many more.[18]

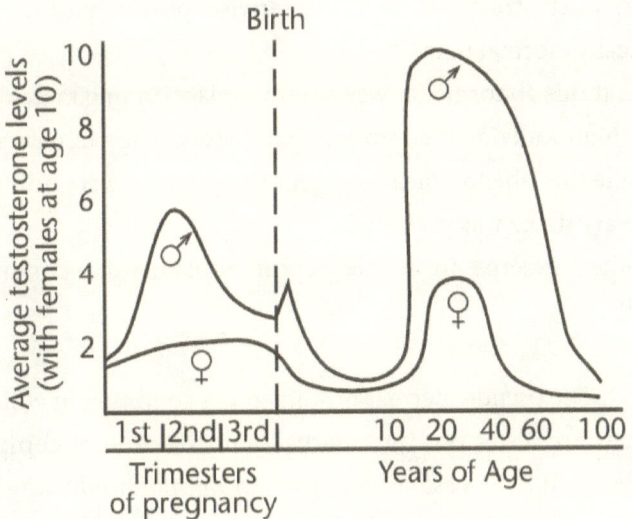

Average testosterone levels in a ratio (with females at age 10 = 1) from 1st, 2nd, and 3rd trimesters, and from birth to death for human males and females.

---

[17] (2019). Typical testosterone levels in males and females. *Medical News Today.*
[18] Marrocco, J., McEwen, B. (2016). Sex in the brain: hormones and sex differences. *Dialogues in Clinical Neuroscience, 18,* 373-383.

Different levels of these sex hormones also affect gender identity, sexual orientation, sex-typed activity interests, sex-typed cognitive abilities, sex-related behavior problems, and forms of psychopathology.[19] Sex hormones are so important that transsexuals take these hormones to alter their physiology to more closely resemble the opposite sex, but more on that later. Even if it was true that hormone levels within and between males and females existed on a flat spectrum with no averages, this would still not prove the category of male and female is unreliable and defunct, as a third gamete would have to exist.

Lastly, after the argument for intersex and hormones fails, the social constructionist will point to case studies with near supernatural levels of uniqueness to prove that the male-female dyad is not a reliable category. For example, in 2014, surgeons discovered a womb in a 70 year-old man, a father of four, and a hysterectomy was performed to remove the uterus and fallopian tubes. The surgeons soon discovered the man suffered from a rare form of male pseudo-hermaphroditism, characterized by "the presence of Mullerian duct structures in an otherwise phenotypically, as well as genotypically, normal man."[20]

Even if this abnormality was commonplace among males, it does not mean such an individual comprises both sexes. The uterus and fallopian tubes inside this phenotypical and genotypical male were non-functional, and no ovary tissue was present.

Another excerpt from the report explains the condition more specifically:

> "Male pseudo-hermaphroditism is a condition in which the gonads are testes but the internal genitalia are not completely virilized. It is possible for pseudo-hermaphroditism to be undetected until puberty. PMDS is a rare form of internal male pseudo-hermaphroditism in which Mullerian duct derivatives

---

[19] Berenbaum, S., Beltz, A. (2011). Sexual differentiation of human behavior: Effects of prenatal and pubertal organizational hormones. *Frontiers in Neuroendocrinology, 32,* 187.
[20] Sherwani, A. et al. (2014). Hysterectomy in a male? A rare case report. *International Journal of Surgery Case Reports, 5(2014),* 1285-1287.

are seen in men. It was first described by Nilson in 1939. Subsequently, approximately 150 cases have been reported. The exact cause of PMDS is not known, however it is thought to result from a defect of the synthesis or release of MIF, or from defects in the MIF receptor. Defects in the MIF gene lead to the persistence of a uterus and fallopian tube in males. It is likely that remnant Mullerian structures lead to cryptorchidism by hindering the normal testicular descent mechanism."

A second example of biological hybrids exists in a case study of an older woman with XY cells. She developed normally, underwent spontaneous puberty, reached menarche (first period), menstruated regularly, and had two normal pregnancies. However, because of her chromosomal abnormality of 46-XY, this woman gave birth to a daughter with complete gonadal dysgenesis, meaning her gonads (ovaries) failed to develop properly. In normal individuals, the researchers outline how the process of sexual differentiation is supposed to work:

> "Normal sexual differentiation in 46-XY individuals relies on a complex cascade of numerous genes, many of which have yet to be identified. Defects in these genes can cause disorders of sexual development of varying severity."

Researchers hypothesized that a novel sex-determination gene was at the cause of this abnormality:

> "The family pedigree on the mother's side was notable for the presence of seven individuals over four generations with either sexual ambiguity, infertility, or failure to menstruate, including one individual with documented 45-X/45-XY mixed gonadal dysgenesis. Both the mother and the 46-XY daughter were screened for mutations in a number of genes known to be involved in mammalian testes determination. In all genes screened, the open reading frame was found to be normal. This

suggests that a mutation in a novel sex-determination gene or a gene that predisposes to chromosomal mosaicism may be responsible for the phenotype in this family."[21]

Knowing the context and nature of these two case studies, the fact that rare abnormalities exist in no way questions the reliability of the male-female category. Consistent chromosomal makeup, genital development, a male or female reproductive structure, and functioning gametes in 99 percent of the population reliably sorts humans into male and female categories despite the rare occurrence of chromosomal, hormonal, or genital abnormalities.

While there is truth to the idea that there are variations in biological sex *characteristics*, this does not mean these variations constitute new sexes. To say otherwise is regressive, backwards, and sexist. Just because someone has unusual genitalia, hormones, or chromosomes, does not make them less of a male or female. Thus, social constructionists ignore statistical reality to argue chromosomal and hormonal variation is predominant and widespread, rendering the male-female category unreliable and defunct. This argument is largely motivated by the need to deconstruct the idea of *category* itself, which in the past, has been used to discriminate and marginalize individuals deemed as "Other." Once this category has been dismantled, then perhaps societal transformation could occur, or as one article puts it:

> "If physicians recognized that implicit in their management of gender is the notion that finally, and always, people construct gender as well as the social systems that are grounded in gender-based concepts, the possibilities for real societal transformations would be unlimited."[22]

---

[21] Dumic, M., et al. (2008). Report of Fertility in a Woman with a Predominantly 46,XY Karyotype in a Family with Multiple Disorders of Sexual Development. *The Journal of Clinical Endocrinology and Metabolism, 93(1),* 182-189.

[22] Kessler, S. (1990). The Medical Construction of Gender: Case Management in Intersexed Infants. *Journal of Women in Culture and Society,* 16(11), 25.

SEX IS SUBJECTIVE

Beyond this, the social constructionist often takes the philosophy a step further from the arguments of indefinite variation: *biological sex is not something which is determined objectively but rather decided upon by subjective societal standards.* Because of this subjectivity, society is the sole author of what is male and female, not biology. And what is male and female exists on a spectrum, just as gender exists on a spectrum comprising indefinite identities. Therefore, if sex exists on a spectrum and we cannot reliably sort people into male and female, then there must be no male and female outside of society. This implies that sex is malleable, transient, fluid, and forever changing according to the subject, and that the traditional understanding of biological sex as solid, fixed, unmoving, and unchanging according to the object is 'behind-the-times.'

In the context of this view, it makes sense why Cass Clemmer, our transgender period activist, is such a believer in the idea that all genders can get periods too. Dismantle the oppressive male-female category through the restructuring of language, and you can fundamentally change people's perception. If you change people's perception, you can change their actions. And if you change their actions, then you can end discrimination, injustice, and oppression.

Social constructionists have always been skeptical of the existence of category, from which many historical injustices have been based, arguing that continual study of category (or group) differences in the present perpetuates stereotypes, injustice, and discrimination: "Arguments about innate biological differences between the sexes have persisted long past the time they should have been put to rest," writes the Scientific American editors in 2017.[23] As an example of this phenomenon, the Scientific American mentions one of their articles called "Woman and the Wheel," written in 1895:

> "[The article] raised the question of whether women should be allowed to ride bicycles for their physical health. After all, the

---

[23] Editors. (2017). The New Science of Sex and Gender. *Scientific American.*

article concluded, the muscular exertion required is quite different from that needed to operate a sewing machine. An eminent French surgeon who authored the article answered in the affirmative the question he had posed but hastened to add: 'Even when she is perfectly at home on the wheel, she should remember her sex is not intended by nature for violent muscular exertion…And even when a woman has cautiously prepared herself and has trained for the work, her speed should never be that of an adult man in full muscular vigor."

Such an article about whether women should ride bicycles due to their 'weak muscles' relies, not on actual science, but on an incredible overemphasis of physical sex differences to the point of absurdity, caricature, and sex-based discrimination.

Similarly, research into male and female differences *can* be sexist, but it depends on how the research is conducted, what its methods are, and how it's applied. For example, highly controversial research into racial differences has shown subtle differences in average IQ levels between races. These average differences are often the result of culture and the environment, not biology. The nature of this IQ fact is not discriminatory, but the interpretation of it *can* be. Racists of the past used a mix of science and pseudoscience to justify their conclusions. Therefore, working from the context of historical injustice, social constructionists dismiss the literature of IQ differences as inherently racist. However, research into racial differences is much different than research into sex differences. First of all, and most importantly, racial differences are much more trivial and much more culturally specific than sex differences. For example, the average white man and black man will likely be much more alike in terms of behavior and interests than the average white man and the average black woman. *Differences between the sexes are often much larger and culturally universal than differences between races.*

This critical fact, however, has not stopped social constructionists from critiquing sex difference research in similar ways that they critique research into racial differences. Although sex difference literature has worked in tandem with the fields of evolutionary biology, psychology, neurology, and anthropology for over three decades, social constructionists view sex difference research with skepticism, and more specifically, as a tool to solidify the existence of historical injustice against marginalized identities, namely females. This skepticism is predicated on the idea that scientific researchers, mostly men historically, are inherently biased and that this bias is inescapable. Rendering entire fields of research defunct based on potential category bias is a common practice of social constructionists, as articulated by social scientists of the 1980s and 90s:

> "Sexist or androcentric background beliefs of scientists cause them to generate sexist theories about women, despite their adherence to ostensibly objective scientific methods."[24]

If male scientists cannot escape from their inherent bias, and if people cannot be reliably sorted into the categories of male and female, then any sex difference research is not just prejudicial, but perfunctory and useless. Instead of working with the possibility that the male-female category exists objectively, the social constructionist seeks to dismantle it through explanations of indefinite variation, society, culture, discrimination, injustice, and bias. If male and female are just constructed terms, why can't we construct, mold, and form human beings in the way we see fit?

The idea that biological sex is socially constructed may still seem out of the mainstream, but this has been an integral part of the social constructionist philosophy for decades.[25] In 1999, Gender theorist Judith Butler articulated what it meant for the category of sex to be unreliable and subjective, contrasting those who claimed sex and gender were distinct:

---

[24] (2015). Feminist Epistemology and Philosophy of Science. *Stanford Encyclopedia of Philosophy*.

[25] Joyce, H. (2018). The new patriarchy: How trans radicalism hurts women, children, and trans people themselves. *Quillette*.

> "If the immutable character of sex is contested, perhaps this construct called 'sex' is as culturally constructed as gender; indeed, perhaps it was always already gender, with the consequence that the distinction between sex and gender turns out to be no distinction at all."[26]

The erasure of sex as an objective biological reality foreshadows why a man can become a woman and a woman can become a man, why a man can become pregnant and a woman can impregnate, and why a man can have a vulva and a woman can have a penis; all true statements when the male-female category loses its linguistic meaning.[27] Here a person's sex becomes transient and ephemeral, as soluble as cream or sugar in a cup of coffee (or at least in the mind of the gender theorist). Without the presence of a reliable category, the subjective nature of perception becomes paramount, and we find ourselves in a plurality of perspectives with no objective reference point. We begin to question the nature of objectivity. Perhaps nothing exists outside an individual's subjective perception and preferences.

Ran to its logical intersectional end, we reach an indefinite number of independent and fragmented identities, with no labels, categories, or groupings. The constraining and prejudicial category of sex is dismantled, and the idea of the individual, with all his or her unique perspectives, thoughts, feelings, and desires, is ironically reinstated as the ultimate identity.[28]

But there is more to the social constructionist view than the supposed subjectivity of biological sex and the argument from indefinite variation. A historical examination of social constructionism's development renders a deeper picture as to what's happening under the surface, from contemporary gender theory and the idea of the blank slate to the postmodern critique of hierarchy and language.

---

[26] (2018). Feminist Perspectives on Sex and Gender. *Stanford Encyclopedia of Philosophy*.
[27] Coleman, N. (2017). Trans Man Gives Birth to Baby Boy. *CNN*.
[28] (2015). Feminist Epistemology and Philosophy of Science. *Stanford Encyclopedia of Philosophy*.

2

PHILOSOPHY OF GENDER

Contemporary gender theory assumes that sex and gender are both socially constructed and are ultimately indistinguishable from one another. This claim is supported using evidence of indefinite variation from medical science: that due to chromosomal and hormonal inconsistencies, people cannot be reliably sorted into the male-female category and that the dividing line for sex is based on socially constructed criteria, or subjective preferences, to decide on whether someone is a "true" male or female. This criteria is then used to construct gendered power structures.

Judith Butler represents a main proponent of this view, writing in the 1990s that "sex is as culturally constructed as gender; indeed, perhaps it was always already gender, with the consequence that the distinction between sex and gender turns out to be no distinction at all."[29] The problem with this view, as I have discussed in the previous chapter, is that in almost 100 percent of cases we can reliably sort individuals into a category of male or female despite the existence of chromosomal, genital, or hormonal abnormalities. Even if we could not sort people into male or female, a third gamete type would have to exist for the male-female category to be unreliable and defunct.

---

[29] (2018). Feminist Perspectives on Sex and Gender. *Stanford Encyclopedia of Philosophy*.

The social construction of sex is the key to understanding the contemporary form of gender theory and the activists who support it. Fighting for diversity, equity, and inclusion (moral qualities if not pathologized), we see our institutions, governments, and companies altering policies, packaging, and laws to dismantle the male-female dyad. To understand where these beliefs originate, let's examine how we got here in the first place.

Before the social construction of biological sex was popularized through the works of Judith Butler and others, one of the major philosophers of social constructionism was a woman named Simone de Beauvoir, a French feminist philosopher and social theorist. She grew up in Paris in the early 1900s and died in 1986 at the age of 78. Her works provided the foundational methods and framework for the critiques of contemporary gender theory.

An existentialist, Beauvoir viewed ethical action as not beholden to a transcendent moral standard, but rather an existential issue contingent on the individual and their relationship to the community. In other words, the moral weight and consequences of action begins and ends with the individual, not God. Because of this, freedom comes not from natural or transcendent order, but from within, where ultimately nothing resides: "The nothingness which is at the heart of man is also the consciousness that he has of himself," Beauvoir writes in *Ethics of Ambiguity,* "...There exists no absolute value before the passion of man, outside of it, in relation to which one might distinguish the useless from the useful."[30]

Values, in other words, come only from our choices. There is no appeal to nature for value. No appeal to religion. And no appeal to God. These things, she says, are mere conceptions of the human mind.

Beauvoir's existentialism can be seen in her feminist works. In 1949, she published her famous foundational work for modern feminism, *The Second Sex*. The philosophy of the book can be encapsulated in the existentialist mantra that "existence precedes essence," meaning essential characteristics of a body, for instance, are only acquired through existence

---

[30] De Beauvoir, S. (1947). *The Ethics of Ambiguity*. Marxists.org.

in life. In other words, no one is born with an innate structure, a patterning system, or a code. We are born as we are, which is to say, nothing. And so the most famous phrase in *The Second Sex* is this: *One is not born but becomes a woman.*

Beauvoir was the first to articulate the foundational concept that sex and gender are separate entities. The former is biological, innate, and fixed, while the latter is transient and socially and historically constructed.

The Stanford Encyclopedia elaborates that "before *The Second Sex,* the sexed/gendered body was not an object of phenomenological investigation." Before this distinction, the idea of gender was relegated to the confines of the Romance languages. Spanish, for instance, still assigns genders to objects. Yet it was rarely, if ever, used to delineate between behavior and essence.

With Beauvoir's writings, the sex and gender distinction allows the behaviors of individuals to be critiqued and understood in the context of social conditioning, specifically the nature of women's historical subordination. "The fundamental source of women's oppression…" Beauvoir writes, "…is femininity's historical and social construction as the quintessential Other." Beauvoir uses an idea developed by Hegel about the Master-Slave concept in Christianity and adapts it to her feminist theory. Master is replaced by Subject, the male. Slave is replaced by Other, the female. This is why the title of her famous book is *The Second Sex*, illustrating the idea that women have always been judged in light of men, or in other words, relegated to the role of Other.

Beauvoir's analysis does two things. First, it examines ways that "masculine ideology exploits the sexual difference to create systems of inequality."[31] Thinking back to the Scientific American's 1895 article on "Women and the Wheel," we recall that the author used language about females' inherent lack of "muscular strength" to argue that women were better off using the sewing machine than the bicycle.[32] While it is true that men, on average, have higher levels of upper body strength and muscle

---

[31] (2018). Simone de Beauvoir. *Stanford Encyclopedia of Philosophy.*
[32] Editors. (2017). The New Science of Sex and Gender. *Scientific American.*

mass, such a sex difference was never and will never be as drastic as the author claimed. The practice of justifying discrimination through overemphasizing sex differences was common throughout the later part of the Industrial Revolution and the early 20th century. This is what Beauvoir was critiquing.

The second part of her analysis focuses on the problems with arguments for equality which *erase* sex differences in order to "establish the masculine subject as the absolute human type." Judging women's equality by a masculine subject ignores the nature and essence of womanhood and unrightfully compares females to a male standard. Here Simone de Beauvoir would find herself in direct opposition to Judith Butler's idea that there is no true sex difference. Beauvoir insists that to treat men and women equal does not mean to erase sex differences, as to do so would be to ignore the differences of human body types. Rather, only through validating sex differences can equality be reached. "Equality is not a synonym for sameness," she argues, a position almost unthinkable in contemporary gender theory which views most sex differences as social constructs.[33]

Further expanding her idea on the Subject-Other relationship, Beauvoir draws from common critiques of the concept of hierarchy seen in Marxist writings. Stereotyping, or relegating someone to Other, could also be understood outside of the man-woman relationship. A similar type of oppression by hierarchy exists between other groups of identities which operate in a power struggle of Subject-Other. Marxists, for example, viewed the Subject-Other divide as the bourgeoisie versus the proletariat. This idea, that history can be broken down into a power struggle between numerous identity groups, is the core of contemporary identity politics.

One might ask, however, when reading Beauvoir's writings, what the goal of her writing was. What did she envision as the end of this Subject-Other oppression? The ultimate goal was liberation, which she defined as our mutual recognition of each other, each group, as free and as other. No one group or individual could lay claim to being the Master or the Slave.

---

[33] (2018). Simone de Beauvoir. *Stanford Encyclopedia of Philosophy*.

We are all just as much Subject as we are Other. In this world, subordination would be gone, replaced with an equality of *value*. In a sense, this is exactly what the Western ideal of the sovereignty of the individual has been working towards ever since it grew out of the Middle East.

Beauvoir explains this paradise concisely:

> "In a world which recognized the phenomenological truth of the body, the existential truth of freedom, the Marxist truth of exploitation and the human truth of the bond, the derogatory category of the Other would be eradicated. Neither the aged nor women, nor anyone by virtue of their race, class, ethnicity or religion would find themselves rendered inessential."[34]

To accomplish this liberation, Beauvoir wrote, women must reject the notion that to be free she must be like men. They must be encouraged to engage and socialize in the world and be allowed to discover the uniqueness of their embodiment. In other words, they must be allowed to discover for themselves what it means to be a woman.[35]

The contemporary mind will find comfort in this liberal doctrine, as it allows for sexual difference while championing the equality of the sexes. In this aspect, Beauvoir's view does not differ from contemporary sex difference research, as modern researchers work under the presuppositions that 1) men and women are both equals and 2) the exploration of biological sex differences can improve society's physical, psychological, and sexual health. However, where Beauvoir's view *does* differ is the social construction inherent in Beauvoir's writing: *one is not born a woman but becomes one*, or that *existence precedes essence*. To a certain degree, these are both true statements. A part of being a woman *is* becoming one, whether it's through learning and action or knowledge and language. Part of essence relies on existence, on socialization, and on embodiment. Yet an

---

[34] (2018). Simone de Beauvoir. *Stanford Encyclopedia of Philosophy*.
[35] Ibid.

issue arises when one uses these concepts to argue that the male-female category does not exist, and that the human mind is born without a code.

THE BLANK SLATE

Simone de Beauvoir's existentialist claims on human nature can be traced back to philosophers such as John Locke and Jean-Jacques Rousseau, who believed that human nature was formless, malleable, and ultimately good, until society corrupted it.

John Locke was one of the most influential English philosophers of the Enlightenment. His writings on inalienable rights and social contracts heavily influenced the founding fathers of the United States and the doctrines they espoused in the Declaration of Independence. One of his most famous phrases did not mention the blank slate, something he never said, but used similar phraseology to describe his views on human nature:

> "Let's suppose the mind to be, as we say, white paper void of all characters, without any ideas. How come it to be furnished? ... To this I answer in one word, from Experience."

Locke's quote must be understood from the historical context in which he lived. For millennia the idea that monarchs were divine captured the mind of humanity. A son or daughter of a monarch was viewed as inheritor of this divinity. And yet, philosophically, a problem arose in the minds of the Enlightenment Europeans: if all humans are equal in their value and rights, then the divine right of kings is a lie, a rouse to instantiate and solidify the powerful and the elite. In other words, people were never born as anything. If there was no divine ordainment and no special heredity, then the traditional views on aristocracy and monarchy were baseless. The values of liberal democracies and the abolishment of slavery all have its roots in this hypothesis.[36]

It is unlikely Locke meant his statements to be construed as holistically social constructionist, but his philosophy has had a large impact on the

---

[36] Pinker, S. (2006). The Blank Slate. *General Psychologist*.

social sciences throughout the 20th century. Psychologist Steven Pinker, in his writings on the blank slate, provides a typical example from a prominent anthropologist and public intellectual, Ashley Montagu:

> "With the exception of the instinctoid reactions of infants to sudden withdrawals of support, to sudden loud noises, the human being is entirely instinctless. Man is man because he has no instincts, because everything he is and has become, he has learned, acquired, from his culture, from the man-made part of the environment, from other human beings."[37]

This rejection of human nature can also be seen across social constructionist circles, where sex and gender are thought of as mere linguistic power structures:

> "There is no essential human. There is no human nature. There is no transcendent self. To be a subject is not to share in common a metaphysical state of being with other subjects … Man and Woman do not exist as labels for certain metaphysical or essential categories of being. They are rather discursive, social, and linguistic symbols which are historically contingent. They evolve and change over time; their implications have always been determined by power."[38]

French intellectual Jean-Jacques Rousseau agreed with this blank slate characterization of human nature and theorized that the white paper of the human mind was only corrupted through civilization:

> "So many authors have hastily concluded that man is naturally cruel, and requires a regular system of police to be reclaimed, whereas nothing can be more gentle than him in his

---

[37] Pinker, S. (2006).
[38] Escalante, A. (2016). Gender nihilism: An anti-manifesto. *Libertarian Communist*.

primitive state...The example of the savages...seems to confirm that mankind was formed ever to remain in...this condition...and that all ulterior improvements have been so many steps...towards the decrepitness of the species."[39]

Did you catch that? *All ulterior improvements have been so many steps towards the decrepitness of the species.* In Rousseau's view, the continual march towards progress of the Enlightenment pushes mankind further and further towards the downfall of the species. The natural state of primitiveness and ignorance is the state of wellbeing, harmony, and peace. Despite anthropological, biological, and historical evidence to the contrary, many social constructionists view human nature in this way.

When Rousseau mentioned the authors in the quote above, he was referencing a famous line by a contemporary who opposed his viewpoint:

> "Hereby it is manifest that during the time when men live without a common power to keep them all in awe, they are in that condition which is called war, and such a war is of every man against every man...In such condition there is no place for industry, because the fruit thereof is uncertain: and consequently...no arts, no letters, no society, and which is worst of all, a continual fear and danger of violent death, and the life of man solitary, poor, nasty, brutish, and short."

This is the famous quote from Thomas Hobbes's *Leviathan*, a semi-anthropological and semi-philosophical book about the nature of the social contract, the idea that freedom is ultimately secured through the force of a constraining government, that humans cannot be trusted to regulate themselves in the absence of the state. Hobbes argued conflict was at the core of the human condition; Rousseau argued it was peace.

Rousseau's view has far surpassed the Hobbesian perspective in the theories of social constructionists. If our nature is fundamentally peaceful

---

[39] Pinker, S. (2006).

and malleable, then society can be used to restructure an individual's actions, thoughts, and behaviors to achieve a utopian ideal.

> "The noble savage, like the blank slate," writes psychologist Steven Pinker, "continues to be an influential doctrine. It's behind the widespread respect for everything natural and a distrust of anything manmade. And it's behind the near-universal understanding of our social problems as repairable defects in our institutions, rather than a traditional view that would ascribe them to the inherent tragedy of the human condition."

It is true societal problems are often repairable defects in our institutions, and yet, it also remains true that certain aspects of the human condition are at fault for the injustices we see in our modern societies. Despite this truism, strong social constructionists continue to argue that the human mind comes into the world with no preconditions, structure, or code. Instead, we are born into a blank mind in a pure body. We learn how to act through societal conditioning alone. Our very soul, the peaceful noble savage within us, is corrupted by the constraints, the lies, and the power hierarchies of society, forcing us to alter our behavior, thoughts, and actions in pathological ways.

For the ultimate social constructionist, the society is all there is to explain human action. It follows logically that if the influences of societal forces could be removed from our ways of being, then perhaps humanity could re-enter the peace, harmony, and equality of our primitive ancestors. It is obvious the kind of imagery conjured up by such a belief: a return to the Garden of Eden, to our pre-fallen state. The Rousseauian social constructionist is no different than the Christian who believes our better natures can ultimately be redeemed. And yet unlike Christianity, which holds that human beings have a fundamentally flawed nature, the Rousseauian social constructionist argues that human nature is ultimately good, pure, and noble.

Because of the ideological belief in the blank slate and the noble savage, the social constructionist denies any evolutionary psychology, neuroscience, and biology that points to the contrary: that average differences exist among groups of people, and that some of these differences are innate. Judith Butler surely did not use the biological sciences to explain her belief that sex is a social construct. Instead, she relied on theory, philosophy, and ultimately, her subjective experience to come to her conclusions on sex and gender. Like Simone de Beauvoir, who believed in the blank slate of human nature and the hope of an equal future, Butler agrees with this, yet she takes the Rousseauian view of human nature a step further: *if we could just dismantle the oppressive nature of the category within society, then we could achieve true equality.*

Because of this Rousseauian view, the social constructionist views any form of inequality as an indicator of inequity. Any statistic showing group differences in occupational preference, for example, is not indicative of free choice but societal conditioning, discrimination, or injustice. If every individual is born a blank slate, then fundamental natures do not exist. Logically, this results in the belief that equality of outcome, rather than equality of opportunity, is the correct aim of society. Ignoring the inherent dangers in this way of thinking, the Rousseauian analysis is incorrect at multiple levels of analysis, which I will examine in the coming chapters.

This overly simplistic explanation of human nature, that if we could just get rid of the corrupting societal forces then we could live in perfect equality, is a doctrine wholly accepted by mainstream social scientists in the humanities and acts as the foundation for the philosophy of gender.

GENDER AS A SOCIAL CONSTRUCT

If humans are born as blank slates with no predispositions or foundational neural structure, then human behavior can only be shaped by the forces of society, culture, and environment. Taken to its logical next step, any significant sex difference between males and females is not a product of biology and is instead a product of socialization. And this is where the modern theory of gender takes hold with two major factions.

The first faction is called second-wave feminism (or gender-critical feminism). This view is best espoused by Simone de Beauvoir, where the distinction is made between biological sex and socialized gender while still recognizing the existence of biological sex differences. *Equal and different* is the perfect phrase for this view. Gender-critical feminists reject gender stereotypes by separating biological sex from gender expression. Gender-atypical play among boys and girls is supported, not discouraged, allowing for a wide diversity of behavior within males and females.

The second faction is known as third-wave feminism. As I have covered, this view is best espoused by Judith Butler, who theorizes sex and gender are socially constructed and that no distinction exists. *Equal and same* is the motto of this philosophy.

Butler represents the strong social constructionist perspective, believing that sex and gender are social constructs, whereas Beauvoir represents the *gender-critical* perspective, believing that while sex is biologically defined, gender is socially constructed. For the remainder of the chapter, we'll be focusing on the strong social constructionist perspective of Butler and how it relates to gender socialization.

In modern gender theory, gender is defined as "the different roles, norms, and meanings assigned to men and women and the things associated with them on account of their real or imagined sexual characteristics."[40] Additional definitions come from West and Zimmerman (1987), where they define gender as a "socially scripted dramatization of the culture's *idealization* of feminine and masculine natures."[41] Or: "the activity of managing situated conduct in light of normative conceptions of attitudes and activities appropriate for one's sex category."

Based on these definitions, gender is a social construct, a conduct, an act, and a performance to align oneself with a cultural idealization of masculine and feminine. Because of real or imagined sex differences, men and women are pushed towards different social roles. Most societies, for

---

[40] (2015). Feminist Epistemology and Philosophy of Science. *Stanford Encyclopedia of Philosophy*.

[41] West and Zimmerman, (1987). Doing Gender. *Gender and Society, 1(2)*, 130.

example, relegated men to the military and politics while placing women with childrearing responsibilities. Society punishes and rewards behaviors which conform or do not conform to established gender roles and norms. For example, men are expected to be aggressive, unemotional, and stoic. A man who exhibits emotion, therefore, is seen as *weak* and *effeminate*. Women are expected to be docile, agreeable, and modest.[42] A women who exhibits aggression or assertiveness is seen as *bitchy* or *controlling*.

According to psychologist of gender Linda Brannon, these role identities between males and females manifest themselves in different ways. For males, a stigma is attached to feminine characteristics. Success and status are seen as the ultimate pursuit. Toughness, confidence, self-reliance, aggression, and violence are encouraged and rewarded. For females, weakness, dependence, and timidity are seen as virtuous, preferably relegated to the domestic life. A woman's place is seen as located in the home as daughter or sister, but most of all, as wife and mother.[43]

Psychological and behavioral traits such as the ones above are assigned to the categories of masculine and feminine as gendered stereotypes. According to gender theory, these traits do not arise from biological and neurological processes but rather are socially constructed and defined through society. *They are not innate.* Because they are viewed by society as biological in origin, a man exhibiting feminine traits is punished through disapproval and discrimination. As a woman, exhibiting masculine traits is just as discouraged. These rewards and punishments then help solidify stereotypical behavior.

Wanting to further re-conceptualize gender, the third-wave gender theorists took the ideas of Simone de Beauvoir and pushed them into the sphere of institutions in the 1990s, applying their theories to how socialization reinforces systemic sexism and how gender is a performative act. Sociologists Candace West and Don Zimmerman said as much in their famous 1987 article:

---

[42] Brannon, L. (2016). *Gender: Psychological Perspectives.* Routledge Press, 162.
[43] Brannon, L. (2016). 162.

"The purpose of this article is to advance a new understanding of gender as a routine accomplishment embedded in everyday interaction. To do so entails a critical assessment of existing perspectives on sex and gender and the introduction of important distinctions among sex, sex category, and gender. We argue that recognition of the analytical independence of these concepts is essential for understanding the interactional work involved in being a gendered person in society. The thrust of our remarks is toward theoretical reconceptualization... **Our purpose in this article is to propose an ethnomethodologically informed, and therefore distinctively sociological, understanding of gender as a routine, methodological, and recurring accomplishment.**"[44]

First, on the meaning of *ethnomethodological*: a common practice in the social sciences, this methodology comes from the word ethnography, which is the practice of studying peoples and cultures through observation. Ethnographers, often times working in the field of sociology, will spend considerable time observing a group of people to understand their behaviors in the context of society. After the observation is complete, the ethnographer will study his or her notes and compile them together for a report, often times to prove or disprove a hypothesis. Analysis on the observations will be done alongside the gathering of other research to support the hypothesis. Ethnography, therefore, consists of observation, notation, analysis, and reflection.

Because of the nature of this practice, ethnography lends itself to being a highly subjective and observational form of research. One ethnographer's analysis, due to the way they see the world, may be entirely different from another ethnographer even when the people or culture being observed are the same. In the absence of a scientific method of consistent and repeatable experimentation, ethnography can often be an unreliable form of research and should be taken lightly.

---

[44] West and Zimmerman, (1987). Doing Gender. *Gender and Society, 1(2),* 125, 126.

Second, on understanding gender as a *routine and recurring accomplishment*: gender is not simply a role or trait assignment. It's also an act. And this is a major aspect of modern gender theory.

"Doing gender…" write West and Zimmerman, "involves a complex of socially guided perceptual, interactional, and micropolitical activities that cast particular pursuits as expressions of masculine and feminine 'natures.'"[45]

Inverting the biopsychosocial model of masculine and feminine traits being defined by existing sex differences, the writers argue that existing sex differences only appear innate due to the mapping of masculine and feminine identities onto human *action*. In other words, masculine and feminine traits are not the *cause* of certain actions but rather the effects. For example, a feminine personality is not seen as the cause, or precondition, for women preferring to work with people rather than things. Instead, women being largely people-oriented is a product of *association* with femininity. Femininity, therefore, assigns people to gendered roles based on their biological sex. It is not a pattern of traits which exists objectively. Thus, if being people-oriented was not stereotyped as a feminine trait, the gender theorists say, then the sex difference would not exist. These traits of masculinity and femininity, rather than being innate personality structures, are the social constructs we see in behavior.

Perhaps West and Zimmerman believed that masculine and feminine traits exist biologically but are often ill-defined or their applications too narrow. On the other hand, they could also be arguing that there is no objective meaning of masculine and feminine, and that all meanings which arise from them are constructed onto the performance of gender. In my opinion, the latter option is more likely.

When discussing the attribution of masculine and feminine traits to male or female action in an effort to encourage or discourage certain behaviors, West and Zimmerman argue that as long as this *gender assessment* is enacted, the constraining and oppressive nature of the male-female category will continue to exist:

---

[45] Ibid., 126.

> "While it is individuals who do gender, the enterprise is fundamentally interactional and institutional in character, for accountability is a feature of social relationships and its idiom is drawn from institutional arena in which those relationships are enacted. If this be the case, can we ever *not* do gender? Insofar as a society is partitioned by 'essential' differences between women and men and placement in a sex category is both relevant and enforced, doing gender is unavoidable."[46]

Invoking the reasoning of West and Zimmerman, the social constructionist views sex difference research as an attempt to place males and females into separate categories, from which the performance and reinforcement of gendered actions, roles, and behaviors are enacted. Like the writers say: *if only we could just move past arguments for "essential" sex differences, then the shackles of gendered action would be eliminated.* Sex and gender differences would then cease to exist, and with it, the utopia of perfect gender equity would be solidified. Here social constructionists find themselves returning to the ideology of the blank slate.

For the believer in the Rousseauian view of human nature, there is no room for biological, neurological, or evolutionary evidence to explain even *some* of the differences between groups of people, let alone differences between males and females; or as West and Zimmerman note:

> "Doing gender means creating differences between girls and boys and women and men, differences that are not natural, essential, or biological. Once the differences have been constructed, they are used to reinforce the 'essentialness' of gender."[47]

While it is true aspects of gender are socially constructed, such as the type of clothing men and women wear, there are other aspects of gender,

---

[46] West and Zimmerman, (1987). 137.
[47] Ibid., 138.

such as toy preferences or interests, which have more of a mixture of biological and psychological origins. Thus, one might think it would be helpful to understand what West and Zimmerman think are natural, essential, and biological differences between men and women, if any. Some social constructionists will acknowledge the existence of minor sex differences, but many downplay or ignore them completely to theorize that both sex and gender are holistically constructed categories. This logic is then used to argue that all societal disparities between men and women, such as differential representation in occupations, are due to socialization, not innate interest. If all differences between boys and girls are socially constructed, then what's stopping individuals from molding boys and girls in ideological ways?[48]

We will explore this question in future chapters, but first, to understand how the essentialness of gender may be constructed, let's breakdown the process of gender socialization and performance from birth through adulthood, using blank slate ideology as our foundation.

CONSTRUCTING GENDER 'ESSENTIALNESS'

Intuitively, you might think of sex as the male and female reproductive capacities required for procreation, while you might think of gender as the roles and behaviors exhibited through these capacities. And yet, as I have shown, strong social constructionists reject this differentiation of sex and gender using arguments of outlier such as intersex, pseudo-hermaphroditism, and phenotypic and genetic abnormalities.[49] For the social constructionist, sex and gender are *both* constructed categories which exist on a spectrum. Thus, when males and females are unable to be categorized because of apparent indefinite genetic variation, the social constructionist theory leaves us with no male-female category and no evolved mind with predispositions, structures, and patterning. Instead, we are left with an unstructured mind connected to a blank nervous system waiting to be molded by potentially corrupting environmental forces.

---

[48] See chapters, "Development of Gender Identity" and "The Gaslighting Ideology."
[49] See Chapter One, "Sex as a Social Construct."

According to social constructionism, the fetal mind is like an amorphous computer code waiting to be activated and built upon. Underlying structural systems do not exist. Instead, the mind exists in a pure, unadulterated state before environmental sensory input creates impressions on the mind after birth, where they begin to form a structure. These impressions influence behavior as the structural system is formed. And, inside an indefinite feedback loop, the structures reinforce each other until behaviors, thoughts, and attitudes are so solidified they seem innate and predetermined.

This social constructionist theory of mind rejects the idea that the brain has a predetermined and adaptive coding structure in the womb, and that the genes of this coding structure are turned on and off by environmental and genetic factors. Instead, social constructionists posit that the nature of the fetal neural structure is fundamentally without a code and that behaviors must be constructed through interaction with the environment. Differences between men and women exist only for reproductive capacities at best and represent stereotyped category differentials at worst. Any other sex difference must be socially constructed.

This philosophy is known as the *theory of gender neutrality*, the idea that one's gender develops from social learning in early childhood and that it can be changed and molded with appropriate behavioral interventions.[50] Or, put another way: "If boys and girls are different, they are not born, but *made* that way."[51]

Sexologist John Money was the main proponent of this *theory of gender neutrality* in the 1950s, when the philosophical distinctions between sex and gender first became prominent. Money theorized that individuals were *psychosexually* neutral at birth, meaning that newborns were born with an undifferentiated brain structure and no differences in innate gender identity; or, in the words of Money and his researchers in the 1950s:

---

[50] Colapinto, John (2001). *As Nature Made Him: The Boy Who Was Raised as a Girl*. New York: HarperCollins. 33-34.

[51] Thorne, B. (1993). *Gender play: Girls and boys in school*. New Brunswick, NJ: Rutgers University Press. 2.

> "Erotic outlook and orientation is an autonomous psychological phenomenon independent of genes and hormones, and moreover, a permanent and ineradicable one as well. It is more reasonable to suppose simply that, like hermaphrodites, all the human race follow the same pattern, namely, of psychological undifferentiation at birth."[52]

Because gender's biological roots were theorized to be non-existent, healthy development depended solely upon the appearance of the genitals. In light of the idea that one's gender-based behavior is independent of genes, hormones, and body morphology, Money proposed a medical model for the problem of intersex infants: if the appearance of the genitals is crucial to psychosexual development, then infants with ambiguous genitalia should be surgically altered to resemble the stereotypical male or female. Money espouses this view in 1955:

> "It is easier to make a good vagina than a good penis and since the identity of the child will reflect upbringing, and the absence of an adequate penis would be psychosexually devastating, fashion the perineum into a normal looking vulva and vagina and raise the individual as a girl."[53]

For Money, the absence of normal genitalia meant that infants with intersex conditions had no chance of healthy psychosexual development and must be surgically mutilated. If one's gender development is simply socially constructed and independent of genetics and prenatal hormones, then giving a genetic male a normal looking vulva and vagina and raising him as a girl should present no issues. This *theory of gender neutrality* became widespread by the 1970s, especially in sociology, psychology, and women's studies. In 1973, based on the experiments done by John Money

---

[52] Diamond, M., Keith, H. (1997). Sex reassignment at birth: a long term review and clinical implications. *Archives of Pediatrics and Adolescent Medicine, 151,* 2.
[53] Diamond, M., Keith, H. (1997). 1.

on raising healthy males as girls, Time magazine famously stated that the theory of psychological sex differences was now bankrupt:

> "This dramatic case provides strong support that conventional patterns of masculine and feminine behavior can be altered. It also casts doubt on the theory that major sex differences, psychological as well as anatomical, are immutably set by the genes at conception."[54]

Time magazine was able to say this, in large part, due to Money's experiments on a set of twin boys in the 1960s, one of whom was damaged by a failed circumcision.[55] The boy's genitals were surgically removed and replaced with a pseudo-vulva and vagina. In consultation with Money and other physicians, the boy was to be raised as a girl. For the next few decades, the experiment was met with reports of great success, proving to social constructionists that gender identity and gender-based behavior was completely independent of genetics, hormones, and chromosomes and was instead socially constructed.

However, as the years went on, these reports on the successful nature of the boy's transition to a normal and healthy girl were all found to be complete fabrications.[56] The boy never accepted his girl identity and continually insisted he was a boy, rejecting all physicians' and psychiatrists' attempts at socializing him as a girl, despite his fake vulva and vagina. It turns out that genetics, hormones, and body morphology *do* directly influence one's behaviors, interests, and preferences, irrespective of the appearance of one's genitals. And it turns out that boys and girls *are* often different in their behaviors, interests, preferences, and even body morphology. Any attempts at *making* them something they are not, no matter how hard physicians, psychiatrists, and parents try, always results in catastrophic consequences for the child. This was known as the

---

[54] Diamond, M., Keith, H. (1997). 2.
[55] Gaetano, P. (2017). David Reimer and John Money gender reassignment controversy: The John/Joan Case. *The Embryo Project Encyclopedia.*
[56] Gaetano, P. (2017).

John/Joan Case, which I will discuss in detail throughout Chapter 9 (Development of Gender Identity).

Ironically, social constructionists argue for the dismantling of the male-female category while at the same time agreeing with John Money that gender identity is socially constructed. By castrating the genitals of intersex infants who deviated from normal conceptions of male and female genitalia and assigning them to the opposite sex, Money reinforced sexist stereotypes.[57] Because of this, he was not a fighter for dismantling the category of male and female. *He was one of the gender binary's most fervent supporters.* In Money's view, there should be no issue with surgically mutilating an abnormal infant if gender is socially constructed, malleable, psychosocial, and based on the mere *appearance* of the genitals. This fundamental agreement between Money and social constructionists (that infants come into the world *gender undifferentiated* and *psychosexually neutral*) is the foundation for conceptualizing gender as a social construct. If infants are born without a predetermined structure in the brain, then their biological sex becomes merely a psychosocial construct where socialization begins molding them the moment they are born.[58] And this is where the construction of gender 'essentialness' begins.

Entering the world for the first time means entering a world of gender stereotypes. The constraining category assignment does not just happen the moment the infant comes home, but right at birth. According to strong social constructionists, a doctor will examine the genitalia of the baby and "assign" him or her to the male or female category. Social constructionists say this is not based on objective criteria of sex but rather socially

---

[57] In reality, intersex infants are born with an innate gender identity through genes and hormones. Because of this, any attempt at surgically altering the ambiguous genitalia to conform to a binary model is often harmful. Instead of performing surgery, parents and doctors should wait for the child to grow and develop, allowing them to explore their gender identity. Once the child has matured, their gender identity will almost always solidify as male or female.

[58] Bullough, V. (2003). 232.

constructed definitions which reinforce normative conceptions of what sex is, marginalizing those who do not fit into the category.[59]

After being arbitrarily assigned to a rigid category, the infant comes home to a world of oppression: blue or pink nurseries filled with cute yet stereotypical toys of trucks and action figures for boys and dolls and kitchens for girls, while the infant's parents begin to interact with them in gendered ways.[60]

For example, the girl may be given extra affection if she smiles more and gazes into her parent's eyes, while a boy may be given extra affection if he pays more attention to things and objects. The parent means no harm of course, but they cannot escape their internal biases. The reinforcement of stereotypical behavior through the biased parent increases as the child grows into the toddler age and begins to catch on to the systems of rewards and punishments which characterize their behavior. For example, boys will be encouraged to play with stereotypically male objects such as trucks, action figures, and even toy guns, while girls will be encouraged to play with stereotypically female objects such as dolls, kitchens, and houses. Perhaps they will even be given a brush to groom the dolls' hair in a caring and nurturing way. Feminine toys are stereotyped to nurturance, care, attractiveness, and beauty, whereas masculine toys are stereotyped to technology, competition, aggression, construction, and action.[61] Furthermore, masculine toys such as trucks are associated with movement, spatial skill, construction, and science, whereas dolls are associated with nurturance.[62]

The point of contention among social constructionists is not that children prefer toys which are stereotyped to their gender but rather that this preference is a construction based in social conditioning: "Parental toy

---

[59] See Chapter One, "Sex as a Social Construct."

[60] Boe, J., Woods, R. (2017). Parents' Influence on Infants' Gender-Typed Toy Preferences. *Sex Roles,* 2.

[61] Kollmayer, M., et al. (2018). Parents' Judgments about the Desirability of Toys for Their Children. *Sex Roles.*

[62] Boe, J., Woods, R. (2017). 12.

selection and responses to toy play are particularly important factors in children's acquisition of gender stereotypes."[63] This preference for masculine or feminine toys occurs due to parents purchasing stereotyped toys and awarding their children when they are played with appropriately.[64]

The parents give their girl, not just attention for doing gender in normative ways, but affirmation and encouragement. This reinforcement of gender is enacted 1) by parents acting as role models for symbolic play, 2) by rewarding desired toy choices and removing undesired ones, and 3) by communicating gendered knowledge and expectations in regard to toy choice and play.[65] Once these processes are instantiated, internal social control becomes a major mechanism for reinforcement. In other words, the norms of society are internalized and accepted as valid regardless of their innateness, or in the words of researchers Boe and Woods: "It is these internalized schemas that lead to the creation of gender differences."[66]

This early childhood socialization goes beyond toy preferences. A study by Serbin et al. (2002) found that "toddlers looked longer at photographs that depicted men and women participating in activities that were inconsistent with gender stereotypes, indicating that the toddlers recognized and expected the associations and were surprised when they were violated."[67] Toddlers also looked longer at dolls if they were female and trucks if they were male, providing evidence that a child's gender-typed toy preferences emerge as early as the first year.

In regards to parental play, evidence shows a difference in the quality of play with male versus female infants. Play with male infants is observed to be highly physical in nature, such as rough and tumble play, whereas play with female infants is observed to be more gentle and female gender-role oriented in nature. These parental play differences reinforce the "strict

---

[63] Ibid.
[64] Boe, J., Woods, R. (2017). 2.
[65] Kollmayer, M., et al. (2018).
[66] Boe, J., Woods, R. (2017). 2.
[67] Boe, J., Woods, R. (2017). 3.

gender binary, placing certain characteristics at extreme ends of a perceived continuum of feminine and masculine."[68]

Parents can play an enormous role in socializing their children to act in stereotyped ways. For example, sociologist Emily Kane writes in her 2006 article that "parents are not simply agents of gender socialization but rather actors involved in a more complex process of accomplishing gender with and for their children."[69] Echoing the characterization from West and Zimmerman that gender is a performative act, Kane argues that gender is accomplished and constrained within the context of accountability, the driving motivating factor for stereotypical gender expression. This accountability comes largely from parents.

Considering all this, learning of stereotyped behavior at such an early age is theorized to occur because of the blank slate of the mind. Stereotypical behaviors such as masculine and feminine toy preferences, interests, and personality are not a mix of complex biological and social factors but rather evidence of social conditioning alone; or, as Professor of Sociology Dr. Heather McLaughlin notes: "[Gender socialization] is a process of social interaction by which a person acquires the knowledge, values, attitudes, and behaviors essential for effective participation in society."[70] The agents of this socialization come from all aspects of society, including family, peers, religion, teachers, and the media. Because stereotyped gender behaviors become solidified in childhood, occupational career choice must be affected significantly as men and women sort themselves into careers which mimic the roles they practiced, learned, and enacted as a child.[71]

The dark side of this socialization is seen in the reactions to nonconformity, where parents or peers punish a child for behaving in non-normative ways. For example, a boy playing with stereotypically feminine

---

[68] Ibid.

[69] Kane, E. (2006). 'No Way My Boys Are Going to be Like That!' Parents' Responses to Children's Gender Nonconformity. *Gender & Society, 20(2)*, 152.

[70] McLaughlin, H. (2018). Gender Socialization: How and When do We Learn the Rules? *Sociology of Gender, Oklahoma State University.*

[71] Kollmayer, M., et al. (2018). 2.

toys may be seen as effeminate or even gay, and because of this nonconformity, he may be discouraged or ostracized by his peers or parents. From parents, this discouragement occurs because they often consider themselves accountable for their child's nonconformity, as Kane writes:

> "Most parents made efforts to accomplish, and either endorsed or felt accountable to, an ideal of masculinity that was defined by limited emotionality, activity rather than passivity, and rejection of material markers of femininity. Work to accomplish this type of masculinity was reported especially often by heterosexual fathers... Many parents invoked biology in explaining their children's gendered tendencies... But one of the things that was most striking to me in the analyses presented here is how frequently parents indicated that they took action to craft an appropriate gender performance with and for their preschool-aged sons, viewing masculinity as something they needed to work on to accomplish."[72]

As you can see, much of gender-critical feminism focuses on this construction of gender differences as 'innate' traits, and here, the feminists are partially right. There are a variety of microbehaviors, values, beliefs, and ways of acting in the world which are influenced by societal conditioning. Gender stereotypes are real and often applied inaccurately; boys and girls are often socialized as completely opposite beings, when in reality, there are many more similarities between them than differences; and caricatures of the masculine and feminine proliferate. Here social constructionists and gender-critical feminists champion an important truth: that parents, the media, and society can have a crucial impact in the child's concept of self. And yet, while this is true, social constructionists extend this philosophy into the realm of existentialism, where 'existence' precedes 'essence.'

---

[72] Kane, E. (2006). 'No Way My Boys Are Going to be Like That!' Parents' Responses to Children's Gender Nonconformity. *Gender & Society, 20(2)*, 172.

Instead of understanding that many of the differences between boys and girls may have genetic, hormonal, and psychological origins, social constructionists believe that the differences between male and female bodies are constructed and institutionalized. Thus, sex differences are not the result of innate essence, but are rather the result of constructed gendered bodies.

It is here, in the realm of body morphology, physiology, and biology, where strong social constructionists reject any innate differences between the bodies of boys and girls. Karin Martin, Professor of Sociology and Women's Studies at the University of Michigan, discusses how the curriculums of preschools actually *construct* the bodily differences between boys and girls:

> "Many feminist scholars argue that the seeming naturalness of gender differences, particularly bodily difference, underlies gender inequality. Yet few researchers ask how these bodily differences are constructed. Through semi-structured observation in five preschool classrooms, I examine one way that everyday movements, comportment, and use of physical space become gendered. I find that the hidden school curriculum also turns children who are similar in bodily comportment, movement, and practice *into* girls and boys--children whose bodily practices differ. I identify five sets of practices that create these differences: dressing up, permitting relaxed behaviors or requiring formal behaviors, controlling voices, verbal and physical instructions regarding children's bodies by teachers, and physical interactions among children."[73]

By ignoring the sex differences in skeletal structure (which allows us to determine sex from a mere skeleton), Martin argues that even the way girls and boys move must be socially constructed. Similar to West and

---

[73] Karin, M. (1998). Becoming a Gendered Body: Practices of Preschools. *American Sociological Review, 63(4),* 494.

Zimmerman's concept that gender is reinforced through practice and action, Martin's critique looks at how classrooms and teachers do not just reinforce bodily differences between girls and boys, but *construct* them. Martin is correct that feminist literature sees the belief in innate gender differences as the cause of gender inequality, and yet, pushes the concept further into how innate properties such as bodily difference are *constructed* onto the bodies of girls and boys. This construction of gendered action in preschool from the curriculum reinforces stereotypical behaviors, making physical differences appear and feel natural in adulthood.

Martin argues these curriculums are more than just harmless structures:

> "Hidden curriculums are covert lessons that schools teach, and they are often a means of social control. These curriculums include teaching about work differentially by class, political socialization, and training in obedience and docility... This curriculum demands the practice of bodily control in congruence with the goals of the school as an institution. It reworks the students from the outside-in on the presumption that to shape the body is to shape the mind... While the process ordinarily begins in the family, the schools' hidden curriculum further facilitates and encourages the construction of bodily differences between the genders and makes these physical differences appear and feel natural."[74]

Martin provides multiple examples for how this construction of bodily difference occurs. One of her examples focused on the controlling of voice. Boys were more likely to be told to quiet down as a group than as individuals, compared to girls, who were told to be quiet three times more often as individuals. When boys were told to quiet down, they were often engaged in behavioral play such as warrior narratives, compared to girls, who were often engaged in a loud emotive response about a person or

---

[74] Karin, M. (1998). 495.

object. This limiting of the girls' emotive responses also affected their bodily movements:

> "By limiting the girls' voices, the teacher also limits the girls' jumping and their fun. The girls learn that their bodies are supposed to be quiet, small, and physically constrained."[75]

When told to quiet down, the girls transformed their behavior into a fun type of whispering:

> "Thus, the girls took the instruction to be quiet and turned it into a game. This new game made their behaviors smaller, using hands and mouths rather than legs, feet, and whole bodies. Whispering became their fun, instead of jumping and humming. Besides requiring quiet, this whispering game also was gendered in another way: The girls' behavior seemed to mimic the stereotypical female gossiping. They whispered in twos and looked at the third girl as they did it and then changed roles. Perhaps the instruction to be quiet, combined with the female role of 'helping,' led the girls to one of their understandings of female quietness--gossip--a type of feminine quietness that is perhaps most fun."

Despite its seemingly innocuous nature, Martin argues this "constructed" quietness hurts girls when it comes to being assertive:

> "Finally, by limiting voice teachers limit one of girls' mechanisms for resisting others' mistreatment of them. Frequently, when a girl had a dispute with another child, teachers would ask the girl to quiet down and solve the problem nicely. Teachers also asked boys to solve problems by talking, but they usually did so only with intense disputes and the instruction to

---

[75] Karin, M. (1998). 504.

talk things out never carried the instruction to talk *quietly*...We know that women are reluctant to use their voices to protect themselves from a variety of dangers. The above observations suggest that the denial of women's voices begins at least as early as preschool, and that restricting voice, usually restricts movement as well."[76]

If you're a woman reading this, does this describe you? Are you afraid to use your voice, to speak up in the face of dangers, and to let your opinions be heard? Were you socialized this way?

While this form of social control of preschoolers reinforces specific behaviors as innate sex differences, it also transforms the actions of gender from mere action to holistic *embodiment*:

"[Embodiment] becomes deeply part of whom we are physically and psychologically. According to Connell, gender becomes embedded in body postures, musculature, and tensions in our bodies. 'The social definition of men as holders of power is translated not only into mental body-images and fantasies, but into muscle tensions, posture, the feel and texture of the body. This is one of the main ways in which the power of men becomes naturalized.'"[77]

Once again, the social constructionist view rejects and inverses the biopsychosocial model that says gender differences are constructed onto biological sex. Instead, *biological sex is what's constructed onto gender differences*. Stereotyped embodiment seen in preschools' differential treatment of girls' and boys' bodies directly contributes to the naturalization of men's power--the patriarchy--through affecting the physiology of boys. It makes them feel powerful, confident, and assertive through bodily action while making girls feel timid, gentle, and weak. In

---

[76] Karin, M. (1998). 504.
[77] Ibid., 504-505.

other words, socialization does not just make bodily differences seem natural; *it constructs them* and *institutionalizes them* into the structure of reality. Muscle tensions, posture, and even the *feel and texture of the body* are actually *created* from social definitions of men as "power-holders."

Literally, the preschool is the first place where boys learn how to enact their embodiment of power over girls, eventually reinforcing their patriarchal status as they grow into adulthood.

Martin concludes with a description of preschool as a *gendered institution:*

> "Sociological theories of the body that describe the regulation, disciplining, and managing that social institutions do to bodies have neglected the gendered nature of these processes (Foucault 1979; Shilling 1993; Turner 1984). These data suggest that a significant part of disciplining the body consists of gendering it, even in subtle, micro, everyday ways that make gender appear natural. *It is in this sense that the preschool as an institution genders children's bodies.*"

If schools are institutions of social control which construct and reinforce the performance and embodiment of gender, further dividing boys and girls along gendered lines of power, and if parents help construct these differences with toy choice and stereotyped-behavior rewards, then what happens after these *socially constructed* children become young adults and enter the workforce and society at large? The answer isn't good.

At this point, the social constructionist argues that these socialized gender differences, which occur from the moment of birth and into school, stratify society into, not a hierarchy of competence and merit, but of power and domination. Labor is divided along lines of subordination and social control. Any type of inequality in the workforce, such as the disparity seen in the STEM fields for example, is not a byproduct of free choice but rather the result of an imbalance of power resulting from gender socialization. This inequality reveals itself through multiple strata of society, from the imbalance of men and women in positions of power all the way down to the gendered products lining our store shelves.

Gendered products are ubiquitous across Western society, from gendered toys and machines to gendered body washes and ear plugs. These unnecessary products are yet another influential factor which contributes to the oppressive gender hierarchy. Gendered products affirm the gender binary (a form of category assignment), reinforce stereotypes, tell us explicitly that women should be subordinate to or dependent on men, cost women money, and ultimately, are tools of structural power.

This unequal stratification of society subordinates women, controls their voices and bodies, and relegates them to lower-paying jobs at best and domestic housewives at worst. It is in this way that the performance and action of masculinity reinforces and solidifies the hierarchy of power as innate and natural based on socially constructed notions of sex and gender, placing women as subordinate and inferior and men as dominant and superior. Through these constructed differences, society structures itself along lines of power, instantiating men as the sole political leaders, owners of capital, and heads of the family unit. This is what social constructionists call *the patriarchy*.

Patriarchal societies, therefore, ultimately operate using a hierarchy of power. Here, individual merit and competence go nowhere, and the people in the society find themselves confined to categorical and stereotypical roles based on social constructs. Gender inequality becomes a social norm, and power becomes the only way of climbing the patriarchal hierarchy. *Is this why we still see large gender differences in the most gender-equal societies? Are men and women in Scandinavia still constrained by patriarchal social practices?*

If human minds are born as blank slates with no differences to other minds, then differences between people must be problematic. And if they *are* socially constructed, rather than biological, then that means these differences can also be dismantled through language. If biological sex, gender differences in behavior and interests, and bodily differences between boys and girls are simply the result of power structures which reinforce these differences through language, then what can we do about it? And can these sex and gender distinctions be eliminated if we restructure the oppressive society, as the gender theorists say, along more equitable lines? Can we, in a sense, use the power of language to dismantle the male-female category, and with it, the patriarchal hierarchy?

3

THE POSTMODERN TYRANNY

Ever since the dawn of civilization, hierarchies have been the ultimate constraints on the creation of a human utopia. A hierarchy, by its very nature, is discriminatory. It categorizes people into groups based on socially constructed criteria, creates a near infinite number of stratifications, and utilizes those at the upper strata to oppress those below them. If only we could dismantle the hierarchy, the arbitrary differences constructed onto us by society, then the utopia of Jean-Jacques Rousseau could transform our age of modernity back into an age of pre-history, where oppression and injustice were not even concepts of the human mind, where our ancestors lived in peace and harmony, and where the sexes lived as equals. This is a paradise constructed by the postmodern mind. It never existed in history, and it will never exist, and yet, this has not stopped social constructionists from positing their visions for a great utopia devoid of inequality.

If humans are born as blank slates, then gender inequality (and gender differences) must arise from socialization alone. The socialization of individuals into male and female categories forms a hierarchy between normative expressions of masculinity (hegemonic) and their differences. Hegemonic masculinity uses power to marginalize other identities while acquiring resources and capital. This occurs not only through physically violent means but through constructed myths, or narratives, which claim the hierarchy is innate and natural. Through the expression of hegemonic masculinity, suppression and greed become indefinitely cycled and continually constructed.

The system is solidified through continual use of language, or rhetoric. This rhetoric utilizes a certain structure to reinforce the power hierarchy. For example, the use of reason, evidence, and logic to argue that gender differences are partly biological is not a form of objective inquiry but rather a subjective rhetorical device of power. To use reason is to utilize just another form of rhetoric to justify the existence of the socially constructed hierarchy. Reason, therefore, is a tool of power, a language strategy of Western civilization to justify discrimination, oppression, and injustice.

If the claims of the social constructionists are correct, that concepts, ideas, material reality, and even the biological sexes are social constructs, then all we are left with is language, which is inherently an act of categorizing, of labeling, of discrimination. Language, therefore, is a system which can be used to structure or restructure reality. If we change language, the very nature of categorizing itself, then perhaps we can restructure societies, and ultimately the hierarchy itself, in the image we see fit. This is where social constructionism reaches its logical conclusion, in a philosophy known as *postmodernism*.

THE THEORY OF REJECTING THEORIES

If social constructionists use the rhetorical framework laid out by postmodernism, then what exactly is the philosophy of postmodernism? To understand this, we first have to look at the presuppositions the philosophy was challenging.

Postmodernism, as its name suggests, is a reaction to modernism and its presuppositions about reality. Modernism grew out of the Age of Enlightenment of the 17th and 18th centuries, where the concepts of the scientific method, logic, and reason saw a rebirth and renewal, and where progress was seen as not just something that was virtuous, but something that was inevitable. History then became a *continual march towards progress.*

Modernism, therefore, can be defined in the four axioms of Truth, Knowledge, Reason, and Progress:[78]

1) **Truth**: there are absolute truths independent of any individual mind, and these truths are universal.
2) **Knowledge**: it is possible to have objective knowledge of these truths.
3) **Reason**: reason is the best method to achieve and justify such knowledge.
4) **Progress**: acting rationally in response to objective knowledge improves our chances of achieving our aims.

In the contemporary world, these axioms are still dominant in Western culture. Science is championed as the ultimate exercise of objective knowledge-seeking, and progress is seen as an ultimate good. And yet at the same time, the idea that there are absolute truths independent of any individual mind is continually under attack from its own image-bearers. The plurality of Western culture and experience, and the method of reason itself, has called into question the four fundamental axioms. The past few centuries we've been championing the idea of progress, of scientific enlightenment, and of the acquisition of objective knowledge, and yet paradoxically, these very methods have led us to question our beliefs in the methods themselves.

Our fundamental axioms have been under siege since the beginning of World War I, where the march towards progress halted, and where we seemed to regress back to the murderous tribalism which characterized the pre-Enlightenment. Our belief that reality was made up of material such as atoms, molecules, and compounds, and that our thoughts and actions were simply the result of material biological processes like the firing of neurons, led us to wonder what the point of it all was. Our religious systems, the

---

[78] Bonevac, D. (2013). Postmodernism, *Philosophy Lecture at the University of Texas, Austin.*

meta-narratives of the past, had collapsed, and in the wake of a failed metaphysics of belief, we found ourselves awash in nihilism and despair. Or, as Friedrich Nietzsche famously declared in *The Gay Science*: "God is dead, and we have killed him."[79]

The death of God brought forth a collapse of fundamental belief systems in metaphysics and in its place the rise of totalitarian states of the 20th century. People put their faith in government, in ideologies, and in collectivist ideas of a grand utopia instead of religion. One could argue, however, it is not that they abandoned religion, but that they converted to something else: *ideology*.

Because our basic axioms of Truth, Knowledge, Reason, and Progress had collapsed and had been shown to be subjective and ultimately meaningless, a response to the rotting philosophy of modernism was needed. Postmodernism was this response, the zeitgeist of its time. Growing out of the writings of French intellectuals from the 1960s, postmodernism challenged the four axioms of modernism and the Enlightenment:

1) **Truth**: there are no absolute truths. Truth is subjective and relative. To say there is such a thing as truth is an untrue statement itself.
2) **Knowledge**: it is not possible to have any objective knowledge of truth, since truth does not exist.
3) **Reason**: reason is a rhetorical method, a system of rationalizations, which reinforces hierarchical structures. It is not a method to achieve knowledge.
4) **Progress**: there is no such thing as progress, as objective knowledge cannot be obtained. Progress, therefore, is subjective.

In these four categories, postmodernism can be defined as a philosophy of relativism, skepticism, and cynicism which rejects the concepts of truth, knowledge, reason, and historical progress. Epistemic

---

[79] Nietzsche, F. (1882). *The Gay Science*.

certainty and objective meaning are thrown out the window in exchange for an air of uncertainty and subjectivity in regards to every aspect of reality.

Postmodernist philosopher Jean-Francois Lyotard defines postmodernism as "incredulity towards meta-narratives," or in other words, a skepticism of *stories about stories*.[80] For example, the idea that reason can be used to gain objective knowledge and universal truths about the world is a meta-narrative, an overarching story about stories. Patterns, therefore, can be described inside the overarching pattern of reality. The method of reason is a meta-narrative describing the structure of the world and how to act within it. Reason tells us that we can obtain objective knowledge and that we can progress using this knowledge.

In this aspect, a meta-narrative describes a pattern structure of the world. It tells you how the world is made up and how to act and interact within it. A meta-narrative, therefore, could be anything from a philosophy of how to act to a comprehensive religious structure. Postmodernists view these narratives with suspicion.

Thus, postmodernism can also be defined as a *theory of rejecting theories*, a *judgment rejecting judgments*, or a *truth rejecting truths*. These are obvious paradoxes, and postmodernists are well aware of the contradictions. Because the structure of reality is ultimately non-objective, there is an indefinite number of structural interpretations to reality, and since objective knowledge cannot be ascertained, communication between groups of people is impossible. In the absence of objectivity, reason, evidence, and the ability to communicate with people of differing views, all we are left with is power--power over other structural interpretations of the world through aesthetic rhetoric. Reason itself, argue the postmodernists, is just one of multiple aesthetic techniques used to consolidate and reinforce the hierarchy of power in the West--the continual oppression of marginalized groups. This is where social constructionism uses the philosophy of postmodernism and its axioms for their critique and analysis of male and female differences.

---

[80] (2015). Postmodernism. *Stanford Encyclopedia of Philosophy*.

For example, the idea that 1) *sex is a biological reality* and 2) *male and female are the only sexes* is viewed as a meta-narrative through which a hierarchy of power is constructed. It is not that essential characteristics influence the structure of society, but it is rather the structure of society which influences essential characteristics. Existentialists such as Simone de Beauvoir and Jean-Paule Sartre would point to the famous statement "existence precludes essence." Social constructionists under the philosophical framework of postmodernism view essences such as sex differences as myths and believe that these myths bring about false essences:

> "Myth exists in a state of tension. It is not really describing a situation, but trying by means of this description to <u>bring about</u> what it declares to exist."[81]

When social constructionists apply these philosophical beliefs into sex and gender research, such that differences between individuals are constructions of a meta-narrative which reinforces hierarchies of power, statements such as this logically follow:

> "To see the world through patriarchal eyes is to believe that women and men are profoundly different in their basic natures, that hierarchy is the only alternative to chaos, and that men were made in the image of a masculine God with whom they enjoy a special relationship. It is to take as obvious the idea that there are two and only two distinct genders; that patriarchal heterosexuality is 'natural' and same-sex attraction is not; that because men neither bear nor breast-feed children, they cannot feel a compelling bodily connection to them; that on some level every women, whether heterosexual or lesbian, wants a 'real man' who knows how to 'take charge of things,' including her;

---

[81] Johnson, A. (1997). Patriarchy, The System. *The Gender Knot: Unraveling our Patriarchal Legacy*, 95.

that females can't be trusted, especially when they're menstruating or accusing men of sexual misconduct.

To embrace patriarchy is to believe that mothers should stay home and that fathers should work out of the home, regardless of men's and women's actual abilities or needs. It is to buy into the notion that women are weak and men are strong, that women and children need men to support and protect them, all in spite of the fact that in many ways men are not the physically stronger sex, that women perform a huge share of hard physical labor in many societies (often larger than men's), that women's physical endurance tends to be greater than men's over the long haul, that women tend to be more capable of enduring pain and emotional stress. And yet such evidence means little in the face of a patriarchal culture that dictates how things *ought* to be and, like all cultural mythology, will not be argued down by facts."[82]

In other words, if you believe that male and female exist objectively, and that these categories represent innate differences which present themselves in the world, then you're embracing patriarchy, you're embracing the oppression of women, and ultimately, you're embracing the hierarchy of power, the meta-narrative, which constructs the patriarchy itself. As a dynamic organism, this hierarchy uses rhetorical strategies to reinforce its legitimacy: truth, knowledge, reason, and progress are tools of the meta-narrative through which power is secured. There is no such thing as objectivity.

On the broad scale, the loss of an agreed-upon meta-narrative, or structural interpretation of the world, breaks the subject into heterogeneous moments of subjectivity which do not cohere to any identity, showing us that everything which exists is the subjective nature of perception. Because of this subjectivity, an aesthetic rhetoric, rather than objective reason, becomes the only effective way to play the game of society. If we cannot operate using reason, then rhetorical strategies based on

---

[82] Johnson, A. (1997). 95.

aesthetic and feeling are the only ways through which we can influence others, and they are also the only ways through which we can destroy the tyranny of the Western patriarchy and the category of biological sex. This is where the postmodernists' emphasis on language begins.

While language can categorize objects and real-world patterns into a structure, it cannot explain the nature of that structure, or in other words, what the makeup of reality actually is. Because of this, the postmodernists view language as not a descriptor of the world as it actually is but rather a tool through which perception can be shaped and re-shaped with the use of rhetoric. If all that exists is subjective perception, then all we have is rhetoric, not objective knowledge. All hierarchies, therefore, are solidified through the use of a specific type of rhetoric championed by the West: *reason*. Rhetoric, with tools of language operating through the process of reason, utilizes power and aesthetic to shape our perception. Here reason itself is the tool which constructs male and female differences and which maintains and perpetuates male's domination over women. As far as society is reason-based, hierarchies of power defined through a patriarchal lens will always follow.

For example, gender socialization uses rhetoric to construct an aesthetic (a myth, a meta-narrative) that males and females have innate differences. This rhetoric, whether it be through scientific papers, scientists, journalists, professors, or even parents and peers, helps solidify the patriarchal hierarchy as innate and biological.

There is some truth to this subjectivity of perception and the use of rhetoric to construct meta-narratives. The postmodernist claim that there are an indefinite number of possible interpretations to reality is true. There are, technically speaking, all kinds of ways we can perceive the world. However, while there is an infinite number of ways to interpret situations, texts, and material reality, there is only a finite set of effective interpretations. In other words, only a few answers work. For example, I can come up with a plurality of interpretations as to what might happen to me if I jump off the Brooklyn Bridge. And yet, my interpretations about my mentally assured survival will prove false the moment I hit the water and die. Or, let's say I'm flying a passenger jet and a warning light comes on in the cockpit. There are an infinite number of possible interpretations to the situation, and yet, choosing the correct interpretation may mean life or death for myself and my passengers.

Postmodernism is also partly true in its skepticism of meta-narratives. Meta-narratives are not a perfectly accurate description of reality's structure. Parts of them may be false. Parts of them may not fully describe reality at the highest resolution. They may neglect some aspects and completely reject others, even when these aspects of reality may be true. Meta-narratives, therefore, must be flexible if they have any chance of long-term success.

Where postmodernists get it wrong is the rejection of meta-narratives altogether. While there are falsities to meta-narratives, they also can be incredibly effective at describing the structure of reality. Patterns of the world may play out inside the meta-narrative. When acting upon the axioms of a meta-narrative, such as reason, we can see the idea's immediate effects. With the Enlightenment, reason proved to be a useful tool for operating and manipulating the world for humanity's technological, societal, and scientific benefit. Based on the incredible material prosperity arising from these beliefs, perhaps some of the Enlightenment's axioms, such as Truth, Knowledge, Reason, and Progress were true. And yet, despite this, both postmodernists and social constructionists reject this Enlightenment meta-narrative as not just untrue, but oppressive, hierarchical, and patriarchal.

Lyotard critiques the concept of reason using this logic:

> "Reason depends upon the unity of the subject for the identity of concepts as laws or rules de-legitimizes its juridical authority in the postmodern age. Instead, because we are faced with an irreducible plurality of judgments and 'phrase regimes,' the faculty of judgment itself is brought to the fore."[83]

---

[83] (2015). Postmodernism. *Stanford Encyclopedia of Philosophy.*

From this, the postmodern era can also be defined as a questioning of the legitimacy of judgment itself, because there is no objective standard, no meta-narrative, and no common sense of perception, revealing the nature of our being: a construction of language. Language as a subjective categorical tool which can create meta-narratives, instantiate myths in the mind of society, and construct a hierarchy based on ideas of 'innateness,' sees its ultimate application in the work of Jacques Derrida, a French postmodernist philosopher.

At the heart of Derrida's work is a special flavor of postmodernism known as *Deconstruction*. For Derrida, language is a written mark or signifier that represents function, not meaning. The word *female*, for instance, functions as a signifier which allows us to differentiate from its opposite: *male*. The erasure of female and male as categorically different, for example, is a symptom of the movement to erase the concept of differential opposites altogether. If we can eliminate the signifiers of *male* and *female,* then perhaps the 'constructed' differences between males and females can be dismantled, and with it, maybe oppression and injustice can be put to an end. Another example of differential opposites can be seen in the world *walk*. *Walk* can only be understood in the context of its opposite: *non-walking.* These words, absent of their differential opposites, do not have meanings on their own. The word woman is only defined through its differential opposite: man. The reverse is also true.

For the postmodernists, written marks or signifiers do not arrange themselves within natural limits, as the modern mind views them, but instead form chains of signification that radiate in all directions, creating infinite interpretations. Based on this logic, Derrida famously states, "There is no outside-text...Every referent, all reality, has the structure of a differential trace." In other words, the words male and female cannot be understood outside of each other. They are constructions which define each other; without one, you do not have the other. Thus, strong social constructionists would say that the signifiers *male* and *female* are not innate categories used to describe an objective world but rather linguistic power structures used to oppress and marginalize.

Derrida takes the idea of *no outside-text* to its logical conclusion with the concept that "a text is not a book, and does not, strictly speaking, have an author. It has a plurality of authors to infinity."[84] Deconstruction, therefore, is not a theory *about* texts but rather about the *practice* of reading and transforming texts, where tracing the movements of *difference* produces other texts interwoven with the first. This idea is seen in the contemporary humanities, where historical texts are not read through the lens of the author, but rather through the lens of differential trace: that words can have a multiplicity of meanings outside the author's original intent.

For example, Shakespeare's famous plays are not texts of great importance to the Western canon and the English language itself, but rather are texts which solidify the hierarchy of power of his time. His words, however innocuous they may seem, radiate themselves outwards to not just the contextual fabric of his era, but to the contemporary politics of modern times. That's why, as a white male who lived in a hierarchy of power, Shakespeare's work is viewed with skepticism and contempt in much of the contemporary humanities. Studying them for their potential merit may only help to perpetuate the patriarchy and colonialism of their time into the present.

Because of the plurality of perspectives and indefinite interpretation, the author may not know the underlying meanings of his own work. To *deconstruct* the text, therefore, means to trace the movements of difference, the radiation, to see how the work is connected to the fabric of contemporary society. And so this idea plays out in how the social sciences often view Western literature: if much of the Western canon was written by white males inside a patriarchal hierarchy of power, then their texts must be taken as suspect.

Derrida uses the concept of Deconstruction to critique reason itself. He identifies the West's meta-narrative as *Logocentric*. *Logos-* meaning 'logic' and *-centric* meaning 'center of,' Derrida rightly labels the axioms of the Enlightenment as logocentric. The West sees the ability of logic and

---

[84] (2015). Postmodernism. *Stanford Encyclopedia of Philosophy.*

reason to transform chaos into order as the central axiom of its civilization. It's the capstone of the pyramid, the highest value in the hierarchy of values. To engage in logical discourse is to operate on the highest value of Western culture. It is no mistake that Derrida uses the word *logos* in his critique, pointing to another central meta-narrative in the West: that the word became flesh and brought order from chaos, the God-centric value of the Enlightenment's metaphysics. This is the core of Derrida's critique.

The Western meta-narrative of Logocentrism is the foundation from which the tyranny of the patriarchy is based; reason is its rhetorical device to reinforce this tyranny. Reason presupposes the objectivity of the world, but takes shape as the confluence of knowledge and power, or as postmodernist philosopher Michel Foucault notes, "Reason is a power that defines itself against an other, a faculty seen operating outside of reason is not allowed to speak for itself and is at the disposal of a power that dictates the terms of their relationship."[85]

Reason, in other words, is a tool through which marginalization and discrimination occurs. Or put another way: reason is a tool by means of which certain empowered groups retain their hegemony, oppressing other groups. Because reason is used to oppress, the emotions and experiences of such marginalized groups are to be valued over rational argument.[86]

In the social constructionist view, the idea of reason is synonymous with the concept of hegemonic masculinity and the hierarchy it produces by relegating females to Other.[87] Therefore, this oppression must be fought by exposing the categories and meta-narratives by which the empowered retain hegemony, valuing instead authenticity.[88] In other words, we must dismantle the linguistic categories of male and female and turn to our authentic inner feelings to destabilize the power structure; such emphasis on inner feelings and authenticity can be seen among social constructionist activists and gender theorists who believe the only way to end oppression is to eliminate the male-female category.

---

[85] (2015). Postmodernism. *Stanford Encyclopedia of Philosophy.*
[86] Bonevac, D. (2013). Postmodernism. *University of Texas, Austin.*
[87] (2018). Simone de Beauvoir. *Stanford Encyclopedia of Philosophy.*
[88] Bonevac, D. (2013).

But how exactly can the oppressive nature of reason, and the patriarchy itself, be subverted? The answer is through what French sociologist Jean Baudrillard describes as the introduction of something outside the system of reason:

> "Direct opposition within the system of communication and exchange only reproduces the mechanisms of the system itself. Strategically, capital can only be defeated by introducing something inexchangeable into the symbolic order, that is, something having the irreversible function of *natural death*."[89]

Baudrillard calls this irreversible function *fatal strategies*, or strategies used to make the hierarchical system suffer *reversal* and *collapse*. A prime example of this is in the actions of graffiti artists who "experiment with symbolic markings and codes in order to suggest communication while blocking it, and who sign their pseudonyms instead of recognizable names."[90] Or, in Baudrillard's words:

> "They are seeking not to escape the combinatory in order to regain an identity, but to turn an indeterminacy against the system, to turn indeterminacy into *extermination*."[91]

Fatal strategies are seen among ideologues who utilize non-communicatory techniques to subvert the symbolic order of society and its hierarchies. Banning speakers you don't like from your college campuses, shouting down points of dissension, and even assaulting someone you disagree with are forms of fatal strategies meant to ultimately destabilize the system. If truth is subjective, then communication between groups cannot occur. Thus, for a radical activist to engage in reason is to use the very mechanism which keeps the tyranny of the patriarchy alive. What's

---

[89] (2015). Postmodernism. *Stanford Encyclopedia of Philosophy*.
[90] Ibid.
[91] Baudrillard, J. (1976). *Symbolic Exchange and Death*, Ian Hamilton Grant (trans.), London: Sage Publications, 1993.

left, however, is rhetorical devices of power, feeling, and indeterminacy. If you use strategies and operational tools such as these, tools which are alien to the modernist system, then perhaps our patriarchal society can be destabilized for its eventual extermination and restructuring.

If you think this is a straw-man of the postmodern argument, think again. Italian postmodernist Gianni Vattimo says that the ideas of modernity must be overcome:

> "What is to be overcome is modernity, characterized by the image that philosophy and science are progressive developments in which thought and knowledge increasingly appropriate their own origins and foundations. Overcoming modernity, however, cannot mean progression into a new historical phase."[92]

When describing what he means by *overcoming modernity*, Vattimo characterizes it as a *twisting or distorting* of modernity itself, rather than a progression beyond it, further expressing that modernity's innate tendencies must be radicalized for it to be dissolved. While he agrees that historical change will continue after the collapse of modernity, he clarifies that only *local histories* are possible. The idea that all of humanity is progressing into a universalism is, in his view, false:

> "We no longer experience a strong sense of teleology in world events, but, instead, we are confronted with a manifold of differences and partial teleologies that can only be judged aesthetically."[93]

In other words, progress can only be judged through the mind of the subject. There is no objective knowledge about what progress is or how it can be developed. The end of modernity laid out by Vattimo shows an inherent flaw in modernity itself: that it results in nihilism. Truth,

---

[92] (2015). Postmodernism. *Stanford Encyclopedia of Philosophy*.
[93] Ibid.

knowledge, reason, and progress collapse under critical examination. In other words, without a strong meta-narrative and underlying metaphysics, modernism reasons its very beliefs out of existence.

Therefore, knowing the theories of indefinite interpretation, deconstruction, and the application of fatal strategies, the postmodern experience can be best realized in aesthetic rhetoric, not reason. For example, when asked whether he considers himself a postmodernist, French sociologist Baudrillard said, "I have nothing to do with postmodernism." Yet such a statement uses the same strategic intent as the graffiti artist of indeterminacy. Insofar the label "postmodernism" has become a theory in and of itself, Baudrillard would want nothing to do with it. In this aspect, it is not entirely clear that postmodernists' externally expressed goals match their internal reality. Because of their rejection of theories, judgment, and truth, the words of the postmodernists should be viewed with great skepticism.

Using the postmodernist framework of subversion, fatal strategies, and rhetoric, strong social constructionists working under the Butler tradition characterize Western society as a hierarchical patriarchy of power. They claim this hierarchy makes women subservient through categorical language use, and that restructuring of language is necessary to dismantle the categories of male and female. If words are just tools of power, not descriptors of an objective reality, then those same words can be molded to restructure the tyranny of the patriarchy. Such is the ultimate goal of erasing the category of male and female.

If the function of male and female can be changed conceptually, integrating an infinite number of potential identities into the binary category, then perhaps our society can finally escape the injustice of gender inequality. Ultimately, this doctrine can only be achieved through equity, or equality of outcome, as I will discuss in future chapters.[94] Inside the postmodernist framework, *equity* is the logical application of social constructionist ideology and the final solution to all discrimination and disparities between men and women.

---

[94] See chapters, "Discrimination and Disparities" and "Interests and Inequality."

PARADOXES OF POSTMODERNISM

To say that postmodernism has contradictions is an understatement. It is *filled* with paradoxes which undermine its credibility as a reliable philosophical framework. For instance, postmodernists Derrida and Foucault commit a *performative contradiction* in their critiques of modernism by "employing concepts and methods only modernism can provide."[95] Or, in other words, they reject the concept of critique while performing the very act of critique itself, a concept which lies at the heart of modernism's axioms.

Furthermore, postmodernists claim that all truth is relative, while claiming their philosophy *tells it like it really is*.[96] They reject meta-narratives, while constructing a meta-narrative of their own. They claim all cultures are equal while they condemn the West as oppressive and destructive. They claim values are subjective, and yet simultaneously claim that sexism, racism, and inequality are *truly* evil. They join Rousseau in agreeing that technology is inherently corrosive to society, and yet claim that it is unfair some people have more technology than others. They claim tolerance is good and dominance is bad, and yet, when postmodernists gain political power, the ideology of political correctness follows. In the words of philosopher Stephen Hicks, these paradoxes of postmodernism can be consolidated into one descriptive statement: "Subjectivism and relativism in one breath; dogmatic absolutism in the next."[97]

The postmodernists are well aware of the contradictions inside their philosophy, and yet, like sociologist Baudrillard, they don't care. It is unlikely most postmodernists truly believe in the paradoxical axioms of their philosophy. What's more likely, however, is that the use of these apparent paradoxes is symbolic of their favorite form of discourse: rhetoric.

It is in the practice of rhetoric where postmodernism resides. Claims such as *all truth is relative* are not statements of pragmatic belief but, in

---

[95] (2015). Postmodernism. *Stanford Encyclopedia of Philosophy.*
[96] Hicks, S. (2004). *Explaining Postmodernism: Skepticism and Socialism from Rousseau to Foucault.* Scholargy Publishing, Inc.
[97] Ibid.

their own words, *rhetorical devices* to destabilize the legitimacy of the present system. Rhetoric, therefore, is utilized as a tool to restructure society through aesthetic appeals. If reason is simply a rhetorical tool to consolidate power, then perhaps a more aesthetic device (something which *feels* authentic) can be used to overthrow the power of reason.

Drawing from the postmodernists, the social constructionist utilizes narrative, not evidence, to argue their points. It doesn't sound so good to say that some societal disparities between males and females are partially the result of biological tendencies, or that male and female exist objectively and that we can reliably sort people into categories. Saying Western society is a patriarchal hierarchy which constructs sex differences as innate sounds much better. That way, if there is such a disparity present, the inequality can be judged as an injustice or a fundamental problem with society. And in that sense, the social constructionist has a lot of rhetorical power. In the social constructionist view, there is no room for the possibility that a specific disparity may be the result of an individual's free choice. Individual interest and freedom of choice is not as important as *equity across all hierarchies*. If equity is the goal, then individual choice must be controlled from the top-down. Ultimately, the postmodernist framework is a fascist way of thinking at its core: we cannot use reason to settle differences and groups of people cannot talk to each other if reason is a tool of oppression, leaving us with only one thing, or as Nietzsche famously stated: "Where there is no truth, there is only *power*."

It is in the final paradox where postmodernism aligns with social constructionism: in the attempt to restructure the supposed corrupting Enlightenment axiom of the sovereignty of the individual, social constructionism actually reinforces it through the proliferation of near infinite individual identities and a postmodern 'plurality of perspectives,' almost limitless in each individual's assertion of 'their truth,' which must be taken as objectively real through the mind of the subject. Taken to their logical end, both philosophies find themselves surrounded by an indefinite number of independent and fragmented identities, and in an ironic twist of the social constructionist theory, the individual becomes the ultimate identity.

THE TRUTH ABOUT HIERARCHIES

For the postmodern social constructionist, all hierarchies are based in power. This belief, however, is like looking at a vein in a leaf and claiming that the vein *itself* is the leaf. In reality, the leaf is comprised of thousands of veins, so small and complex, all working together to produce the structure. A hierarchy is too like the leaf. Hierarchies are comprised of a near infinite set of variables, thousands of veins, which construct them and support them. Power is only one of these variables. In some hierarchies, power is more important than other variables, while in other hierarchies, power is a negligible variable. In Enlightenment societies based on the axioms of Truth, Knowledge, Reason, and Progress, power is only a negligible variable inside its most fundamental hierarchies.

The first truth about hierarchies is this: hierarchy in the West is based mostly on competence, on the idea that whoever does the best and most efficient job in their field will rise to the top, or, in other words, be rewarded. The hierarchy of competence is not a hierarchy of tyranny, but rather a hierarchy of merit, skill, determination, and some luck for sure. Individual interests, preferences, and values largely determine how the contemporary West structures itself.

This is not to say that certain hierarchies within the system are not corrupt. But when they do become corrupt, such as relying solely on power, they became rigid, stale, and inefficient. Revolution, figuratively and literally, becomes a prime avenue for change. For example, a business based solely on a hierarchy of power, of subordination and domination, will eventually collapse under its own ideological rigidity. If it isn't willing to hire the people who have the most competence, who speak truth, and who work honestly, then its structure will eventually disintegrate.

This is the second truth about hierarchies: they are transient. Hierarchies are in constant motion, moving people up and down their strata and even into different hierarchies. The 1% income bracket in the United States, for instance, is highly fluid. A statistic from the Panel Study on Income Dynamics highlights this basic fact about hierarchies:

> "By age 60, 70% of the population will have experienced at least one year within the top 20th percentile of income; 53% of the population will have experienced at least one year within the top 10th percentile of income; and 11.1% of the population will have found themselves in the much-maligned 1% of earners for at least one year of their lives."

Therefore, the idea that all hierarchies are 1) comprised solely of power and 2) remain in a fixed state indefinitely is anathema to the data. So the question is not "*should hierarchies be dismantled?*" but rather "*what are they comprised of, how do they change,* and *which ones are good and which ones are bad?*" That is a more accurate line of questioning.

The gender hierarchy of the social constructionists, too, has been in constant flux. Statistics on gender inequality are constantly changing across the world, especially in Enlightenment societies. People aren't just becoming more prosperous through free market economies and free trade, but their societies are also becoming more gender egalitarian. The patriarchal hierarchy described by the social constructionists seems to be a symptom of the past at best, and a complete fabrication at worst. Power, as it turns out, is not the driving force.

The truth is individuals can be categorized in a near infinite number of ways. Each person has unique traits, interests, preferences, desires, and even genetics, epigenetics, personality, neurology, and family history. All of these variables come together to form the amazing biological and social creation known as *you*. The idea that power is the sole driver of inequality among individuals in the most gender egalitarian societies is asinine.

Here social constructionists run into a major break in their chain of logic. They claim concepts such as the male-female category are social constructs which form a hierarchy of power. The problem, however, is in the existence of the objects themselves. While a concept may be a social construct, the object it's describing certainly can't be! For instance, the *concept* of a tornado is a social construct, but tornadoes themselves cannot be social constructs. When you see a tornado barreling towards you, wishing it away as a "construct of society" won't do much to change its damage path.

The same goes with male and female. The existence of reproductive capacities which allow for procreation is not a social construct, *but the words themselves are.* And here the social constructionist has hope, for there is always language. If the concept of male and female can be dismantled through language, then the utopia of perfect gender equity can finally be realized. *Language and power, that's all there is.*

The postmodern world is full of paradoxes. Postmodernists claim that all truth is relative, that differences between males and females are constructs of society, that all hierarchies are based in power, that the West is a sexist and racist patriarchy, that reason is a tool of oppression, that the male-female category should be abolished, and that modernity should be exterminated.

Postmodernists are ideologues who believe power is the only game in town. They wish to restructure society along equitable lines, flatten out all hierarchies, use the power of language to reshape perception, and ultimately, destroy the concept of male and female itself. They claim to have all the answers for the oppressed, the marginalized, and the downtrodden, and yet they use *fatal strategies* to destabilize the very system which has brought the individuals they claim to be fighting for out of poverty.

Like the Marxists of the past, the postmodernists are no liberators of the oppressed, no comfort for the marginalized, and no great saviors of the downtrodden. Where their ideas go, destruction and chaos follow. In their contempt and resentment of the prosperity the Enlightenment brought each one of us, they wish to tear it all down--to twist it, distort it, and ultimately, exterminate it. They don't love the prosperity afforded to them from their ancestors of the Enlightenment; they hate it. Drowning in their nihilism as they cling to the hope of a great utopia, they find themselves alone in a society which has long forgotten their ideas even existed. They do not have reason, logic, or evidence to save them. All they have is words.

It is only through language where the postmodern social constructionist has power, where they can craft the aesthetic of reality they want *you* to believe in. *Do not give them that power.*

4

DISCRIMINATION AND DISPARITIES

Despite the postmodernist claim that all of history can be compressed into hierarchies of power, the truth is most of human history was focused on one thing and one thing only: *survival*. Men and women both worked together by harvesting the land and taking care of children as they kept the wolves and marauders outside the gates at bay. Famines, infant death, and disease were not just problems, but the common realities of life. And slavery, one of the great evils of history, was practiced by every society. Because of these things, civilization was often not much better than living alone in the wilderness. There was no such thing as leisure, and certainly no such thing as time to think about how *oppressed* one was. If you happened to not be oppressed, you were among the luckiest men and women in all of history.

While most peasants and commoners worked similar jobs, men and women were still divided in some major areas. Men could be sent off to a war at a moment's notice, while the absence of the birth control pill for women meant their sexuality was a major controlling factor of their lives. When they became pregnant, their commitments forever changed.

The difference in these areas can be explained through the most important biological fact: women have to bear children; men do not. Big gametes such as eggs are produced much less often than small gametes such as sperm. The nature of large gametes makes them more valuable, and therefore, they are the controlling factor for most of the sexual selection in mammals. Knowing this, it makes sense why women are often the sexually selective force in reproduction, not men.

The Industrial Revolution further segregated men and women's jobs, as many people moved away from farmland and into the cities. Men became the primary factory workers and bosses, while women often remained at home. As the revolution progressed, more women entered the workforce, especially in factories, where conditions were beyond imaginable for men and women both.

As technology progressed, the Enlightenment axioms began to take hold. The Declaration's statement that *all men are created equal* began to make people question the status quo. If all men *are* created equal, then women and minorities are just as deserving of rights. Soon women saw their rights expand into all spheres of life such as voting and property.

The watershed moment came when the birth control pill was invented and distributed in the 1960s. It is almost impossible to fully understand just how revolutionary such an invention would be. For the first time in human history, women's sexuality was not controlled by the possibility of pregnancy. This allowed them to push further into the different strata of society and enter all types of professions. Suddenly the average woman could focus on a career and personal interests rather than taking care of a child. The full effects of the birth control pill are still beyond understanding across evolutionary and socialization theories.

However, neither the birth control pill nor voting rights suddenly liberated women from their historically unequal status. Gender socialization and stereotypes continued, producing discrimination and disparities of all kinds across the strata of society. Yet with the explosion of gender-critical feminism into the mainstream culture, gender inequality began to decrease as more and more women moved into the workforce in the 1970s, 80s, and 90s.

While the average person may look at these statistics and be encouraged by the increasing number of women participating in education, business, and politics, a major problem arises for the social constructionist: *the statistics are still not equal!* Despite decades of programs, activism, and policies constructed to reach equitable amounts of men and women across all fields of society, the statistics on gender inequality have not panned out as the social constructionists expected.

Enlightenment nations, more than ever in history, are some of the most gender *diverse* countries on the planet, not through law but through individual choice. It seems as though, with our civilization's immense material prosperity and gender egalitarian policies, that men and women would become *more the same* in their interests, preferences, and choices, not less. And yet in a twist of the postmodern social constructionist framework, the exact *opposite* has happened: **the more gender egalitarian a country is, the more gender differences there will be.**[98] This is the *Gender Equality Paradox I discussed in the introduction.* Aren't gender differences supposed to decrease as societies become more gender-equal?

Here social constructionists run into perhaps the single greatest defeat their theory has ever witnessed: individuals are not exercising their equality in predicted ways. Or, in other words, in ways the social constructionists *wanted*. In one of the greatest discoveries of the social sciences, researchers found that *if you minimize sociocultural effects on people's choice, you allow for innate differences to maximize.* This discovery reveals the biggest flaw in social constructionist theory: *people are not blank slates.*

The cynic, however, views these gender differences with a high degree of suspicion: perhaps the system is *still* sexist, racist, and oppressive. Here the free choice of individuals is not accounted for, and any inequality which doesn't match with the ideologue's framework must be viewed as suspect. Writing in the 1990s, social constructionist Allan Johnson has one word to describe this gender inequality: *oppression*.

> "Since gender oppression is, by definition, a system of inequality organized around gender categories, we can no more avoid being involved in it than we can avoid being male or female. All men and all women are therefore involved in this oppressive

---

[98] See chapter, "Solving the Gender Paradox."

system, and none of us can control whether we participate, only how…"[99]

If we cannot avoid being involved in this oppressive system, then the male-female category must be eliminated. Johnson's cynicism about the system reveals itself in how he views inequality. To sensible minds, the presence of inequality does not always mean injustice, while at the same time, some inequality *can* be indicative of a broken system.

For the social constructionist, however, there is no room for freedom of human action in a system they define as oppressive, sexist, and patriarchal. And why would there be? For ideologues who wish to see *equity across all hierarchies* with no differences between groups of people, it makes perfect sense why social constructionists continue to blame the system for all inequality. Again, this is where their power resides: if they can draw on past injustices and argue these injustices continue into the present, they have a lot of rhetorical power. It is here where the social constructionist crafts their most effective arguments, using statistics on gender inequality to reveal the effects of the male-female category.

GENDER INEQUALITY IN LIBERAL DEMOCRACIES

Data on gender inequality makes up some of the most well-documented group statistical analyses we have. From workforce participation, income earnings, and representations in the different strata of society, to the distribution of college bachelors, masters, and doctorates, the data is clear: men and women are not equal across multiple levels of analysis. While they are more evenly distributed in certain areas, many aspects of society remain gender unequal. Why is this? Social constructionists have the answer: discrimination and socialization. Before we analyze the causes for gender inequality in liberal democracies, let's look at the data.

---

[99] Johnson, A. (1997). Patriarchy, The System. *The Gender Knot: Unraveling our Patriarchal Legacy*, 96.

Economically, women continue to be underrepresented in high-level, highly paid positions and are overrepresented in low-paying jobs.[100] Inequality.org and other activist groups claim this is due to gender discrimination and sexual harassment, not individual interest and choice.

> "The global trend towards extreme wealth and income concentration has dramatically strengthened the economic and political power of those individuals, overwhelmingly male, at the top...Gender discrimination and sexual harassment in the workplace contribute significantly to these persistent economic divides."[101]

This statistic of underrepresentation is illuminated through the wage gap. As of 2017, women earn about 82 cents for every dollar a man earns. This is called the female-to-male earnings ratio, which is calculated by taking women's median full-time annual earnings and comparing it to men's.

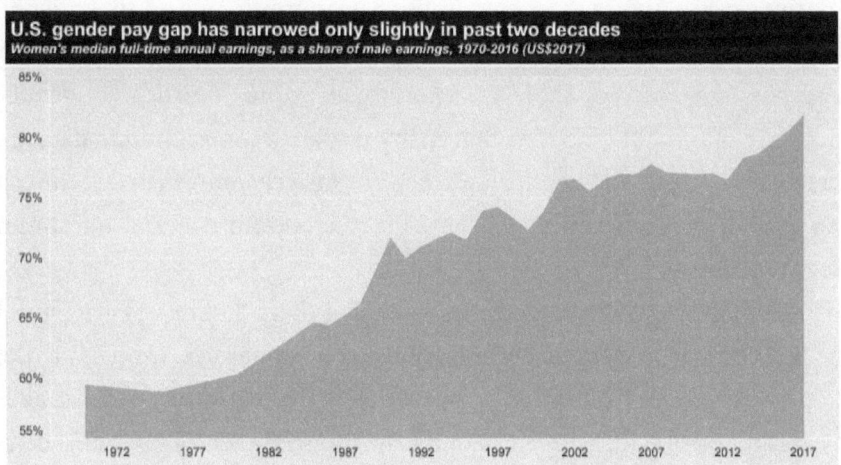

U.S. gender pay gap has narrowed only slightly in the past two decades. Source: Institute for Women's Policy Research.

---

[100] (2019). Gender Economic Inequality. *Inequality.org*.
[101] Ibid.

Social constructionists argue this is due to discrimination, that employers pay their women workers less for doing the same work as their men workers. The problem, however, is that the wage gap considers only one variable: *gender*. This low-resolution, univariate analysis is used to argue that women are unfairly paid and undervalued by their patriarchal bosses. And yet, this argument falls apart as we analyze the wage gap using a high-resolution, multivariate model. If we take into account the type of job, hours worked, time taken off, overtime, and a handful of other meaningful variables, we find that the wage gap is reduced to only a few percentage points of the near 20 percent gap.[102] It turns out that most employers are not paying women less for the same job. Rather, it is generally the women who decide they want to work less, work in a specific type of job they're interested in, take time off to care for their children, or not work full time. Most of these reasons relate to an individual's choice, not discrimination.

In fact, when we analyze the full-time salaries of young women in 147 of the 150 largest cities in the United States, we find that these women's salaries are 8% higher than men in their peer group.[103] And so, knowing this, the social constructionist should be met with a revelation: there's no true *pay* gap; instead, there's a *choice* gap. More statistics on gender inequality illuminate the idea that many disparities in economic data are due to life choices. For example, women are scarcely represented at the top strata while being heavily represented in the bottom strata of minimum wage workers.

The social constructionist will argue that these differences are caused by two things: 1) gender socialization, which relegates women to the domestic role, or 2) outright discrimination and harassment, which makes it near impossible for women to enter the top strata. Our biopsychosocial model proposes, instead, that much of this disparity relates to women's life choices, interest, and priorities. In other words, most women are not

---

[102] Horwitz, S. (2017). Truth and Myth on the Gender Pay Gap. *Foundation for Economic Education.*
[103] Heritage Foundation. (2019). Why There is No Gender Wage Gap. *YouTube.*

interested in working 70, 80, or 90 hours a week to be at the highest economic strata such as those in the Fortune 500. To their credit, women tend to have a better work-life balance than men. This is not solely due to sexism, but is rather due to women having different priorities when it comes to life and work. And this idea remains true across cultures, as the disparity in the upper economic strata of income earners remains mostly dominated by men. In most liberal democracies, women make up only about 20% of the top 1% income group.

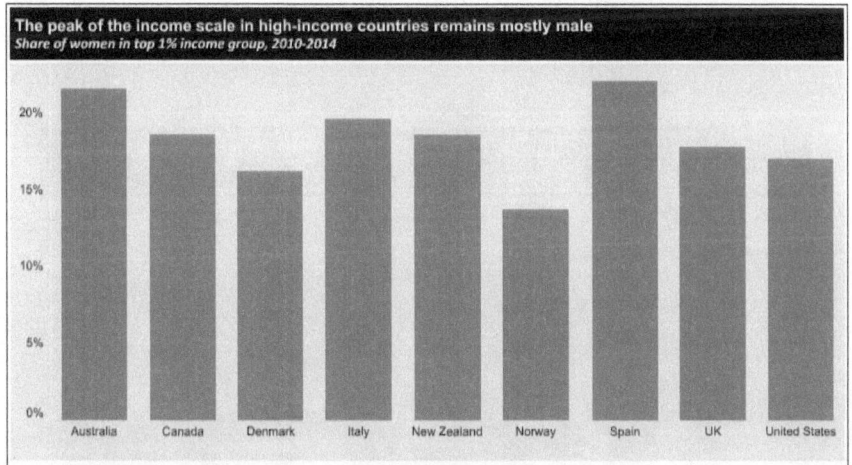

The peak of the income scale in high-income countries remains mostly male. Source: National Bureau of Economic Research.

The gap in pay and income earnings is not relegated to the United States. Other gender egalitarian countries, such as South Korea, have larger wage gaps, with men earning 37 percent more than women, on average.[104] Comparatively, small nation states such as Luxemburg have almost no gap, with men earning 3.4 percent more than women on average. Women also tend to do more unpaid work such as caring for children and the elderly, doing housework, or other people-oriented jobs. Thus, in the most gender-equal societies, gender differences in high income earnings are largely the result of women's own choices when it comes to life and work.

The second analysis of gender disparity in economics relates to the distribution of wealth, or the sum of one's assets minus debts.

---

[104] (2019). Gender Economic Inequality. *Inequality.org*.

"In 2018, only 256 women ranked among the world's 2,208 billionaires. 77 of those women hail from the United States, more than double the number in any other country."[105]

Again, this economic disparity is not evidence of discrimination, injustice, or oppression. Rather, with an understanding that women tend to prioritize their work-life balance more and take more time off than men, this statistic is not surprising.

When it comes to debt, women share the greater burden of student loan debt. This also is not surprising, as women make up the majority of college graduates.

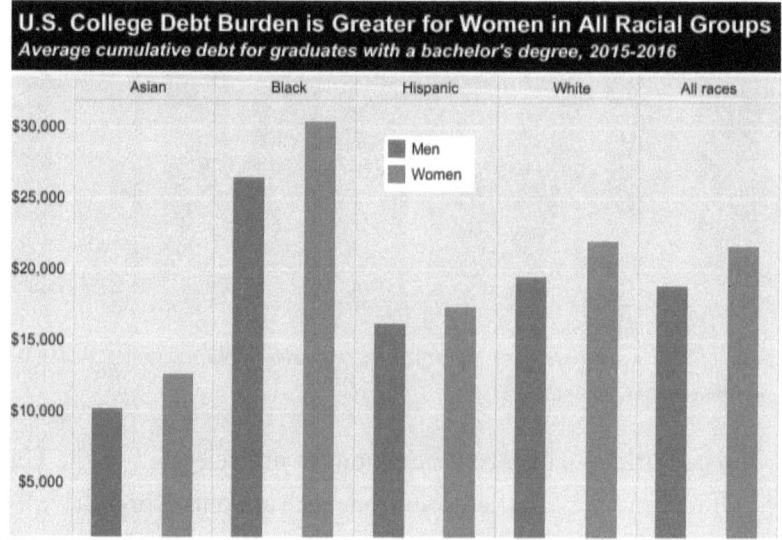

U.S. college debt burden is greater for women in all racial groups. Source: American Association of University Women.

For those not up-to-date on the statistics of gender inequality, it may come as a surprise that women, not men, comprise the majority of college students. According to the U.S. Department of Education, women comprise more than 56 percent of students on campuses nationwide. By 2026, the figure will rise just one percentage point to 57 percent. In the 1970s, this ratio was reversed.[106]

---

[105] (2019). Gender Economic Inequality. *Inequality.org*.
[106] Marcus, J. (2017). Why Men are the New College Minority. *The Atlantic*.

When it comes to bachelor's degrees, women are the majority degree holders in some fields and the minority in others. For example, the percent of bachelor's degrees earned by women in psychology, the social sciences, biosciences, and the physical sciences has been steadily on the rise since 1991. The bachelor's degrees in psychology, the social sciences, and biosciences are all majority women. Percent of bachelor's degrees earned by women in computer sciences and mathematics have declined slightly since 1991. Men are the majority in the computer sciences and engineering, and hold only a slight majority in mathematics.

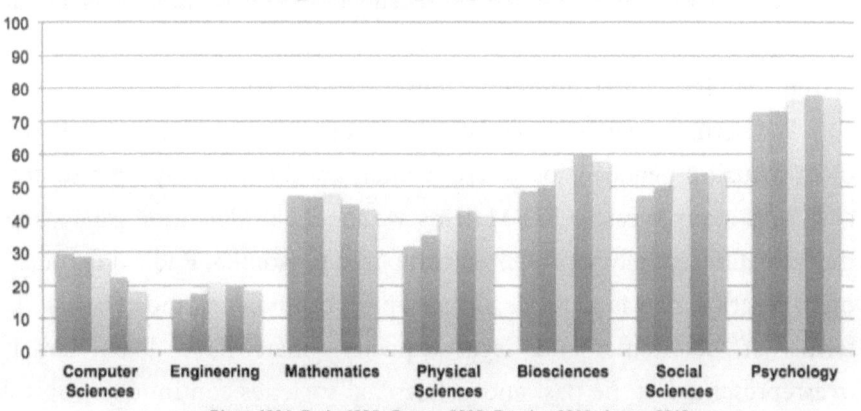

Percent of bachelor's degrees earned by women, from 1991-2010, left to right. Source: National Science Foundation, National Center for Education Statistics.

These disparities in bachelor's degrees may explain a very simple fact: that men and women, on average, have differing interests when it comes to their career preferences.[107] Data from the United States Department of Labor illustrates this fact of different career choices. Dental hygienists, for example, are almost exclusively women (98%), while brick and cement masons are *exclusively* male (near 0% women). Women constitute large majorities as therapists, licensed and vocational nurses, registered nurses,

---

[107] Su, R., & Rounds, J., et al. (2009). Men and things, women and people: a meta-analysis of sex differences in in interests. *Psychological bulletin, 135(6),* 859-884.

elementary school teachers, and mental health counselors. Men constitute large majorities as carpenters, mechanics, truck drivers, software developers, and sales representatives. The distribution of men and women is more equal in physician assistants, accountants, marketing specialists, medical scientists, postsecondary teachers, database administrators, and physicians.

All of this is not to say that discrimination or injustice does not occur. It would be an absurdity to claim that our society is perfect or free of immorality. But claiming, as the social constructionists do, that all inequality is the result of discrimination is a dangerous presumption. It assumes that all people's choices, especially ones which differ, must be the result of socialization, discrimination, and injustice. If such an assumption is true, then individual differences must be equalized.

Furthermore, while social constructionists and gender theorists will often focus on gender gaps where women are differentially represented, such as in many Western STEM fields, we should also mention gaps where men are differentially represented compared to women. Radical feminists often describe men as society's ultimate privilege-holders, and of course, in terms of economic and legal equality, men have been largely overrepresented in the past and patriarchal societies continue to oppress women across the world. While all of this is true, women share many advantages over men in Western societies, and it would be unfair to only focus on women's inequality with men; to gain a full understanding of gender inequality in the West, we must also look at men's inequality compared to women. To do this, I am going to present a table of statistics produced by Thomas Mortenson, a senior scholar at the Pell Institute for the Study of Opportunity in Higher Education. Mortenson, an independent higher education policy analyst, compiled and published a large list called *"For Every 100 Girls..."* which elucidates gender disparities between girls and boys in education, disability, and many other dimensions. Such a table allows us to see many areas where boys and men are at a disadvantage compared to their female counterparts. This is not to say that women do not have disadvantages or do not face discrimination, but we must explore the disadvantages of boys and men as well.

FOR EVERY 100 GIRLS...

Mortenson's Gender Inequality Table (United States).[108]

For every 100 girls/women...	There are this many boys/men...
Who take AP/Honors courses in Art/Music	54
Who earn an associate's degree	63
Who take AP/Honors courses in English/Language arts	64
Enrolled in US graduate schools	73
25 to 29 years who have at least a master's degree	73
Who earn a bachelor's degree	74
Who earn a master's degree	74
Enrolled in US colleges	77
Who are in the top 10% of their high school class	79
Who take AP/Honors courses in Natural Sciences	79
25 to 29 years who have at least an associate's degree	80
25 to 29 years who have a first professional or doctor's degree	81
Who take AP/Honors courses in math	82
25 to 29 years who have at least a bachelor's degree	85
Who take the SAT test	89

---

[108] Mortenson, T. (2011). For every 100 girls. *Pell Institute for the Study of Opportunity in Higher Education.*

Who earn a doctor's degree	90
25 to 29 years who have at least some college but no degree	94
Whose entry into kindergarten is delayed	139
Who repeat kindergarten	145
Who are homeless	154
Ages 3-17 years diagnosed with communication disorders	168
Who abuse illicit drugs and alcohol	180
Who have problems with alcoholism	200
In K-12 and classified as having a specific learning disability	207
Who die by opioid overdose	212
25 to 34 years old who die	232
4-17 years diagnosed with ADHD	237
Suspended from school	240
Who are homeless and unsheltered	242
15 to 24 years who die	280
Expelled from schools	291
Ages 15 to 19 who commit suicide	293
Who receive services in public schools for autism	300
In public schools classified as having emotional disturbance	355

Ages 20 to 29 who commit suicide	450
Ages 20-29 who die of homicide	648
Under age 18 who are in a correctional facility	770
In adult correctional facilities	1,000
Who die on the job	1,294
In a federal prison	1,333

From Mortenson's table, men are overrepresented for homicide, suicide, diagnoses of ADHD, problems with alcoholism, suspension from school, and expulsion from school; they are ten times more likely than girls to die on the job, to be in adult correctional facilities, and to be in a federal prison. Everyone, no matter their beliefs, should be willing to explore men's disadvantages just as much as they are willing to explore and fight against women's disadvantages. It is only when we begin treating men and women equally, putting as much emphasis on men's needs as women's needs and women's needs as men's, that we will see some of these disparities minimize. Other disparities, such as a tendency towards physical aggression and low Agreeableness in men, may be more difficult to reduce because of the influence of more biological and psychological factors. For instance, high physical aggression and low Agreeableness on the extreme end for men tends to predict the probability that one will be in imprisoned. Regardless, these disparities must be separately analyzed to see how biology, psychology, and society affect such differences, and whether such differences can be reduced through socialization.

As we analyze these gender gaps between men and women, we should champion freedom of choice. While some disparities may be due to socialization, discrimination, or injustice, many other disparities are also the result of the free choices and innate interests of individuals. Many women share different preferences for work-life balance than men. Perhaps these differences can be changed through a shift in how our culture views men's and women's roles, and perhaps we can minimize gender disparities between men and women through more socialization. However, while eliminating gender disparities can be possible, such a practice depends on which disparities need to be eliminated and which ones do not, because some gender differences are the result of free choice, while others are the result of discrimination. Understanding the difference requires we look at both sociocultural and biological factors for each disparity. Postmodern social constructionists, however, will only focus on the former (sociocultural).

THE LIE OF EQUITY

If the social constructionist theory is correct, then gender disparities such as these are the result of discrimination, injustice, and ultimately, socialization. From this disparity, the social constructionist concludes that equity policies must be enacted so that an equal distribution of men and women in each field can be achieved. How can this be done? Through individual choice? Certainly not. Preferential hiring quotas and equity policies must be instituted for such a disparity to be erased. There is no room for individual freedom of choice in this view, except for the freedom of the social constructionist to dictate that people's choices fit within their postmodernist framework of gender equity. Knowing that equity policy can only be enforced at the expense of individual freedom of choice, it is appalling that such a belief can be seen as something worthy of pursuit.

Even though economic opportunity, economic mobility, and freedom of choice are very high in our liberal democracies, disparities such as these continue to exist between men and women. And yet social constructionists continue to claim this is because of discrimination or the *patriarchy*, without supplying much evidence other than showing that the disparity exists. For decades now, quotas in hiring have been instituted to achieve equal representation of women in certain fields, especially STEM, with varying degrees of success. For example, the percent of computer science bachelor's degrees held by women fell from 30% in 1991 to 19% in 2010.[109]

---

[109] Source: *National Science Foundation* and *National Center for Education Statistics*.

Has sexism in that field increased since then? Or could these disparities simply be a product of men's and women's choices, interests, and preferences?

What if equal representation in *every* field (50% men and 50% women) is unreasonable or even impossible if liberal democracies give people the freedom to make their own choices? For the social constructionist, however, nothing is impossible. Equity *can* be achieved. It's just a matter of *how*. How can we mold individuals, socialize them, and re-educate them, so that their choices fit within our ideology? The logic follows that if only we could remove society's stereotypes, its injustices, and its discrimination, then we could achieve perfect gender equity. The inequities of the past, present, and future would disappear, and the utopia would arrive.

It is not a coincidence that strong social constructionists utilize Marxist concepts inside their philosophical framework: *if only we could redistribute the wealth, take from the bourgeois and give to the proletariat, then we could achieve equality.* Despite the claims of the Marxist leaders, that they knew what was best for the oppressed, the marginalized, and the downtrodden, the great Marxist utopia of equality never came. Instead, hell rose up from the depths. Blood was spilled across entire societies as the Marxists fought to achieve their vision of equity, collectivizing entire industries to redistribute wealth along more equitable lines. Such a task was finally accomplished once hundreds of millions of innocent people were equally *dead*.

No one's arguing the social constructionist theory will lead to hundreds of millions of deaths, but the idea of *equity across all hierarchies* certainly did, and it must be taken with great caution and suspicion wherever it's championed.

Postmodern social constructionism, like Marxism, believes in the blank slate of the human mind. They believe that differences between groups of people are social constructs, and that these constructs arrange themselves into hierarchies of power, domination, and oppression. Unequal distributions of men and women, therefore, must be seen as the result of societal imposed gender roles which are misogynistic, sexist, and bigoted. These roles are often imposed unconsciously through people's biased behavior, words, and actions. Like Allan Johnson said, we are all contributors to the patriarchy whether we realize it or not.

The postmodernist solution, therefore, is to fix all inequality through the use of quotas, equity policies, and so-called "diversity" campaigns which push for equitable representation among different gender and racial groups. For example, equity policies are common in universities, where departments aim to achieve 50% men and 50% women through preferential and discriminatory hiring practices. The most qualified candidate is not the one who is most accomplished, but the one who best fulfills the diversity quota. Here we see a contradiction in the words and actions of the social constructionists: they say they are fighting against "institutional discrimination" while simultaneously instituting policies which discriminate based on sex, gender, and race. They are no dismantlers of institutional discrimination; *they are its damn torch-bearers.*

The problem with achieving diversity through group identity is immense. Consider this question: *Are there more differences <u>within groups</u> or <u>between them</u>?* Both the postmodernists and social constructionists answer by saying, "Between them!"[110] They believe that diversity is achieved through one's immutable characteristics such as sex, race, and gender; such a belief makes people see the world through the lens of trivial factors such as race rather than individual merit, personality, and skill. This assumes that there are more differences *between* these groups than there are *within* the groups. And yet, the exact opposite is true. For instance, a social constructionist might believe that there are more differences *between* white men and black men than there are within white and black men. However, while such a view is certainly bigoted, it's also technically incorrect. There is actually more variation *within* a group of white men and *within* a group of black men than there is *between* the groups. This is the fundamental concept of individual differences, and social constructionists reject this revolutionary idea in favor of outright differential and

---

[110] Peterson, J. (2017). Jordan Peterson on Diversity. *Bite-sized Philosophy, YouTube.*

discriminatory treatment of individuals based on their immutable characteristics, the oldest practice in existence.

Here the social constructionists show that all they have is the power of rhetoric, for their beliefs are filled with contradictions, paradoxes, and lies. They claim to be fighting against the sexism and racism of the West while simultaneously championing the most fundamentally sexist and racist idea: *that individuals should be judged by their group identity, their gender and their race.* They are full of it.

The truth is you cannot get diversity through choosing applicants based on their sex, gender, and race. You get diversity by selecting among *individuals*. Scientific research from both biology and the social sciences prove it, as I will soon discuss.

Categorizing and judging people solely by their immutable characteristics is not just bigoted, it's also technically impossible due to the indefinite number of ways an individual can be categorized. We can be divided into categories of *intelligence, temperament, geography, historical time, attractiveness, youth, age, health, sex, athleticism, wealth, family structure, friendship, education, race,* and thousands more.[111] These traits can be considered the cards of fate, the gifts or curses of God, or the empty arbitrariness of a nihilistic existence. It doesn't matter where these traits come from. What matters is the categories exist. Obviously the world is not fair, or else these categories would all be equal. And yet this problem of practicality doesn't stop social constructionists from categorizing individuals by their group identity.

On a technical level, the category of *intelligence*, which has a strong biological basis despite the objections of social constructionists, is far more important to long-term success than the categories of race or gender. On the other hand, the category of *temperament* may be more valuable in certain occupations which work best with specific temperaments. For example, some people are naturally more extraverted than others. Some are more sensitive to negative emotion, more agreeable, and more open to

---

[111] Peterson, J. (2018). Identity Politics & The Marxist Lie of White Privilege. *Sovereign Nations, YouTube.* 48:00.

experience than others; some are more conscientious, more competitive, and more creative than others. This doesn't mean that one trait is better or worse. All traits have their good and bad, healthy and pathological elements, just as being born as male or female has its good and bad qualities. On the sexual side, women can bear children and even have multiple orgasms, while men on average are physically stronger and yet die eight years sooner. These differences in cost and benefits can go on indefinitely, which shows us that 1) individuals are incredibly complex and 2) that judging them by a single category is both practically and technically incorrect.[112]

Ironically, the concept of *intersectionality* is based in this framework of indefinite intersections of identity categories, and yet intersectional theorists, mostly comprised of social constructionists, fail to recognize the logical conclusion of their arguments: *that the ultimate identity is the individual*, something the West already figured out.

For most liberal democracies, the individual is what is championed and valued. This is evidenced in the fact that most hierarchies in the West are comprised of competence, not immutable characteristics. You're not given rewards for your work due to any concept of your *intrinsic worth, your race, or your gender*. You're given rewards based on your *skill* and the value you contribute. This idea is summarized by Dr. Jordan Peterson, Professor of Psychology at the University of Toronto:

> "The reason you get promoted up a competence hierarchy is because other people want to *maximize* the value you can produce for society; that's why you get paid. It's not an award...The most intelligent thing to do, even if there are arbitrary reasons for competence differences, is to place the most competent people where they can do the best job, because that's best for everyone else."[113]

---

[112] Peterson, J. (2018). Identity Politics & The Marxist Lie of White Privilege. *Sovereign Nations, YouTube.* 48:00.
[113] Ibid., *53:00.*

Knowing these truths, what is the solution to inequality? The answer is this: *allow people to engage in free market systems which champion individual preference, interest, and choice while protecting their rights*. Such a system will be kept as long as we understand the core truth of being human: that individuals differ. They differ in their traits, desires, interests, preferences, and ultimately, their choices, as I will soon show. In a free economic system, you will not get equal outcomes; you will not achieve the *equity* the social constructionists so vehemently desire. But you can achieve *justice* by giving people the freedom to live their own lives.

The attempt at restructuring society in a more equitable way has led to a major paradox in social constructionism: the countries who have the highest gender equality scores from the UN, such as Sweden and Norway, have the largest *gender differences* in personality, interests, and occupational preferences. This is likely because, as the sociocultural effects on individual action were removed, the innate tendencies of men and women *maximized*, not minimized. To understand why, we're going to need a much better framework than social constructionism. We're going to need something which explains this paradox in a high-resolution, multivariate model.

Unlike those who use rhetorical device and language to argue their points, it's time to finally examine sex and gender differences with the scientific tools of reason, logic, and evidence. As we do, the postmodern social constructionists are going to lose any semblance of sanity they had left.

5

NATURE AS FORMWORK

Postmodernism claims that all truth is relative, that meta-narratives are false, that there is no objective way to ascertain knowledge, that reason is a tool of power used to oppress marginalized groups, and that power is the ultimate structure of society.[114] Such claims are impossible to work with. Despite their paradoxes, there is no way to use such a framework to study the nature of objective reality. For the postmodernists, reality only exists in the mind of the subject. This incredibly elitist, hands-off approach to *being* is ultimately useless for any type of serious scientific investigation.

Because of this, I propose an opposite epistemology to postmodernism: that truth is universal and that these truths can be ascertained through the tool of reason to illuminate the objective knowledge underlying them. Reason is not a tool of power, as the postmodernists claim, but a tool which allows us to further understand the reality around us. It allows us to discuss different perspectives with evidence and logic, and it allows us to not just tolerate but support the opinions of others different from our own. Finally, perhaps most important is that the belief in objective truth, knowledge, and reason is the very thing which allows us to have conversations in the first place. Without it, we instead live in a postmodern world comprised of nothing but power. This is not the world we wish to live in.

---

[114] See chapter, "The Postmodern Tyranny."

I agree with the postmodernists that there are an indefinite number of interpretations to reality, but I disagree with their next step. While there are infinite interpretations, there are only a fixed number of *functional* interpretations to reality, ones which allow us to interact effectively in the world both for our benefit and the benefit of others. Using this epistemological framework, I propose a detailed biopsychosocial model for the existence of gender inequality in liberal democracies. I will explore a multivariate framework which places gender differences in the areas of *biology, epigenetics, neurology, endocrinology, evolutionary science, personality psychology, behavioral science, sociology of gender,* and *anthropology*.

Synthesizing this large variety of fields into a single theory, I will show that sex is an objective category which differentiates *functionally different* reproductive capacities. Gender differences, therefore, are both biological and sociocultural in origin. These somewhat dynamic gender differences map onto the biological category of sex and are interpreted in the context of it. **In other words, I define gender as an expression of one's sex category through behavior, actions, interests, and preferences. Therefore, gender is not something which is completely independent from sex, but something that is actualized through the interaction of both biological and sociocultural realities.**

Using this framework for gender as being a biological and sociocultural expression of one's sex category, I will differentiate aspects of sex and gender differences utilizing the fields of research mentioned above. I reject the idea of the blank slate and believe that each individual is born with a predetermined genetic patterning structure which can mold and develop based on exposure to environmental, sociocultural, and epigenetic effects. In this sense, actions of individuals cannot be understood in the absence of a pre-existing structure as social constructionists claim.

Because of this truth, differences between individuals emerge due to different physical characteristics, cognition, brain anatomy, physiological structure, hormone levels, personality, socialization, and interests. Each of these variables, and many others, influence the choices an individual makes across their lifetime.

This theory, that individuals are born with a predetermined neural and physiological structure, is supported by nearly all of contemporary scientific research, disproving both the social constructionists' blank slate theory and the social Darwinists' biological essentialism. I reject the extreme view that all of human action is determined by sociocultural forces, while I also reject the opposite extreme view: that all of human action is determined by biology.

The truth is that all of human behavior and psychology is a complex interaction between biological and sociocultural factors at almost every stage of development. From genomics and heritability, to the study of personality and IQ, both biology and society have been shown to affect our behavior in interconnected ways. Some aspects of our behavior are more affected by biology and other aspects are more affected by society and culture. Simultaneously, there are some behaviors, such as language, that are solely affected by culture, while there are diseases, such as Huntington's disease, which are solely affected by heritability.

The cliché debate on nature and nurture needs to come to an end. Neither side is completely correct, nor is it the case that nature and nurture are balanced entities.

Rather, *nature is the foundation on which the existence of nurture arises*. Nurture cannot exist in the absence of *being*. Therefore, nature can be seen as the formwork onto which nurture grows, expands, and develops. The formwork provides the structure through which the mixture of nurture can be poured. Because of this, the correct question is not *to what degree does nature and nurture affect human action?* The correct question is, instead, two-fold:

1) *What is the formwork of nature?*

2) *How does nurture fill the formwork?*

HUMAN UNIVERSALS

Humans have a genetic patterning structure present in the brain when we're born, which is already set up to interpret, sense, analyze, judge, and react to both internal and external inputs and information. This structure can be compared to a computer program which is written to adapt to new inputs based on pre-existing code. This is the main reason why humans are much more complex than AI.

Our brains, physiology, and senses are already set up to interpret our world from the moment we're born. The context for all of our future actions is already set with the pre-existing code. This is why the development of AI systems has been incredibly difficult. It turns out that intelligence is directly related to action and movement which is able to learn and develop utilizing an existing structure. AIs do not have the existing structure, nor do most of them have a body to interact with the physical world. In the AI world, the postmodern observation that reality has an indefinite number of possible interpretations has become a major problem: how do we get our robots to interpret the world like humans?

Without the ability to use the body and nervous system to interact, play with, and learn from the world, AI development will continue to suffer from this *frame problem.*

The multivariate biopsychosocial model, therefore, begins with this truth claim: that human beings are not born as blank slates but have an existing structure and nature arising from millions of years of genetic evolution. Our model acknowledges the more than 300 universals which define human beings across all cultures.[115] A handful of which are:

- Abstraction of speech and thought
- Aesthetics
- Actions under self-control
- Belief
- Childhood fears

---

[115] Brown, D. (2002). Human Universals, *New York: McGraw-Hill.*

- Classification
- Collective identities
- Critical learning periods
- Culture variability
- Language
- Logic
- Making comparisons

   ...

   All the way to:
- Sexual attraction, jealousy, modesty, and regulation
- Socialization
- Status
- Symbolism
- Time
- Tools
- Trade
- Vowel contrasts
- Weapons
- Worldview

Aside from our psychology, the blank slate theory is also challenged by evidence in neuroscience. Here the complex genetic patterning system is present in the brain from birth. For example, there are more than 50 distinct areas comprising the visual system which interconnect during gestation like a complex computer code beyond any imagination.[116]

The idea that the brain has a predetermined code holds up even in studies on fraternal twins. For example, fraternal twins who are separated at birth have highly correlated measures of intelligence and personality, likelihood of getting divorced, becoming addicted to tobacco, and even

---

[116] Pinker, S. (2006). The Blank Slate, *General Psychologist*, 3.

have a high likelihood of holding similar political beliefs. Behavioral geneticists call this interconnectivity of genes the *First Law of Behavioral Genetics,* which says that all behavioral traits are partially heritable.[117]

These heritable traits even extend to aspects of our behavior such as the tendency to have an antagonistic personality, a tendency toward violent crime, a lack of conscience, and psychopathy.[118] Psychologist Steven Pinker explains that "neuroscience has identified brain mechanisms associated with aggression. And evolutionary psychology and anthropology have underscored the ubiquity of conflict in human affairs."[119]

DANGERS OF BLANK SLATE IDEOLOGY

Despite the near limitless amounts of scientific evidence on the topic of human nature, social constructionists continue to believe in the utopian ideal of the blank slate. Why is this? The answer lies in what the social constructionist viewpoint fears: *inequality.*

The concept of inequality is anathema to the social constructionist ideology. If humans are born as blank slates, then they *must* be equal. If, however, the mind *does* have a pre-existing structure, then "different races, sexes, or individuals could be biologically different," and, as the social constructionists fear, "that would condone discrimination and oppression."[120] This is why we see the vehement push for equity from social constructionists. If humans are all equal in their traits, then equity is the logical fix to the problem of inequality. And yet, the view that the acknowledgement of biological differences between individuals would lead to discrimination is a non-sequitur. Just because someone is different does not mean they are not equal in value or not equal before the law. However, it also does not mean they are equal in *traits.* Individuals differ in all kinds of traits, and this is the beauty of being human.

---

[117] Pinker, S. (2006). 4.
[118] Ibid.
[119] Pinker, S. (2006).
[120] Ibid.

While there are dangers in relying too strongly on the existence of a biological human nature, there is also a danger in the denial of our nature altogether. The ideologies of both the social constructionists and biological essentialists have no place in the scientific study of human beings and should be rejected as unscientific views based on *feeling*, not evidence.

The belief in the flawed nature of human beings immediately rejects the notion that a great utopia can ever be reached. Here the social constructionists have a major problem: no matter what they do or what they say, their ideological beliefs will never match up with reality. The equitable utopia they envision can never be actualized.

Instead, there's a brighter future ahead, one based in reason, science, and liberty. As flawed human beings we can continually strive for more justice, more fairness, and more equality of opportunity for every individual to pursue their own interests. An ideal political system of equality would therefore be defined as policies which treat people as individuals with rights, not members of their group; not blank slates ready to be molded, but as individuals, each with their own structure, framework, and motivations. Knowing all this, the blank slate theory of the social constructionists is one of the most flawed theories in history:

> "The blank slate is a mistake. It's a mistake because it makes our values hostages to fortune, implying that someday, discoveries from the field or lab could make them obsolete. And it's a mistake because it conceals the downsides of denying human nature, including persecution of the successful, totalitarian social engineering, an exaggeration of the effects of the environment, a mystification of the rationale behind responsibility, democracy, and morality, and the devaluating of human life on Earth."[121]

Moving past the social constructionist theory requires that we reject both the blank slate and social Darwinism. The truth is biology and society

---

[121] Pinker, S. (2006). 8.

interact in complex ways we do not fully understand. And yet, lucky for us, this is where the fun begins. For the rest of the book, the goal is to present a high-resolution, multivariate model on why disparities between men and women still exist despite gender egalitarian policies in liberal democracies.

I will begin with the details of the biopsychosocial model, discussing the structure of the theory and how it can be conceptualized. After this, we'll explore the category of sex as a biological mechanism, studying how it's determined, structured, and represented in the world from conception through adulthood. We'll then explore how children are influenced by both biological and social factors, how gender stereotypes can turn pathological, how puberty differentiates the sexes, how personality maps onto sex, how discrimination and disparities reveal themselves in society, and how the Gender Equality Paradox dismantles social constructionism.

In doing so, there are six critical things we will learn:

1) Average sex and gender differences between males and females are real.
2) The source of sex and gender differences can be explained through a high-resolution multivariate analysis.
3) There are more differences *within groups* than *between them*.
4) Males and females are substantially more similar than they are different.
5) Statistically significant differences on the extremes result in degrees of sex and gender differences.
6) Individuals should be judged as *individuals,* not as members of their group.

The social constructionist framework can no longer survive on the doctrine of equity: that all hierarchies must have 50% men and 50% women. Using the multivariate biopsychosocial model to explain the causes of gender inequality and explore possible solutions, I will show that the social constructionists only have one tool at their disposal: like the postmodernists before them, they do not have reason, logic, or evidence; all they have is *words*.

6

EXPLORING GENDER DIFFERENCES

Without the blank slate, social constructionists have nothing but aesthetic appeals to argue their case for the causes of sex and gender differences. Their framework allows for only one variable: *socialization*. Because of this, social constructionist arguments that sex and gender have no basis in biology should be viewed with great suspicion due to their lack of serious scientific research. If we want to understand the causes of gender inequality in a non-partisan and nuanced approach, we have to utilize a high-resolution framework with many variables. We must account for a wide range of theories, data, and empirical research if we are going to get anywhere close to the truth.

Therefore, to explain the causes for sex and gender differences, I propose a variant of the biopsychosocial model: *sex category (male-female) is the biological formwork within which gender differences in behavior, personality, and interests develop.* I argue these gender differences are influenced by a mix of three factors:

1) **Biology** (chromosomes, genes, hormones)

2) **Psychology** (neuroanatomy/brain circuitry, perceptual structure)

3) **Society/Culture** (socialization, cultural practices, social learning)

After male-female behavior is differentiated by these three factors, gender differences become apparent and can be represented through average group distributions, and extreme ends of the group distributions often represent themselves in society as gender inequalities.

Unlike the social constructionist model that says sex and gender differences exist on a flat spectrum, I propose that sex and gender differences exist on a bimodal distribution, which is to say, a set of statistics which have *two modes*, or two averages. For example, the average height for humans can be broken up into two averages, one for males and one for females. Males have an average height that is five inches higher than the average height for women.[122] Therefore, this data can be separated into a graph with two distributions, or a bimodal distribution. Much of the research into sex differences can be represented through these double-bell curve graphs. The closer the averages between men and women are, the greater the overlap is, and the farther the averages between men and women are, the lesser the overlap is.

Graph: A common bimodal distribution of male and female traits.

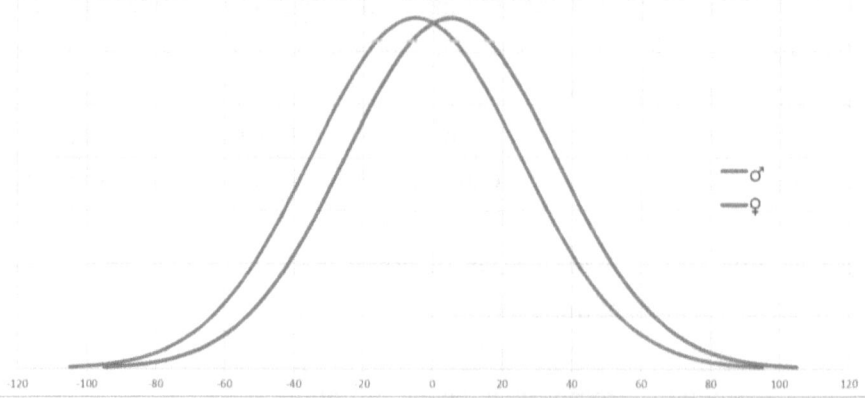

A bimodal distribution (two averages) showing a distribution for males and a distribution for females. While the averages are slightly different, there is immense overlap between the traits. Thus, men and women often share considerable overlap in a given trait. Some sex and gender traits have more overlap than others, while some have little to no overlap.

---

[122] Schmitt, P. (2017). Sex and Gender are Dials (Not Switches), *Psychology Today*.

This is the *law of bimodal distributions*, and it's the most important concept to understand when learning about sex and gender differences.[123] The fact that sex and gender differences arrange themselves into bimodal distributions challenges the idea that every individual sits on a *spectrum*. Differences *within* and *between* groups exist as averages, not as spectral distributions where every color of the rainbow is equally distributed. Because of this, our model utilizes bimodal distribution data to study the composition of sex and gender differences. I recognize that sex and gender should not be conceptualized as binary switches but *dials* with common indicators, interconnected dimensions, and two averages.

The social constructionist view that sex and gender exist as spectral distributions is represented in the infamous child-like diagram known as the *Genderbread Person*. In perhaps the most stunning diagram of the 21st century, the Genderbread Person shows children and adults that, ultimately, everyone is a unique individual whose gender identity, gender expression, anatomical sex, sex assigned at birth, sex attraction, and romantic attraction all vary *independently* from one another. Or, in the words of the non-binary cookie:

> "Gender is one of those things everyone thinks they understand, but most people don't. Gender isn't binary. It's not either/or. In many cases it's both/and. A bit of this, a dash of that. This tasty little guide is meant to be an appetizer for gender understanding. It's okay if you're hungry for more after reading it. In fact, that's the idea."[124]

The Genderbread Person tells us in crunchy-cookie fashion that every individual exists on a continuum, a spectrum, between 0 and 100, in a beautifully crafted and incredibly diverse cooking recipe where all of it's mixed in together.

---

[123] People unfamiliar with sex difference research will bring up anecdotes to argue against the validity of sex differences. In doing so, they fail to understand the basics of bimodal distributions.

[124] The Genderbread Person, *It's Pronounced Metrosexual (Website)*.

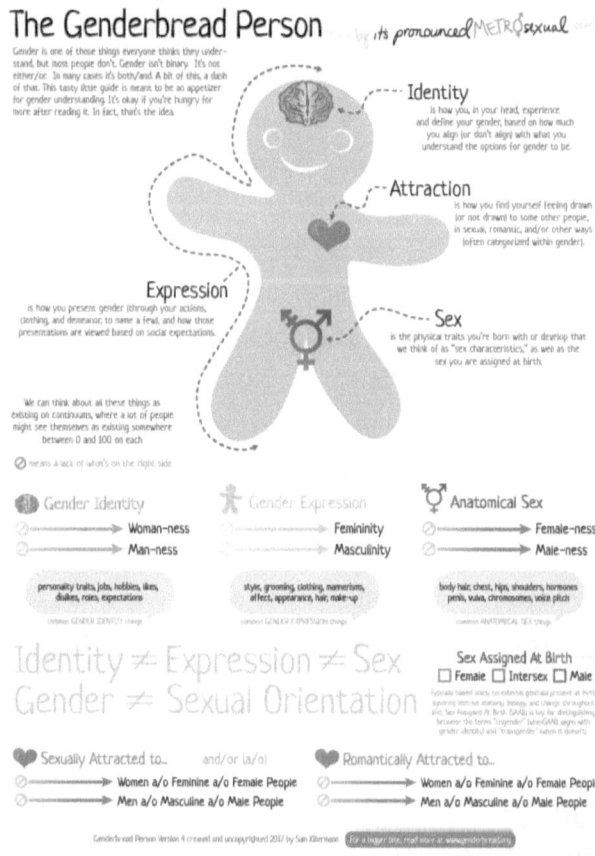

Despite its very inclusive and feel-good nature, the Genderbread Person crumbles into a sad pile of dust upon any critical examination. The variables of gender identity, gender expression, anatomical sex, sex assigned at birth, sex attraction, and romantic attraction vary *dependently*, not independently. These sexual traits are so heavily linked that changing one will most likely change another.

For example, most people with the female sex will be attracted to members of the male sex, and most people with the male sex will be attracted to members of the female sex. In fact, these traits are so interconnected that a person's biological sex predicts sexual attraction, gender identity, and gender expression with a high level of accuracy.

Even though sexual traits are heavily interconnected, it does not mean the traits exist as a binary switch, or *on* and *off*; rather, sex and gender differences vary within interconnected and trait-dependent *bimodal distributions*, with an average for females and an average for males. Because of the bimodal nature of sex and gender differences, most traits have high degrees of overlap between men and women, proving that there is more variation *within* males and females than there is *between* them. In other words, *men and women are not much different on an average level*.

Using the idea that sex and gender can be conceptualized as interconnected dials which form bimodal distributions, not switches or rainbows, we can propose a more realistic model for sex and gender differences.

SEX AND GENDER AS DIALS

Conceptualizing sex and gender as dials causes an immense problem for any scientific research: it requires researchers to be incredibly careful and specific with their words while understanding that sex and gender differences can be analyzed from an indefinite number of possible traits and causes.

For example, sexual attraction is not a binary switch, nor is it a flat spectrum. Rather, just like most sex and gender differences, sexual attraction fills out as a bimodal distribution with two averages (one for men and one for women). Men tend to be attracted to one sex or the other exclusively, with some overlap. Women, however, show a more complex distribution, where same-sex attraction and erotic fantasies towards both sexes are more common.[125] Social constructionists view this disparity in sexual fluidity as a product of societal stigma towards bisexual men. However, in many different mammalian species, bisexuality is an innate dimorphic trait which develops in *only* the male or female of the species through exposure to prenatal androgens (a sex hormone with higher levels in males and lower levels in females). Because of this, evolutionary

---

[125] Ngun, T., et al. (2010). The Genetics of Sex Differences in Brain and Behavior. *Frontiers in Neuroendocrinology, 32(2011)*, 238.

scientists believe women are the more "bisexual sex" due to lower levels of prenatal androgens, whereas men are more likely to be either heterosexual or homosexual exclusively.[126] The influence of prenatal androgen to the differentiation of male and female in the womb and its subsequent effects on sexual orientation is one of the most well-established findings in biology:

> "The possibility that people differ in sexual orientation because of hormonal differences has been the most influential causal hypothesis involving a specific mechanism."[127]

This knowledge should be well-accepted by those who have claimed for decades that being gay or bisexual is innate and biological. *Surely the social constructionists can support this evidence.* Thus, for sexual attraction variation, humans show a much more complex distribution than the highly binary animal kingdom. Using this concept, experts in sex difference research have come up with a model that I will be utilizing in ours: *sex and gender differences as interconnected, dimensional dials.*

> "Rather than thinking about men and women as categorically different (or different along one dimension of masculinity versus femininity), I think it is more useful to think about sex differences as a series of numerous *interconnected, multidimensional dials.* Dials that can be turned up or down (individually, or in combinations) depending on one's genetics, hormone levels, organizational effects *in utero*, activational effects in puberty, and a wide range of social, historical, and cultural factors."[128]

---

[126] Bailey, J., Vasey, P., Diamond, L., et al. (2016). Sexual Orientation, Controversy, and Science. *Psychological Science in the Public Interest, 17(2),* 56.

[127] Bailey, J., Vasey, P., Diamond, L., et al. (2016). 69.

[128] Schmitt, D. (2017). Sex and Gender are Dials (Not Switches), *Psychology Today.*

If we can think of sex and gender traits as *dials,* rather than binary switches or flat spectrums, then our analysis of sex and gender differences will be much more accurate. Through conceptualizing sex and gender traits as interconnected and dimensional dials, we can form a superior replacement to the univariate social constructionist theory which conceptualizes sex and gender differences as flat spectrums with no averages.

To begin, we can break down sex and gender differences into ten main dimensions, or traits, with each dimension being controlled by its own dial. Whether a dial is turned to the more masculine or more feminine end will be determined through a mix of *biology, psychology,* and *society.* Each of these dimensions can be represented through a bimodal graph, with the male-typical extreme on one end and the female-typical extreme on the other. Many males will fall under the typically masculine end of the distribution, while many females will fall under the typically feminine end. However, statistically significant amounts of males will fall under the female-typical side of the distribution, while significant amounts of females will fall under the male-typical side. This does not mean that *average sex and gender differences* do not exist, but that there is a bimodal variety in sex and gender dimensions.

This ten-dimensional *dial-based* model integrates the existence of feminine males and masculine females, an important aspect of any scientific model for sex and gender research.

Exploring each of the ten dimension's male-typical and female-typical ends will help us understand the extremes we're comparing. For example, the male-typical extreme for mental abilities tends to be focused on spatial manipulation and systematizing, while the female-typical extreme for mental abilities tends to be focused on the *location* of objects and verbal categorizing. These ten dimensions are provided by psychologist David Schmitt. Each one has its own dial which can be affected by biology, psychology, and society. There is significant overlap between the masculine and feminine extremes listed for each dimension, as males and females can share many of these traits in common. I have arranged these ten dimensions into a table on the next page.

Schmitt's Ten Dimensions of Masculine and Feminine Traits.[129]

Sex and Gender Dimensions	Masculine extreme	Feminine extreme
PHYSICAL TRAITS	Tall, late puberty	Short, early puberty
MENTAL ABILITIES	Spatial manipulation, systematizing	Location memory, verbal skills
MATE PREFERENCES	Youthful, fertile	Older, high status
SEXUAL DESIRE	High sex drive, paraphilia	Sexual fluidity
PERSONAL VALUES	Power, status, achievement	Benevolence, unity, harmony
SOCIAL INTERESTS	Competitive, Things	Domestic, People
SOCIAL BEHAVIORS	Aggressive, risk-taking	Compliance, agreeable
MENTAL HEALTH	Psychopathy, ADHD	Depression, anxiety
PERSONALITY	Low Agreeableness, Low Neuroticism	High Agreeableness, High Neuroticism
OCCUPATIONAL INTERESTS	Realistic, investigative (Things)	Artistic, social (People)

---

[129] Schmitt, D. (2017). Sex and Gender are Dials (Not Switches), *Psychology Today*.

Each of these dimensions' male-typical and female-typical descriptions are *extremes*, not averages, of scientifically-documented masculine and feminine traits. To understand this, imagine a bimodal distribution for height between men and women. The extreme ends of the height distribution have almost no overlap between men and women, and this is where the height differences between men and women are most apparent. Therefore, because many sex and gender differences conform to bimodal distributions (two averages), *gender differences are most apparent in the extremes* (or the tails of the two averages).

For instance, the shortest people in the world are almost exclusively female, while the tallest people in the world are almost exclusively male. The overlap occurs near the middle, the region of the averages, or peaks. Some dimensions listed above have a higher overlap than others, with men and women sharing near equal levels of a trait. Other dimensions have a lower overlap, with men and women sharing very different levels of a trait. It all depends on what we're analyzing.

Another example is found in the differences between males and females in the fundamental frequency of speech, which has almost zero overlap.

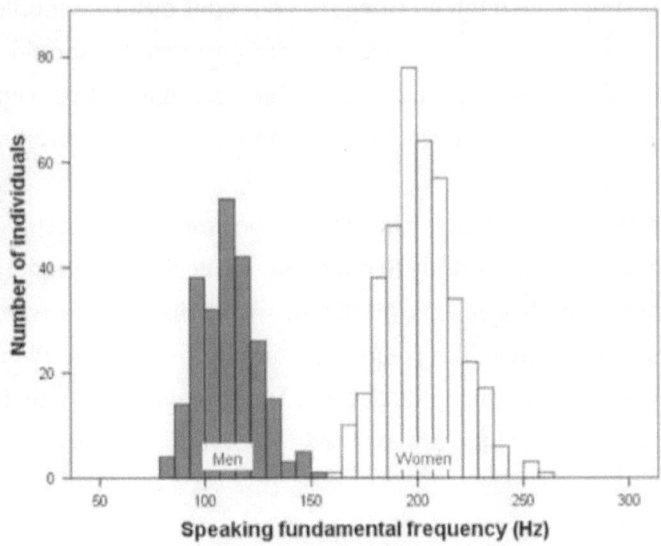

There is almost no overlap between men's and women's mean habitual speaking fundamental frequency in Hz. Source: Puts et al. (2014).

Is this difference in vocal pitch socially constructed? While the social constructionists might find a way to argue so, traits such as vocal range have clear biological origins through exposure to sex hormones. Other traits, however, have clear sociocultural origins. Because of this variation, an analysis of each trait is necessary for a holistic understanding of gender inequality, illuminating the complicated yet beautiful nature of sex and gender research.

Psychologist David Schmitt explains that the first step towards understanding this sex and gender variety occurs when we think beyond binary categories and dichotomous causes of variation and into an understanding of how variations in sex and gender are defined by time and context:

> "What is especially interesting, I think, is how sexual diversity scientists have tracked evidence regarding these different sex/gender dials, evidence accumulating into an integrated, coherent account of how various causes of sex/gender variation are developmentally-timed and domain-specific."[130]

Thinking beyond binary categories does not mean we should accept a social constructionist perspective, nor does it mean we should reject the reliability of the male-female category. Rather, it means that we recognize the *complexity of sex and gender variation within and between males and females.*

For example, prenatal androgen exposure during gestation is not fixed, nor does it always apply to male and female fetuses in equal amounts. The amount of androgen exposure in the womb directly affects male-typed preferences for things like rough-and-tumble play.[131] On average, male fetuses are exposed to higher levels of androgens than female fetuses. Because of this, *girls* who are exposed to prenatal androgens higher than

---

[130] Schmitt, D. (2017).

[131] Ibid.

the female-average tend to exhibit more male-typical psychological traits compared to their unaffected sisters.[132] Therefore, the variability of prenatal androgen levels supports our model that sex and gender traits are not binary switches. Such an interconnected and multidimensional model for sex and gender can be applied to many more traits than just prenatal androgens, as I will show in the coming chapters.

Modeling sex and gender differences as dials also allows for so-called *master-dials* which affect many traits simultaneously:

> "Undoubtedly, there exist some master-dials that affect a bunch of other dials (e.g., both rough-and-tumble play and mental rotation, or both vocal pitch and grip strength), and some dials may have antagonistic effects on other dials (i.e., more masculinity along one dial may cause less masculinity along another dial). Moreover, sex/gender movement along dials at one point in time might affect subsequent sexual expressions, both of which may depend on other direct genetic and activational effects, and so forth. There is also increasing evidence that many sex differences emerge in *middle childhood,* mediated by hormone changes that occur before the onset of adolescent puberty."[133]

Of course, modeling using dials is not a complete explanation, but it does offer perhaps the most holistic understanding we can hope to reach, for it gives us a complex and multivariate framework for how sex and gender differences can be highly variable. This allows us to explore how differences between men and women arise from variations along the dials.

If sex and gender differences can be conceptualized as dials, what about the origins of the dials themselves? Or put another way: **What are the origins of sex and gender differences?** This is the question our biopsychosocial model addresses next.

---

[132] Hines, M. (2016). Prenatal androgen exposure alters girls' responses to information indicating gender-appropriate behavior. *Philosophical Transactions, 371.*
[133] Schmitt, D. (2017).

ORIGINS OF SEX AND GENDER DIFFERENCES

Research into the origins of sex and gender differences has had a complex history ever since its creation. For centuries, humans believed that gender differences in behavior were fixed, binary, and immutable.[134] Religious meta-narratives provided the framework for the belief that gender differences were created *ex nihilo* by a divine creator and therefore were innate. A critique of this meta-narrative has characterized much of the debate on the origins of sex and gender differences: were males and females created with an innate and binary structure or were males and females shaped only by sociocultural factors? As it turns out, neither question is correct.

The emergence of formal scientific psychology in the 1870s created the initial surge of sex and gender research. The debate was contentious from the beginning, with one side arguing for large gender differences and the other side arguing for large gender similarities.

Because of this debate, studying sex and gender differences and similarities naturally causes a question to arise: **What is the purpose of understanding sex and gender differences?** Social constructionists claim that the useless distinction between males and females only serves to perpetuate gender stereotypes, discrimination, and injustice. On the other extreme, biological essentialists claim that the purpose is to sort people into their rightful places based on 'innateness,' though it is hard to find any scientists who hold this view today. Both extremes, however, could not be more wrong.

Sex and gender research is not useless, and it is not used to discriminate against males and females. The purpose, rather, is two-fold, or as Gender and Women's Studies Professor Janet Hyde explains:

> "First, stereotypes about psychological gender differences abound, influencing people's behavior, and it is important to evaluate whether they are accurate. Second, psychological gender

---

[134] Hyde, J. (2014). Gender Similarities and Differences. *Annual Review of Psychology*, 65, 374.

differences are often invoked in important policy issues, such as single-sex schooling or explaining why, in 2005, there were no women on the faculty in mathematics at Harvard; *it is crucial to have accurate scientific information available to evaluate such policy recommendations and explanations."*[135]

Therefore, sex and gender research is useful for 1) combatting and evaluating the accuracy of gender stereotypes and 2) for understanding the causes of gender inequality in society.

Because gender differences form a bimodal distribution, there is significant overlap between men and women in any given trait. This important fact supports Hyde's hypothesis that males and females are similar on most, but not all, psychological variables. Our model agrees with this *gender similarities hypothesis:* that men and women are, for the most part, very similar. Evidence for gender similarities being larger than gender differences comes from a review of 46 meta-analyses of research on psychological gender differences.[136] Hyde summarizes this research:

> "The 46 meta-analyses yielded 124 effect sizes (Cohen's *d* equal to the mean score for males minus the mean score for females, divided by the within-groups standard deviation) for gender differences. Strikingly, 30% of the effect sizes were between 0 and 0.10, and an additional 48% were in the range of a small difference, between 0.11 and 0.35. That is, 78% of the gender differences were small or very close to 0. The gender similarities hypothesis provides important input."[137]

---

[135] Hyde, J. (2014). 374.

[136] A meta-analysis is a statistical technique that allows a researcher to combine the results of many research studies on a given topic. Meta-analyses, therefore, utilize data from many countries and hundreds of thousands of people.

[137] Hyde, J. (2014). 375.

78% of gender differences being small or close to 0 supports our model that within-group variation is larger than between-group variation among males and females; or, in other words, males and females are *more similar than different* across a given trait. The effect sizes found in the meta-analyses are used to compare the *two averages* of male and female scores. *Effect sizes* are the most common measure of the magnitude of gender differences in psychology research. Most effect sizes in psychology are between 0 and 0.35 (zero to small). The larger the effect size, however, then the larger the gap between men and women in a given trait. The fact that most effect sizes are small or close to 0 is indicative of the immense overlap in the bimodal distributions across traits, as the *dials* theory explains.

These effect sizes have been shown to be incredibly accurate due to the methodology of a meta-analysis, which is comprised of three important factors:

> "1) An assessment of whether multiple studies find the same result.
>
> 2) An examination of whether there is a gender difference in a particular psychological attribute by estimating the magnitude of the gender difference.
>
> 3) A systematic exploration of moderators, such as socialization, that may contribute to the presence or absence of gender differences."[138]

Once the reliability of the meta-analysis can be understood, we can move past the questioning of methodology and into the causation of sex and gender differences. If 78% of gender differences have an effect size of 0.35 or *smaller*, that means 22% of gender differences have an effect size of 0.35 or *higher*; or, in other words, **22% of gender differences are statistically significant**. It is in this 22% where the causes of gender inequality are of particular importance.

---

[138] Hyde, J. (2014). 375.

Sex and gender researchers use three unique theories to explain the sources for this 22% variation, which are:

1) Evolutionary Theory
2) Cognitive Social Learning Theory
3) Sociocultural Theory

I will briefly explain these theories below and work to integrate aspects of each into our biopsychosocial model for the origins of sex and gender differences and similarities.

THREE UNIQUE THEORIES

Evolutionary Theory proposes that gender differences are products of natural selection. We will define *gender* as how individuals express their biological sex through behavior, interests, actions, and preferences. In Evolutionary Theory, differences in gendered behaviors arise due to adaptive strategies for males and females which become genetically patterned and heritable. Evolutionary Theory is comprised of two major components: sexual selection and parental investment. The first, sexual selection, relies heavily on two processes:

> "1) Members of one gender (usually males) compete among themselves to gain mating privileges with members of the other gender (usually females), and 2) members of the other gender (usually females) have preferences for and exercise choice in mating with certain members of the first gender (usually males). Sexual selection can be invoked, for example, to explain gender differences in aggression."[139]

The process of sexual selection differs depending on the animal being studied. For most mammals, however, sexual selection is often controlled and mediated by female choice. Females exert immense selective pressures

---

[139] Hyde, J. (2014). 376.

on males due to the value of their gametes; they are produced much less often and require greater bodily investment than male gametes. Because of this, parental investment becomes the second major component of Evolutionary Theory:

> "Gender enters the picture because human females generally have substantially greater parental investment in their offspring than do human males. Women invest a precious egg (compared with the millions of sperm that men can produce every day) and then invest nine months of gestation, which is costly to the body. At birth, then, women have greater parental investment than men do, and it is to the advantage of the person with the greater investment to care for the offspring, making sure that they survive to adulthood."[140]

Because of the high degree of parental investment necessary to take care of offspring, women's interests and life choices tend to reflect this parental investment structure which adapts the nervous system to the handling and care of infants. Caring for infants requires high degrees of Agreeableness (compassion and politeness) and high degrees of Neuroticism (volatility and withdrawal).[141] This tendency for higher Agreeableness and higher Neuroticism influences women's average interests and preferences.

The theory that men and women's psychological gender differences arise from sexual selection and parental investment has considerable merit, as these differences can also be seen in mammals. However, Evolutionary Theory does not explain all of the difference. These adaptive strategies likely evolve as societies and cultures change, allowing for a possibility that gender differences are not fixed but *dynamic*. Therefore, many evolutionary psychologists who study gender differences have concluded, like Hyde, that most gender differences in psychological traits are moderate

---

[140] Hyde, J. (2014). 376.
[141] See chapter, "Personality and the Big Five."

to small. Rather than proving gender differences are mostly large, Evolutionary Theory indicates the universality of gender *similarities* while acknowledging important differences through sexual selection and parental investment.

Cognitive Social Learning Theory proposes that human behavior is shaped by reinforcements and punishments. Children and adults model their behavior after role-models in their environment, and their behaviors are reinforced and/or punished by society. Just like Evolutionary Theory, there is an abundance of evidence for the merit of this theory. Hyde summarizes this process of internalization:

> "As children grow, control of their behavior shifts from externally imposed reinforcements and punishments to internalized standards and self-regulation. In particular, children internalize gender norms and conform their behavior to those norms. Self-efficacy, another cognitive component, refers to a person's belief in her or his ability to accomplish a particular task. Self-efficacy may be important in explaining certain gender effects. For example, although girls' math performance is equal to that of boys, generally there is a wider gender gap in math self-efficacy ($d = 0.33$, Else-Quest et al. 2010). Self-efficacy is important because of its power in shaping people's decisions about whether to take on a challenging task, such as majoring in mathematics."[142]

Belief in one's ability is a critical factor in shaping men's and women's choices that should not be ignored. Considerable evidence shows that girls' self-efficacy in math ability versus verbal skills combined with boys' high levels of self-efficacy in math ability predicts the existence of the gender gap in mathematics. In other words, because boys tend to lack in verbal skills, they prioritize math-based ones.[143] Therefore, society's expectations may

---

[142] Hyde, J. (2014). 377.
[143] See chapter, "Interests and Inequality."

have a large effect on human behavior, and this is where social constructionists are partially correct. Social learning is a real process, and it affects multiple aspects of gender inequality.

Lastly, the **Sociocultural Theory** proposes that a society's division of labor by gender drives all other psychological gender differences which are not the result of biological sex. This theory acknowledges the existence of biological differences between men and women and says that gender differences are mapped onto biological sex through the culture's division of labor. In other words, the way the society decides to divide labor is also the way in which gender differences are produced. This allows for a flexible origin story of gender differences that takes into account cross-cultural variations which are then applied onto sex. The Sociocultural Theory is perhaps the one theory which integrates both biology and culture into a single mechanism, a mechanism which relies on both to function. Gender differences, therefore, may be a byproduct of the culture's division of labor:

> "The theory acknowledges biological differences between men and women, such as differences in size and strength and women's capacity to bear and nurse children. These differences historically contributed to the division of labor by gender. Men's greater size and strength led them to pursue activities such as warfare, which gave them greater status and wealth, as well as power over women. Once men were in these roles of greater power, their behavior become more dominant, and women's behavior become more subordinate. Women's assignment to the role of child care led them to develop qualities such as nurturance and a facility with relationships."[144]

Agreeing with our biopsychosocial model, the Sociocultural Theory for gender differences emphasizes the effects of epigenetics (gene adaptation to the environment). Women's assignment to more child care roles created an environment for certain psychological traits to develop for

---

[144] Hyde, J. (2014). 377.

facilitation with these roles. In this sense, gender differences may also be the result of biological genetics being altered by environmental context.

The **Evolutionary, Cognitive Social Learning**, and **Sociocultural** theories all express important truths for the origins of sex and gender differences. **Evolutionary Theory** rightfully ascribes certain gender differences to sexual selection and parental investment, which originate from genetics, chromosomes, hormones, and different reproductive capacities. However, it cannot explain all of the variation. **Cognitive Social Learning** rightfully ascribes other gender differences to reinforcement and punishment. Humans are social creatures, and much of our behavior is expressed within the context of societies which shape us, mold us, and constrain us. However, just like the Evolutionary Theory, taking the Cognitive Social Learning Theory to its extreme provides no truthful insight into gender differences. Lastly, the **Sociocultural Theory** rightfully ascribes gender differences to the division of labor while acknowledging the existence of male and female. It accurately describes gender differences as varied across cultures to a degree and accurately explains that gender differences may evolve as society changes to accommodate different roles.

THE BIOPSYCHOSOCIAL MODEL

Our multivariate model does not ascribe to a single theory but rather analyzes gender differences through a mix of these three colorful lenses. Because of this, I reaffirm that gender differences are the result of:

1) **Biology** (chromosomes, genes, hormones)
2) **Psychology** (neuroanatomy/brain circuitry, perceptual structure)
3) **Society/Culture** (socialization, cultural practices, social learning)

Using these factors and the three theories discussed above, I propose that *gender is the behavioral expression of biological sex, influenced by biological, psychological, and sociocultural factors*. For example, a child's interest in male-typical toys may be from an interaction of higher levels of prenatal androgens combined with the reinforcement and encouragement by parents of male-typical play. Likewise, and perhaps more importantly, a girl's interest in male-typical toys may also be from the interaction of higher levels of prenatal androgens which produce changes in the visuospatial system for an interest in mechanical and moving objects.[145] Such differences may be further influenced and molded through epigenetic factors.

From this three-factor model, gender differences in behavior, personality, and interests can be conceptualized as *dials*, not switches, which are dynamic and not static.[146] These gender differences produce bimodal distributions, where males and females have two averages of a given trait, and where these traits have significant overlap. This overlap signifies that men and women are *more similar than different*, and that the differences occur mostly at the extreme ends of the distributions. For example, the most agreeable people across cultures are almost exclusively women, and the most disagreeable people are almost exclusively men. The gender difference in Agreeableness is only a 60-40 ratio, making this a *small* gender difference in the averages and a *large* gender difference on the extremes.[147] The extreme ends, therefore, represent much of the gender differences we see across gender-equal societies.

Using our biopsychosocial model requires that a given trait is analyzed in the context of the three factors: *how does biology, psychology,* and *society affect a given trait?* Our model does not allow for univariate explanations to gender differences such as those seen in social constructionist theory. Instead, our model utilizes a multivariate, high resolution, and integrative approach to the origins of gender differences and the causes of gender inequality in liberal democracies. Thus, the key to sex and gender research is found in integration: that nature and nurture interact in complex ways, and that nature is the formwork within which nurture develops and evolves.[148] And neither arises out of context with the other. I agree with the

---

[145] See chapters, "Conception to Birth" and "Development of Gender Identity."
[146] Schmitt, D. (2017).
[147] Peterson, J. (2017). Jordan Peterson on Diversity. *Bite-sized Philosophy, YouTube.*
[148] Eagly, A., Wood, W. (2013). The Nature-Nurture Debates: 25 Years of Challenges in Understanding the Psychology of Gender. *Perspectives on Psychological Science, 8(3),* 351.

description by researchers Eagly and Wood, who explain the importance of this *interactionist* approach:

> "We believe that the future of science pertaining to gender and sex differences lies in overcoming ideological and identity biases and formulating theories that effectively integrate principles of nature and nurture into interactionist approaches. Yet, the complexity of such theories presents intellectual challenges for psychological scientists who try to model the intrinsic dependence of nature on nurture and vice versa.
>
> Perhaps as a result, research has tended to focus on one or the other type of cause, yielding a muddled scientific voice in public discourse. Adding further difficulties, the media and public need simplifying frameworks that facilitate using scientific evidence to reason about gender in daily life. Excellent communication is essential because any messages from psychological science on gender issues compete with robust informal reasoning based on ideology, everyday observation, and cultural traditions. Among the competing information sources on sex and gender, science may not be winning."[149]

It's true: science is under attack from the outside by the public and media who wish to dismantle critical research into sex and gender differences, and it's under attack from within by ideologues who wish to alter the discourse through championing univariate theories such as social constructionism and biological essentialism. Like Eagly and Wood said, if we wish to get anywhere with the discussion on sex and gender differences and similarities, we must use excellent communication and a clear framework to combat the ideologues. I hope the biopsychosocial model can provide that framework.

---

[149] Eagly, A., Wood, W. (2013). 351.

7

SEX AS A BIOLOGICAL MECHANISM

The foundational structure of our biopsychosocial model begins by understanding the formwork within which gender differences evolve. This formwork we call *biological sex*. Social constructionist theory proposes that biological sex is subjective, determined by criteria agreed upon by society, and arbitrarily pigeon-holed into a binary category differential.[150] While it is true that the *terms* male and female are linguistic constructs, there is still a biological reality to which the words are referring. The words male and female are not signifiers with an indefinite number of interpretations, as Jacques Derrida might argue. Rather, the words *male* and *female* are written marks used to signify two unique reproductive capacities which allow sexual reproduction to occur. This is why, unlike the social constructionist theory which conceptualizes sex as a social construct, biological sex refers to the two reproductive structures used for procreation.

A simple definition search for male and female provides us with clarity on what *sex* actually is:

> **Female:** *of or denoting the sex that can bear offspring or produce eggs, distinguished biologically by the production of gametes (ova) which can be fertilized by male gametes.*

---

[150] See chapter, "Sex as a Social Construct."

> **Male:** *of or denoting the sex that produces small, typically motile gametes, especially spermatozoa, with which a female may be fertilized or inseminated to produce offspring.*

Social constructionists claim that male and female exist on a spectrum, using arguments of indefinite genetic variation, intersex infants, pseudo-hermaphroditism, and a host of hormonal, chromosomal, and genetic abnormalities.[151] While it is true that a certain amount of variety exists when it comes to sexual characteristics, near 99% of infants are born with healthy sex organs and genitals. Almost all girls are born with a vagina, vulva, clitoris, ovaries, fallopian tubes, and a uterus, and almost all boys are born with a penis, testes, prostate, seminal vesicles, and Cowper's gland. Such a consistent appearance of these two distinct reproductive structures allows us to construct a category we call male and female.

This does not mean, however, that infants born with genital abnormalities should be surgically altered when they are born. Each infant with a disorder of sexual development must be individually analyzed to see what is best for them. For many who have abnormal genitalia, it is often times more humane to wait on surgical reassignment and to allow the individual to understand their sense of self. Some cases, however, require surgical alteration or hormone treatment for the proper functioning of the body, but such cases must be taken at an individual basis.[152]

Now that we know how to define sex, how is this unique biological formwork determined? Social constructionists point to the cases of intersex infants to argue that a baby's sex is *assigned* at birth through unreliable and subjective criteria, not objective observation. Is this true?

---

[151] See chapter, "Sex as a Social Construct."
[152] Diamond, M., Keith, H. (1997). Sex reassignment at birth: a long term review and clinical implications. *Archives of Pediatrics and Adolescent Medicine, 151.*

SEX DETERMINATION

Just a few decades ago, you may have gotten away with claiming that sex is assigned, not determined. Social constructionists in the 1990s relied on subjective definitions of sex and observations of how medical professionals often placed intersex infants into binary categories. While it is true that intersex infants were relegated to overmedicalized views of what constitutes "healthy" genitals, this does not mean that these infants were new sexes. Intersex infants of the 1990s were still male or female; they were just males or females with developmental abnormalities, many of whom had issues with the production of and sensitivity to sex hormones. The problem for social constructionists, however, is that contemporary technology has allowed us to see how sex is determined on the genetic and molecular level in extraordinary detail, making social constructionist claims about the subjectivity of male and female asinine.

It may come as a surprise to many that specific genes which produce sexual differentiation were not discovered until the end of the twentieth century. In the 1950s, the mammalian chromosome Y was found to determine male sex. The specific sex-determining region of the Y chromosome was then discovered in the 1980s, and finally, in the 1990s, the specific sex-determining *gene* inside this region was found, known as the SRY gene, which opposed the network that regulated female development.[153]

Therefore, sexual differentiation is not determined through the environment, socialization, social learning, or culture. Sex is established *in utero* through many different mechanisms which form phenotypic structures required for the production of sex-specific hormones.

We each begin as an undifferentiated zygote at the moment of conception, which is the fertilization of two gametes each with their own genetic information. One zygote may have two X chromosomes, while another may have an X and a Y. (Sometimes abnormalities arise which produce unusual combinations of chromosomes as in the cases of intersex infants. Despite the existence of these abnormalities, most people have XX

---

[153] Gamble, T., Zarkower, D. (2012). Sex Determination. *Current Biology, 22(8)*.

or XY.) As development progresses, both sexes possess equivalent internal structures. However, because of the presence of the Y chromosome in about half of all fetuses, the sexes begin to diverge. The presence of a gene known as the *sex-determining region Y* (SRY) on the Y chromosome causes the development of testes in males, which then release sex-specific hormones.[154] These hormones affect the entire body and the brain, causing further sexual differentiation between males and females.

The SRY gene, therefore, encodes proteins which create sexual differentiation, forming sex-specific phenotypic structures such as testes. It is in the presence of testosterone produced from the testes where male phenotypes develop further. The absence of testes, however, allows for *feminization* of the body and the brain. This means that the default pathway for sex development is female, not male.[155]

The mechanisms are different depending on the species, but they often relate to a class of transcriptional regulators known as the DM domain proteins.[156] The DM domain genes regulate sexual development in most vertebrates, including humans. A gene inside the DM class known as DMRT1 is universally required for male gonadogenesis (development of male gonads). DMRT1 is found in the genital ridge of both sexes before sexual differentiation occurs. It is likely that sexual differentiation occurs when the SRY gene activates DMRT1 in males, which then encodes proteins for use in the sex-specific DM domain.[157] These proteins then promote testes development from the default *female* structure. Because of sex-specific encoders such as SRY and DMRT1, sex is both chromosome and gene specific.

Concisely put, the SRY gene activates the formation of testes, which then produce male-specific hormones which communicate this sex

---

[154] Gamble, T., Zarkower, D. (2012). 4.

[155] Yang, C., Shah, N. (2014). Representing Sex in the Brain, One Module at a Time. *Neuron, 82,* 263.

[156] Gamble, T., Zarkower, D. (2012). 1.

[157] Zarkower, D. (2013). DMRT Genes in Vertebrate Gametogenesis. *Current Topics in Developmental Biology, 102,* 327-356.

determination decision to the rest of the body. Gonads are so important to sexual differentiation that removing the male's testes during fetal development causes XY embryos to develop as *females*.[158]

> "Manipulations of gonadal sex hormones in other vertebrates also have confirmed the critical importance of the gonad in sexual development...There is good evidence that the sex chromosomes also act directly on the fetal mammalian brain to cause sexually dimorphic gene expression."[159]

Thus, sex is not determined through societally-defined criteria but rather through a gene-defined, dimorphic development structure which begins to form as early as the 8th week of gestation.[160] This claim is supported by contemporary scientific research from a variety of fields and methodologies:

> "A wide range of experimental approaches have confirmed that this is so, including experiments involving surgical removal of the fetal gonad and conditional deletion of sex determination genes in mammals, temperature shifts in reptiles with TSD [temperature-dependent sex determination], and temperature shifts using conditional alleles of sex-determining genes in worms. In each case sex reversal required intervention during embryonic development and did not alter phenotypic sex after that time."[161]

Because males and females have different gonads which produce different levels of hormones, neuroendocrinologists have produced a

---

[158] Gamble, T., Zarkower, D. (2012). 4.
[159] Ibid.
[160] Cohen-Bendahan, C., Beek, C., Berenbaum, S. (2005). Prenatal sex hormone effects on child and adult sex-typed behavior. *Neuroscience and Biobehavioral Reviews, 29,* 355.
[161] Gamble, T., Zarkower, D. (2012). 5.

specific field of experimental research showing how these sex hormones affect the brain.[162]

For instance, sex hormones produced in the gonads of males and females directly affect the neural structure of male and female brains in the womb. High levels of testosterone released from the testes "induces masculine patterns of neural and behavioral development, while preventing feminine patterns of differentiation."[163] The details of this hormonal sex differentiation will be discussed in later chapters, where I will cover how gonads and sex-differentiated gene expression produce sex and gender differences in males and females.[164]

EFFECTS OF SEX HORMONES

Sex-specific hormones do not simply produce differences in genital development. Different levels of them in males and females produce differences in brain structure, bone structure, sexual desire, sexual arousal, mood, cognitive function, blood pressure regulation, motor coordination, sensitivity to pain and stress, ovulation in females, sperm production in males, memory retention, spatial-affinity, interest in people versus things, and many more.[165] Different levels of sex hormones also affect gender identity, sexual orientation, sex-typed activity interests, sex-typed cognitive abilities, and sex-related behavior problems and forms of psychopathology.[166] Some of these differences form prenatally with exposure to different levels of testosterone and estrogen.

For example, as mentioned previously, a female fetus exposed to male-typical levels of prenatal androgens will exhibit male-typical psychological

---

[162] Neuroendocrinology is the study of how hormones affect the brain.
[163] Dewing, P., Shi, T., et al. (2003). Sexually dimorphic gene expression in mouse brain precedes gonadal differentiation. *Molecular Brain Research, 118,* 82.
[164] See chapters, "Conception to Birth" and "The Activational Stage."
[165] Marrocco, J., McEwen, B. (2016). Sex in the brain: hormones and sex differences. *Dialogues in Clinical Neuroscience, 18,* 373-383.
[166] Berenbaum, S., Beltz, A. (2011). Sexual differentiation of human behavior: Effects of prenatal and pubertal organizational hormones. *Frontiers in Neuroendocrinology, 32,* 187.

traits such as an interest in *things* rather than *people* and will have a slight affinity for manipulating objects in three-dimensional space.[167] She will also have a higher predisposition to be sexually attracted to both males *and* females.[168] This effect was found through studying cases of Congenital Adrenal Hyperplasia (CAH) in females, a genetic condition which produces elevated levels of prenatal androgens. A meta-analysis on this topic showed that spatial ability in CAH females compared to their unaffected sisters was significantly elevated.[169]

Other traits differentiate more in puberty, where a spike in testosterone in both sexes, but a much higher spike in males, further differentiates men and women in physical and neurological dimensions.[170] Our biopsychosocial model reaffirms that these traits, just like most sex and gender differences, arrange themselves into bimodal distributions, creating a significant overlap between males and females.[171] *It is in the tails of these distributions, however, where neurological differences are most apparent.*

In summary, the standard development path for XX and XY embryos is as follows: XX chromosomes have no sex-determining Y region, which allows for the development of ovaries. Ovaries then produce estrogen and progesterone, which affect the structure of the brain at multiple levels. This sex-differentiated brain structure forms a feminized brain, which then affects the expression of female-typical behavior in childhood and adulthood.[172] On the other hand, XY chromosomes have a sex-determining Y region which promotes the development of testes. Testes then produce higher levels of testosterone which affect the structure of the brain as well. This sex-differentiated brain structure forms a masculine brain, which then

---

[167] Hines, M. (2016). Prenatal androgen exposure alters girls' responses to information indicating gender-appropriate behavior. *Philosophical Transactions, 371.*

[168] Bailey, J., Vasey, P., Diamond, L., et al. (2016). Sexual Orientation, Controversy, and Science. *Psychological Science in the Public Interest, 17(2),* 45-101.

[169] Berenbaum, S., Beltz, A. (2011). 192.

[170] Schmitt, D. (2017); See chapter, "The Activational Stage."

[171] See chapter, "Exploring Gender Differences."

[172] Berenbaum, S., Beltz, A. (2011). Sexual differentiation of human behavior: Effects of prenatal and pubertal organizational hormones. *Frontiers in Neuroendocrinology.*

affects the expression of male-typical behavior. These sex-determining events influence how both the genes and the environment interact to influence gender-specific behaviors in children and adults.[173] Gender-specific behavior, therefore, is often influenced by the gonadal sex hormones:

> "The unique hormonal, genetic, and epigenetic environments of males and females during development and adulthood shape the neural circuitry of the brain. These differences in neural circuitry result in sex-typical displays of social behaviors such as mating and aggression. Like other neural circuits, those underlying sex-typical social behaviors weave through complex brain regions that control a variety of diverse behaviors…In addition, the actions of estrogens and androgens produce sex differences in gene expression within these brain regions, thereby highlighting the neuronal subpopulations most likely to control sexually dimorphic social behaviors."[174]

Furthermore, neuroscientists have found a common model for the genetic control of sexually dimorphic behaviors through sex hormones:

> "Sex hormones control a sexually dimorphic transcriptional program in the nervous system such that individual dimorphically expressed genes control one or a few components of a sex-typical behavior. This model is supported by work showing that genes downstream of sex hormone signaling are required for the normal display of sexual or aggressive displays. Many genes downstream of sex hormone signaling still remain to be identified."[175]

---

[173] Yang, C., Shah, N. (2014). 263.
[174] Bayless, D., Shah, N. (2016). Genetic dissection of neural circuits underlying sexually dimorphic social behaviors. *Philosophical Transactions B, 371.*
[175] Yang, C., Shah, N. (2014). 269.

Diagram: Sex determination and differentiation in mice and humans.

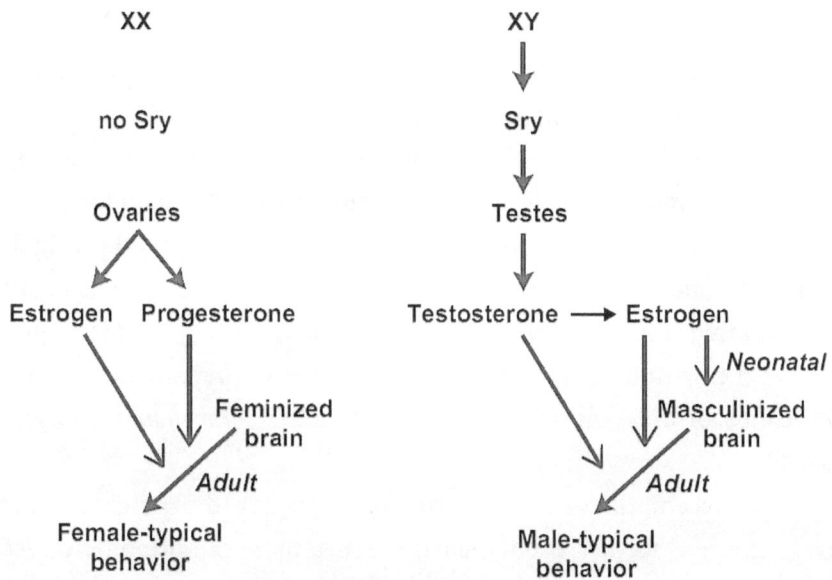

This is a simplified diagram of the sex determination and sexual differentiation process in mice. Such a process occurs in humans as well. The presence of the SRY gene initiates testes development from the bipotential gonads. Testes produce testosterone; some of this testosterone is converted into estrogen which helps masculinize the fetal brain. Without the presence of the SRY gene, the bipotential gonads develop into ovaries, and estrogen and progesterone help produce a feminized brain. Thus, sex determination is binary.[176]

---

[176] Yang, C., Shah, N. (2014). Representing Sex in the Brain, One Module at a Time. *Neuron, 82,* 263.

Sex hormones directly affect the expression of sex-specific genes, which then affect gender-specific behavior. The sex-specific gene *Brs3*, for instance, has been documented to directly affect levels of male aggression. Testosterone is the primary activator of the receptors which encode Brs3 activation. On the other hand, maternal aggression behavior has been mapped to the sex-specific gene Irs4, which is directly affected by exposure to sex hormones from ovaries.[177]

To understand the complexity of how sex-specific genes work in the brain, take the trait of *empathy*, for example. Empathic action tends to be activated in different regions of the brain for men and women. Functional magnetic resonance imaging (fMRI), commonly used to map the activity of the brain, revealed the activation of different brain regions, showing that men and women tend to use different approaches when showing empathy. The amazing thing, however, is that men and women performed *equally* well with empathic action. Therefore, the true sex difference occurs in the *utilization* of different brain regions, not in the *performance* of a given task.[178]

The recent discovery of more than 50 sex differences in gene expression has further strengthened the idea that sex hormones directly influence certain genes, which cause sex differentiation to occur in cellular and molecular properties of neurons.[179] Sex-differentiated brain regions are therefore comprised of sex-differentiated *neurons*.

These discoveries have been largely the result of technological advancements in genetics and molecular biology, which has further elucidated the miraculously complex mechanisms driving human sexual development:

> "Coupled with modern neural circuit mapping tools, the recent dramatic advancements in systematic gene expression profiling, genetic manipulations, and potential deorphanizing of many pheromone receptors will lead to rapid progress in

---

[177] Yang, C., Shah, N. (2014). 269.
[178] Marrocco, J., McEwen, B. (2016). 379.
[179] Ibid., 372.

understanding how sex is represented in the brain and transformed into gender-typical behavior in mammals (Isogai et al., 2011; Luo et al., 2008; Mardis, 2008; Wang et al., 2013; Xu et al., 2012)."[180]

These recent technological advancements have also illuminated the complex interaction between genes and the environment. Unlike animals, human genomes are affected by environmental factors in a sociocultural context. Genes, therefore, are not insulated capsules of heredity but rather are dynamic and adaptive coding mechanisms which are altered by both heredity and the sociocultural environment.

For example, there is reasonable evidence to suggest that if women's sex roles were altered to less parental investment in infants, environmental changes would alter not just personality, but gene expression, which would then affect neuronal structures in the brain. This evolution on the genetic level may be passed down to future offspring so that the affected genes adapt to the new environment. This is the conceptual framework for *epigenetics*, the theory of how gene expression is altered through environmental influences, and it may be able to explain how certain gender-specific behaviors are formed. Some gender-specific behaviors may be *encoded* into our nervous systems through an interaction of socialization, environment, and genetics, all of which evolve and develop inside the formwork of biological sex.

Understanding how sex hormones affect the expression of certain genes is not just important for understanding gender differences; it's also an important field of neuroscience which can facilitate the development of more effective treatments for patients who have sex-differentiated psychiatric illnesses.[181]

Because of the plethora of advancements in molecular biology, genomics, and neuroscience, research into sex and gender differences and similarities when it comes to the brain has never been more profound:

---

[180] Marrocco, J., McEwen, B. (2016). 372.
[181] Yang, C., Shah, N. (2014). 272; Xu et al. (2012).

"These sex differences, and responses to sex hormones in brain regions and upon functions not previously regarded as subject to such differences, indicate that **we are entering a new era in our ability to understand and appreciate the diversity of gender-related behaviors and brain functions.**"[182]

From the activation of the SRY gene and the effects on the gonads to sex hormones and sex-differentiated gene expression, the social constructionist theory is left with not scientific evidence but rather *rhetoric* and *aesthetic arguments* to argue their position that all gender differences are the result of social forces. Social constructionists cannot explain these incredibly diverse findings on sex hormones and their effects on gene expression through the lens of mere socialization; or, in the words of neuroscientists Sheri Berenbaum and Adriene Beltz: "*There is now little question that hormones exert permanent and powerful effects on human sex-typed behavior.*"[183]

While genes are influenced by environmental factors through epigenetics, this does not disprove the fact that genes are expressed through biological mechanisms such as proteins, DNA, chromosomes, and sex hormones. These sex hormones influence the structure of the male and female brain, affecting which parts of the brain activate during specific tasks, as well as modulating numerous traits ranging from brain structure, and bone structure, to sensitivity to pain and stress, memory retention, spatial-affinity, and interest in *people* versus *things*.[184]

Sex is not a social construct. It's a biological mechanism, the formwork within which gender differences evolve, develop, and are also constructed. Gender-specific behaviors are influenced through differentiated sex hormones originating from the gonads during prenatal development, epigenetic factors of modified gene expression to

---

[182] Marrocco, J., McEwen, B. (2016). 373.

[183] Berenbaum, S., Beltz, A. (2011). 197.

[184] Su, R., Armstrong, P. (2008). Men and things, women and people: a meta-analysis of sex differences in interests. *Psychological Bulletin, 135(6),* 859-884.

environmental input, and social learning of ideal masculine and feminine traits which are often overemphasized.

Findings such as the equal performance in empathic action between men and women should give social constructionists a breath of fresh air when it comes to sex difference research. Nuances such as this show that sex and gender research is a complex field worthy of debate, discussion, and inquiry--a field which allows us to gain a better understanding of how nature and nurture interact to form gender-specific differences and similarities. It allows us to dismantle inaccurate and constraining stereotypes, evaluate the usefulness of accurate ones, and provide a solid framework for understanding the origins of sex and gender differences.

Social constructionists can offer a more robust and testable framework if they embrace scientific inquiry and reject aesthetic appeals and ethnographic arguments, but such an abandonment of ideology is unlikely. Social constructionism cannot last forever, however, as its decaying beliefs are stuck in the 1990s. Thirty years of scientific research into the brain, genetics, and molecular biology has far surpassed the aesthetic appeal of strict social constructionist arguments.

In the coming chapters, I will further challenge the postmodern social constructionist theory by integrating research from a variety of fields. And for the rest of the book, we will be exploring how gender differences arise through the complex formwork of biological sex, and how the three factors in our biopsychosocial model (biology, psychology, and society) influence disparities between men and women in liberal democracies.

To understand how important genetics and hormones are to the determination of sex and gender differences and similarities from childhood through adulthood, we need to further elucidate the details of prenatal development.

8

CONCEPTION TO BIRTH

For the next phase of our biopsychosocial model, I will build the biological formwork for sex and gender differences by following the life of a fictional baby girl from conception to birth, birth to adolescence, and adolescence through adulthood. Doing so will help us understand the origins of sex and gender differences. Along the way, I'll integrate research from a variety of fields and provide narrative-based thought experiments to form a high-resolution picture of what sex and gender are, how they interact, and ultimately, how we should treat every individual. Whether you're a parent, a teacher, a sibling, or someone's close friend, you'll gain a better understanding for the true diversity of individual differences.

Unlike social constructionists, we're not looking to explain sex and gender differences through a single variable. Using a univariate explanation to such a complex problem is evident of an incredible lack of intellectual curiosity and is indicative of either complete ignorance or ideological possession. Either way, such a practice must be avoided.

Instead, I will build upon our biopsychosocial model by continuing to integrate the three frameworks of Evolutionary Theory, Cognitive Social Learning Theory, and Sociocultural Theory. Because there's so much to discover, I recommend that everyone enjoy the journey and keep an open mind. There's no telling what you might learn, what beliefs might be changed, and what new perspectives you might gain.

DEVELOPMENT OF THE ZYGOTE

Meet Julia, our fictional baby girl. Right now, she's just a small zygote. Having been conceived just a minute ago, Julia is a completely new person with half her genetic material from her dad and half from her mom; she's unlike any human in all of history, a completely new combination of DNA. The moment she was conceived her chromosomal sex was set.[185] She was given one X chromosome from her mom's egg and another X from one of her dad's sperm cells. Each chromosome is carrying around 2,000 genes. Her XX chromosomes determine her sex, but the other 44 determine many other physical characteristics, such as her hair color, eye color, blood type, height, and build.[186] Barring any hormonal or genetic abnormality, Julia will develop as a healthy baby girl with functioning reproductive organs and near-average levels of sex hormones.

With the genes in place, it's time to get to work. She's traveled all the way from her mom's fallopian tubes to the uterus, which is quite the trek for a zygote measuring only a tenth of a millimeter, or just about 12 to 15 cells.[187] Over the course of a few weeks, Julia begins to multiply and grow into a blastocyst, where she then implants into the wall of the uterus. This blastocyst is comprised of an inner and outer layer. The inner layer will develop into the embryo (Julia's body), while the outer layer will develop into the placenta, which will protect and nourish her as she grows. Implanted into the wall of the uterus, the pregnancy now officially begins.

Around the third or fourth week, Julia begins to differentiate into three layers: the ectoderm, the mesoderm, and the endoderm. The *ectoderm* develops into the nervous system and skin. The middle layer, the *mesoderm*, develops into the muscles, bones, kidneys, heart, and her reproductive system. The inner layer, the *endoderm*, develops into the liver, lungs, certain glands, and the digestive system.[188] The umbilical cord soon

---

[185] Todd, N. (2018). Conception & Pregnancy, Ovulation, and Fertilization. *WebMD*.
[186] Smith, L. (2018). *Your pregnancy at week 3*. Medical News Today.
[187] Ibid.
[188] Jerome, B. (2010). Reproduction and Development. *Visual Learning Systems, In Amazing Human Body, Narrated by Nina Keck.*

develops along with these parts, connecting Julia and her many layers to the placenta. At this point, Julia's mother finds out she is pregnant and stops her nightly intake of Mango White Claw, as alcohol has been shown to have adverse effects on embryotic development.

Soon, Julia's reproductive organs begin to develop from the intermediate mesoderm layer. They start as purely embryonic structures known as the Wolffian and Mullerian ducts. If she had an X and a Y, Julia's Wolffian duct would remain, forming a male reproductive tract, while her Mullerian ducts would degenerate. However, having two X chromosomes means that male sexual development will not be initiated, so Julia will follow the default development pathway of female and her Mullerian ducts will remain.

For about nine weeks after conception, her external genitalia cannot be distinguished from a male or female because of their common development structure. For example, the clitoris in females is the analogue to the penis in males, while part of the labia in females is the analogue to the scrotum in males.[189] It is the presence of the Y chromosome's SRY gene which initiates the Wolffian duct structure's development into the male reproductive system such as the penis, scrotum, testes, and the vas deferens. Once the gonads have differentiated into the testes, they produce testosterone and the anti-Mullerian hormone, causing the Mullerian ducts to degenerate and the Wolffian ducts to remain.[190] The persistence of vestigial Mullerian ducts in genotypic and phenotypic males is rare, but it can happen. For instance, if you remember from Chapter One, the case of the 70 year-old man with a uterus and fallopian tubes is an example of such an abnormality.[191]

For Julia, however, her development will continue as normal, and her Wolffian ducts will atrophy. Throughout the coming months of development, her Mullerian ducts form her fallopian tubes, uterus, cervix, and the upper two-thirds of her vagina.[192]

---

[189] Development of the Reproductive System. *Boundless Anatomy and Physiology, Lumen.*

[190] Sperling, M. (2014). *Pediatric Endocrinology 4th Edition.* Elsevier, Inc. 107-156.

[191] See chapter, "Sex as a Social Construct."

[192] Development of the Reproductive System. *Boundless Anatomy and Physiology.*

PRENATAL ANDROGEN EXPOSURE

By the 9th week, her major body systems are developed, and by 14 weeks, her hands, arms, legs, feet, nose, eyes, and ears are developed.[193] Between 8 and 24 weeks of gestation, male fetuses are exposed to about six-times more androgens than what Julia is exposed to thanks to the development of the testes, which produce increased amounts of testosterone (or androgens).[194]

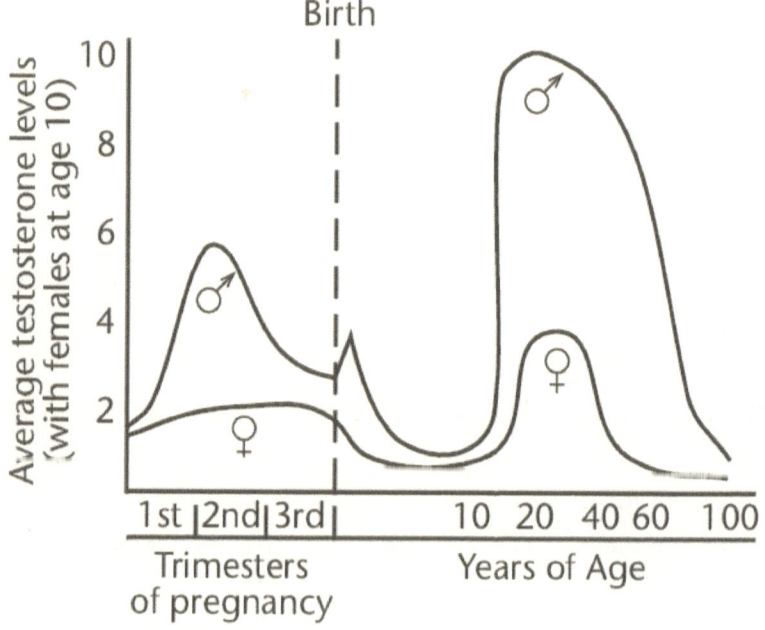

Average testosterone levels in a ratio (with females at age 10 = 1) from conception to death for human males and females.
[Ellis, L. (2011). Identifying and Explaining Apparent Universal Sex Differences in Cognition and Behavior. *Personality and Individual Differences, 51,* 552-561.]

---

[193] Jerome, B. (2010).
[194] Cohen-Bendahan, C., Beek, C., Berenbaum, S. (2005). Prenatal sex hormone effects on child and adult sex-typed behavior. *Neuroscience and Biobehavioral Reviews, 29,* 355.

This increase in androgen doesn't just initiate male sex development, it alters the brain structure by modifying the expression of sex-specific genes, which then produce sex differentiation in cellular and molecular properties of neurons.[195] All of this is initiated by the presence or absence of male-typical androgens in the womb:

> "Once the differentiation of these sexual organs is settled, sexual differentiation of the brain happens, by permanent *organizing* effects of sex hormones on the developing brain. During puberty, the brain circuits that have been organized in the womb will be *activated* by sex hormones."[196]

Because of the presence of sex hormones such as androgens, sex-differentiated brain regions become comprised of sex-differentiated *neurons*.[197] For example, the hypothalamus region of the brain, which controls the release of hormones, body temperature, sexual behavior, and emotions, has large amounts of sexually dimorphic cells which are produced through exposure to sex hormones in the womb.[198] Other structural sex differences have been found in other parts of the brain, such as the anterior commissure, the interthalamic adhesion, the corpora mamillaria, the amygdala, and the cortex.[199]

There are two major effects of prenatal androgen on the brain worth noting. The first is known as *suboptimal arousal,* which refers to the tendency to grow bored of one's environment relatively quickly. High androgen exposure promotes suboptimal arousal. The higher the

---

[195] Marrocco, J., McEwen, B. (2016). 372.

[196] Bao, A., Swaab, D. (2011). Sexual differentiation of the human brain: Relation to gender identity, sexual orientation and neuropsychiatric disorders. *Frontiers in Neuroendocrinology, 32,* 215.

[197] Yang, C., Shah, N. (2014). 261.

[198] For a list of traits and behaviors which are affected by prenatal androgen, see chapter, "Sex as a Biological Mechanism."

[199] Bao, A., Swaab, D. (2011). 216.

androgen, the more the person will likely seek new environments and experiences and get bored of old ones, even in the face of risk.[200] Suboptimal arousal also insulates one from feeling pain. This can be seen in the tendency for teenage boys to take higher risks and to be pain-insensitive compared to teenage girls. Therefore, the activation of suboptimal arousal through androgen exposure is one of the major mechanisms underlying this sex difference in risk-taking.

The second major effect of prenatal androgen relates to the functional structure of the brain. High androgen levels initiate what's known as a *rightward shift* in neocortical functioning. Rightward shift does not just cause a lateralization of the brain to the right hemisphere; it's the nature of the brain's connections which change. The connections in one's brain move *within* hemispheres and *between* hemispheres, but there are sex differences in how these connections work. Females tend to have brains whose connections move *between* the hemispheres, thereby creating a higher interconnectivity. However, androgen exposure causes this interconnectivity to become more lateralized, which is to say, to become less interconnected and more specialized in one hemisphere. Males, therefore, show much more connections *within* the hemispheres than between the two.[201]

Androgen exposure produces more lateral emphasis on both hemispheres, especially during development in the womb, which then allows for increased hemisphere-based specialization. Because of this, babies exposed to high levels of prenatal androgens "rely less on language-based reasoning, emphasizing instead reasoning which involves spatial and temporal calculations of risk and reward possibilities."[202] Therefore, a masculinization of the brain, or a rightward shift, means the brain's connections are more lateralized to the right, which then specializes the brain for more specific cognitive tasks. This is, in part, why males and females share a sex difference in verbal skills: Since the average male brain

---

[200] Ellis, L. (2011). Identifying and Explaining Apparent Universal Sex Differences in Cognition and Behavior. Personality and Individual Differences, 51, 554.
[201] Khazan, O. (2013). Male and Female Brains Really Are Built Differently. *The Atlantic.*
[202] Ibid.

is less interconnected between the hemispheres than the average female brain, the side of the brain which focuses on spatial reasoning and arrangement of objects into an organized system (systematizing) can be more specialized and developed. Since female brains tend to be more interconnected between both hemispheres, girls tend to excel at *both* verbal skills and math compared to boys' skills with math *only*.

Because of these two major structural differences in both *suboptimal arousal* and *rightward shift*, exposure to high levels of androgens in the womb predicts male-typical psychological traits in play, behavior, personality, and occupational interests, where as an absence of high levels of androgens predicts female-typical psychological traits in play, behavior, personality, and occupational interests.[203] These sex differences are some of the largest and most robust findings in all of psychology.[204] Because of the huge range of data, androgens' influence on personality and occupational interests will be a topic for later chapters.

The effects of sex hormones on the brain's organizational structure has not just been documented in humans, but in thousands of studies on nonhuman animals, including primates:[205]

> "It is now clear…that the levels of sex hormones present during early development produce long-lasting effects on a variety of behaviors that show sex differences, including sexual behavior but extending well beyond to include, for example, learning and memory, aggression, and play. This evidence comes from experimental manipulations of hormones (e.g., castration of males, androgen treatment to females) and from natural variations (e.g., gestating between animals of the opposite sex)."[206]

---

[203] Hines, M. (2016).
[204] Berenbaum, S., Beltz, A. (2011). 187.
[205] Bachevalier J., Hagger, C. (1991). Sex differences in the development of learning abilities in primates. *Psychoneuroendocrinology, 16,* 177-188.
[206] Berenbaum, S., Beltz, A. (2011). 184.

For example, female animals exposed to high prenatal testosterone showed increased male-typical behavior and decreased female-typical behavior. It also works the other way: male animals who have their testes removed early in development show increased female-typical behavior and reduced male-typical behavior.[207]

Despite the impact of sex hormones on the brain, researchers have also discovered over 50 genes which express themselves at different levels in male and female brains even before these sex hormones come into play. Using microarrays and a laboratory technique known as *reverse transcription polymerase chain reaction* (RT-PCR), which is used to study gene expression, molecular geneticists have discovered over 50 candidate genes for differential sex expression.

Many of these genes show differential expression between the developing brains of male and female mice before any hormonal activation can occur, suggesting that genetic factors may influence brain sexual differentiation.[208] For example, one of the genes, *Eif2s3y*, was found to be expressed nine times higher in males than in females. This suggests that this gene acts in the brain to "promote masculine patterns of neural and behavioral development."[209] Genes such as the one studied, therefore, may contribute to a variety of sex differences in behavior.

Other examples of sex-differentiated genes have elucidated the amazingly complex nature of sex differences in the brain. For instance, stress-induced remodeling of dendrites and synapses in the hippocampus occur near puberty based on elevated levels of testosterone, whereas brain regions such as the prefrontal cortex and primary sensory-motor cortex show estrogen-regulated spine synapse formation.[210] This sex hormone-dependent synapse formation can then produce behavioral differences in

---

[207] Hines, M., Constantinescu, M., Spencer, D. (2015). Early androgen exposure and human development. *Biology of Sex Differences, 6(3),* 2.

[208] Dewing, P., Shi, T., et al. (2003). Sexually dimorphic gene expression in mouse brain precedes gonadal differentiation. Molecular Brain Research, 118, 82.

[209] Dewing, P., Shi, T., et al. (2003). 88.

[210] Marrocco, J., McEwen, B. (2016). 377.

males and females. For example, females exposed to high levels of estrogens can often respond differently to stressors than males exposed to low levels of estrogens.[211] A few of these brain-level sex differences show almost no overlap between males and females. For instance, the nucleus of the preoptic area (SDN-POA) in brains of rodents comes close to complete sexual dimorphism.[212] Females tended to show a higher hippocampus gene activation than males through acute stress, providing evidence that male and female brains activate different genes in response to the environment. This feminization of the brain is maintained by the "active suppression of masculinization through DNA methylation, pointing to epigenetic modifications that promote and maintain sex dimorphic features."[213] *DNA methylation* occurs when DNA is tightly packed into *heterochromatin*, where the DNA cannot be transcribed or "read" to elicit a normal response. Thus, DNA methylation is the process of causing certain genes to be *inactive* and not expressed. In females, DNA methylation suppresses masculinization of certain brain features. The fact that genes modify themselves to maintain sexually dimorphic features shows a curious relationship between sex differences and *epigenetics*, or how genes can be modified based on environmental factors.

Despite this research into mice, many sex differences in the brain between human males and females are subtle and involve complex connection patterns. Such discoveries in neuroscience and neuroendocrinology would not have been possible in the 1990s, when technology for analyzing such neuron differentiation was not available, marking the peak of social constructionist theory.

Based on this research, Julia's brain structure will likely be influenced by two things: 1) organizational sex hormones in the womb and 2) sex-specific genes in the brain. The amount of prenatal testosterone she is exposed to will determine whether she develops male-typical or female-typical genitalia.

---

[211] Marrocco, J., McEwen, B. (2016). 377.
[212] Ibid., 376.
[213] Marrocco, J., McEwen, B. (2016). 377.

PSYCHOLOGICAL EFFECTS OF ANDROGENS

Starting around the 8th week, the absence of high levels of testosterone (due to no testes) allows Julia to continue developing as a female. Her Mullerian ducts remain and her reproductive structures finally begin to form. The absence of high levels of testosterone also allows the clitoris, labia majora, and separate vaginal and urethral canals to develop. However, Julia is also slowly exposed to *higher than average levels of androgens for a female*. Usually a female fetus would be exposed to relatively low levels of androgens, while a male fetus would be exposed to six times as much. And yet, for Julia, her body is met with increasingly higher levels of androgens as she develops. The exact reason is unclear, but current findings suggest a mixture of three sources: the umbilical cord, sex hormones found in the mother's blood during pregnancy, and amniotic fluid. Out of the three, sex hormones from the mother's blood is the most likely candidate for variance of sex hormones in the fetus.[214]

Therefore, we can postulate that the presence of higher androgens for Julia is a result of higher testosterone in her mom. High androgens in the mother often results in masculinized gender-role behavior in female offspring.[215]

> "Consistent with expectations based on findings in Congenital Adrenal Hyperplasia (CAH) girls and studies in other mammals, mothers who had high serum testosterone during pregnancy were more likely than mothers with low serum testosterone to have daughters who were interested in boy-typical toys and activities."[216]

---

[214] Cohen-Bendahan, C., Beek, C., Berenbaum, S. (2005). 365.
[215] Ibid.
[216] Hines, M., Golombok, S., Rust, J., Johnston, KJ., Golding, J. (2002). Testosterone during pregnancy and gender role behavior of preschool children: a longitudinal population study. *Child Development, 73,* 1678-1687.

8 CONCEPTION TO BIRTH

This higher level of prenatal androgen won't affect the development of Julia's reproductive organs since she hasn't reached high enough concentrations of androgens, but a certain level of androgens *can* affect the sex-differentiated regions of her brain, which then would affect the properties of her neurons and the expression of sex-specific genes. As her brain develops in the womb, the higher levels of prenatal androgens begin to produce organizational effects in her brain.

For instance, high prenatal androgen levels are thought to be associated with reduced eye contact and small vocabulary in infancy.[217] Compared to infant girls, infant boys made less eye contact and had a smaller vocabulary, a statistically significant effect size that ranged from small to moderate for both eye contact and vocabulary use.[218] Such differences in behavior are theorized to result from high levels of androgens, which causes sex-differentiated brain structures. This effect can be especially seen in autism, where incredibly high levels of androgens masculinize the brain to such a point where autistic boys struggle to interact effectively in social situations.[219]

Furthermore, research has also shown the effects of androgens on girls. Studies on Congenital Adrenal Hyperplasia further illuminate the effects of prenatal androgens on sex-typed activity interests:

> "Compared to their unaffected sisters or unrelated controls, girls with CAH play more with boys' toys and less with girls' toys, and are reported by their parents to be more interested in male-typed toys and activities and less interested in female-typed toys and activities; when given a toy to keep, they are more likely to pick a boys' toy (such as a toy airplane); girls with CAH make drawings that are more like those of typical boys than typical girls, using masculine characteristics (such as moving objects,

---

[217] Lutchmaya, S., Baron-Cohen, S., Raggatt, P. (2002). Fetal testosterone and vocabulary size in 18- and 24-month-old infants. *Infant Behavioral Development, 24,* 418-424.
[218] Lutchmaya, S., Baron-Cohen, S., Raggatt, P. (2002). 418-424.
[219] Cohen-Bendahan, C., Beek, C., Berenbaum, S. (2005). 367.

dark colors, a bird's-eye perspective) more than feminine characteristics (such as human figures, flowers, and light colors)."[220]

Among girls with CAH, this sex-typed preference does not just end in childhood; women with CAH report that they prefer more male-typical activities and careers:

> "Females with CAH continue to be interested in male-typical activities into adolescence and adulthood, although the evidence here is based on self-reports and not observations. Thus, compared to their sisters, teenage girls with CAH report greater interest in activities like electronics, cars, and sports, and less interest in activities like cheerleading, make-up, and fashion.
> Adult woman with CAH report that they are more interested than their sisters in male-typed activities, such as sports and electronics. Across age (childhood, adolescence, and adulthood), females with CAH also express interest in male-typed careers, such as engineer, construction worker, and airline pilot. These interests manifest themselves in behavior: adult females with CAH engage more in male-typed occupations than do females without CAH. Differences in interest between females with and without CAH are very large, with little overlap between the groups. It is characteristic of girls and women with CAH to be interested in male-typed activities."[221]

One reason for this preference in male-typed activities is in the category of cognition. While males and females have equal levels of overall intellectual ability, males tend to perform better at mental rotation and spatial ability than females, making these two categories some of the largest

---

[220] Berenbaum, S., Beltz, A. (2011). 188.
[221] Ibid.

sex differences in cognition.[222] These differences have been shown to directly relate to the exposure of prenatal androgens.[223] Part of this data has been further solidified through examining females with CAH, who tend to have a higher spatial ability than their sisters throughout all of development, including childhood, adolescence, and adulthood.[224] Spatial ability positively correlates with the preference for male-typical tasks and occupations. While males tend to have an advantage in mental rotation and spatial ability, females tend to be significantly better at *spatial location memory*.[225] Girls with CAH, however, tend to have the male-typical advantage when it comes to spatial ability.

While Julia does not have Congenital Adrenal Hyperplasia, she has been exposed to higher than normal levels of androgens for an average female because of the high levels of testosterone in her mom's blood during pregnancy. This difference may manifest itself throughout her childhood if she is left to freely explore her own interests. She may explore more male-typed activities due to the effects of androgens on her organizational brain structure, giving her a greater interest in mechanical things and objects rather than more relational categories such as people and language.

While the high levels of androgens may affect her play preferences, behavior, sex-typed occupational interests, and the organizational structure of her brain, it will most likely *not* affect her *gender identity* (female or male identified). This may come as a surprise to many, but even CAH doesn't really affect one's gender identity to a great degree. Evidence from females with CAH shows that the overwhelming majority identify as female throughout their life, despite the abnormally high levels of

---

[222] Berenbaum, S., Beltz, A. (2011). 192.

[223] Ibid.

[224] Hines, M., Fane, BA., Pasterski VL., Mathews, GA., Conway, GS., Brook, C. (2003). Spatial abilities following prenatal androgen abnormality: targeting and mental rotations performance in individuals with congenital adrenal hyperplasia (CAH). *Psychoneuroendocrinology, 28,* 1010-1026.

[225] Berenbaum, S., Beltz, A. (2011). 192.

androgens.[226] However, for *sexual orientation*, higher levels of androgens has been shown to increase non-heterosexual orientation in women:

> "For example, of 30 women with CAH, 20 reported exclusively heterosexual fantasies; all 15 control women reported exclusively heterosexual fantasies. Other studies indicate that about 15-25% of women with CAH are not exclusively heterosexual, with nonheterosexuality associated with higher prenatal androgen levels."[227]

Androgens are known to affect more things than just cognition and behavior; they are also known to affect differences in physical traits. For example, the ratio of the length of the second digit (index finger) to the length of the fourth digit (ring finger) is fixed by the 14th week of gestation. This ratio, known as the 2D:4D ratio, is likely affected by prenatal androgen exposure during weeks 8-24. The higher the androgens, the lower the ratio is. In other words, men tend to have longer ring fingers relative to their index fingers, and women tend to have longer index fingers relative to their ring fingers:

> "This sex difference was noted more than 125 years ago, and a standard method for measuring it was developed 55 years later. The finger ratio appears to be stable from age 5 and shows a sex difference across races. In fact, the 2D:4D ratio shows larger ethnic than sex differences. The sex difference in the 2D:4D ratio is found on both hands, although it is somewhat larger on the right than the left hand, d's of 0.85 and 0.74 respectively."[228]

---

[226] Berenbaum, S., Bailey, J. (2003). Effects on gender identity of prenatal androgens and genital appearance: evidence from girls with congenital adrenal hyperplasia. *Journal of Clinical Endocrinology and Metabolism, 88,* 1102-1106.
[227] Ibid., 191.
[228] Cohen-Bendahan, C., Beek, C., Berenbaum, S. (2005). 373.

Let's do an experiment! Hold up your right hand and look at your ring and index fingers. What do you see? If your index finger is longer than your ring finger, or equal, you're likely a female. If, however, your *ring* finger is longer than your index finger, then you're likely a male. On the other hand, if you're a female whose ring finger is longer than your index finger, then this means you may have been exposed to higher levels of prenatal androgens than your female counterparts. And the reverse is true for males with a female-typical finger ratio. Surprised yet?

Amazingly, this 2D:4D ratio is also associated with other sexually dimorphic traits. For instance, body mass index, waist-to-hip ratio, and waist-to-chest ratio are all associated with a low 2D:4D ratio, which corresponds to high androgen levels. In other words, masculine body forms are correlated to a masculine finger ratio. And yet, this association goes further still. Adult sperm number and circulating testosterone levels are negatively associated with right-hand finger ratio. In other words, the lower your 2D:4D ratio is (lower means more masculine), then the higher your sperm count and testosterone levels will likely be. In terms of other associations with 2D:4D, cognition traits such as spatial ability have been correlated with finger ratio.[229] Spatial ability is negatively correlated with finger ratio, meaning the higher your spatial performance is, the lower the ratio will be, but, according to the 2005 research, this has not yet been produced consistently.[230] Such correlations from finger ratio are simply that...*correlations*. Based on the existing research, there is not conclusive causal evidence for prenatal androgen effects on such ratios, only indirect evidence and strong correlations.

Considering the evidence of prenatal androgen's effects on a baby's development, including her brain structure, digit ratio, and later sex-typed behavioral and psychological traits in early childhood, it is clear that prenatal androgens play an important and significant role in sex and gender differences.

---

[229] Manning, JT. (2002). *Digit ratio: a pointer to fertility, behavior, and health.* New Brunswick, NJ: Rutgers University Press.

[230] Cohen-Bendahan, C., Beek, C., Berenbaum, S. (2005). 375.

HORMONES AND SOCIALIZATION

Evidence from girls who were exposed to abnormally high levels of androgens (CAH) showed that prenatal androgens not only affected their sex-typed toy preferences, but influenced their interests and occupational preferences as they grew into adulthood.[231] High levels of androgens in girls have been shown to arise, in part, from the levels of testosterone in the mother's blood during pregnancy. Putting aside the effects of sex hormones, further sex and gender differences have been shown to originate from more than 50 genes in the brain which are expressed differently in males and females before hormonal organization can occur.[232]

The exposure to prenatal androgen seems to induce brain changes during gestation, which then manifest themselves in sex-typed behavioral play in early childhood. However, sex-typed behavioral play cannot be solely the result of prenatal androgen exposure. As social constructionists have rightfully pointed out, utilizing the Cognitive Social Learning Theory, gender-typical behavior is also encouraged by parents through parental-toy choice. Parents will give their boys male-typical toys, and they will give their girls female-typical ones.

> "Once children understand that they are girls or boys, they tend to model, or imitate, the object choices and other behaviors of individuals of the same sex more than those of individuals of the other sex. They also respond to gender labels, preferring objects, including toys and activities that have been labeled as appropriate for their own sex over those that have been labeled as appropriate for the other sex."[233]

---

[231] Berenbaum, S., Beltz, A. (2011). 196.
[232] Dewing, P., Shi, T., et al. (2003). Sexually dimorphic gene expression in mouse brain precedes gonadal differentiation. Molecular Brain Research, 118, 82.
[233] Hines, M. (2016). 2.

Therefore, the process of modeling and labeling gender-appropriate behavior is a common phenomenon which affects gender-development and gender-typed behavioral preferences. However, Cognitive Social Learning Theory does not explain all of the origins of sex and gender differences.

Because both prenatal androgens and socialization have been shown to affect gender-typed behavior, we should conceptualize prenatal androgens and socialization as *working together* in different ways, rather than working apart. For example, it is likely that prenatal androgen exposure directly affects gender-typed behavior through organizational effects on the fetal brain. Male fetuses, having been exposed to high levels of androgens, consistently exhibit male-typical psychological traits and toy preferences in infancy and early childhood.[234]

By recognizing the effects of prenatal androgens, we can develop a better model for socialization, which should be conceptualized as the process by which external modes of gendered behavior become internalized. This internalization, or self-evaluation, is not solely affected by societal expectations, however. The process of socialization is also the process of *maximizing* or *suppressing* the effects of sex hormones on the brain; in other words, biology and culture interact to produce differences in gender expression.

A boy who has lower than average androgen levels, for example, may be more interested in female-typical toys. However, due to societal expectations of male-typical behavior, this innate interest may be suppressed through parental influence. In this sense, his natural predispositions towards female-typical behavior become toned down and less evident, even though he may have been exposed to less androgens than his brothers.

On the other hand, socialization may play a smaller role in children whose parents let them explore their own interests. If parents let their children play with a variety of toys of their choice, the organizational effects of prenatal androgens will maximize. For instance, young CAH girls given

---

[234] Hines, M. (2016). 6.

a choice of male-typical and female-typical toys will mostly show interest in the *male-typical* toys, maximizing the effects of prenatal androgens through free choice rather than minimizing them through socialization.[235] Unlike the strong social constructionists who only allow for one variable (socialization), I account for both biological and cultural factors in the example above.

To deny the effects of prenatal androgens is to ignore an abundance of research from many fields, including newly developed areas of scientific exploration such as *Frontiers in Neuroendocrinology*. Denying the influence of androgens ignores the thousands of studies of prenatal androgen effects on nonhuman animals, and it fails to recognize the possibility that certain behavioral preferences and interests may be innate. Ironically, social constructionists reject the very arguments which can be used to support the fair treatment of others: *namely, that gender atypical behavioral differences, even sexual orientation and gender identity, are partly biological, not mere social constructs.*

At the same time, we must recognize the relationship of prenatal androgens to socialization. Like the relationship between biological sex and gender, prenatal androgen is often the *formwork* upon which socialization will occur. Early androgen exposure doesn't exist on its own; it influences the continual process of self-socialization through childhood. Children will evaluate their own interests and behaviors through modeling of others and interaction with gendered labels. These social evaluations interact directly with the already-present organizational effects of prenatal androgens on the brain. Or, in the words of sex researcher Melissa Hines:

> "Thus, the influences of prenatal androgen exposure on behavior across the lifespan may reflect in part the cascading effects of alterations in processes related to self-socialization of gendered behavior, as well as neural changes induced by androgen prenatally."[236]

---

[235] Hines, M. (2016). 6.
[236] Ibid., 8.

Due to the effects of prenatal androgens, society may never be able to fully *equalize* sex differences in motivations, behaviors, and interests. Despite social constructionist attempts at *equity across all hierarchies,* sex differences in desires, preferences, and interests continue, and these differences are especially large across those nations who have done the most for equalizing the playing field for boys and girls when it comes to education, occupations, and other forms of socialization.[237] In other words, even the most egalitarian societies may never see equality of outcome when it comes to sex differences in choices, interests, and behavior.

Far from the blank slate claims in social constructionist theory, letting people make their own choices based on their own individual preferences will maximize biological innateness, resulting not in *equity* but *inequality.* This inequality would not be the result of socialization pressures, but rather the product of innate tendencies and natural interests. It would show that the blank slate was a farce, that sex and gender differences exist, and that equality of outcome should not be pursued. For the social constructionist, such an inequality through individual choice would be intolerable, as it renders their entire philosophy obsolete.

With few new models at their disposal, the social constructionist theory has become incredibly outdated and unreliable. Their simplistic theories and models which only account for socialization and patriarchal power hierarchies do not match reality. However, to keep their philosophy from atrophying into non-existence, social constructionists shift their attention from scientific inquiry and deny the validity of scientific practice altogether through aesthetic arguments, rhetoric, and fanciful claims of gender equity.

Social constructionists deny the 300 universals of human experience through blank slate ideology; they deny prenatal androgen's effects on the organizational structure of the brain; they deny its effects on male-typical and female-typical psychological traits; they deny biological causes for CAH girls' male-typical behavior; they deny more than 50 sex-differentiated genes in the brain; and they deny entire fields of research,

---

[237] Ellis, L. (2011). 555.

from neuroscience and neuroendocrinology to molecular genetics and evolutionary biology. Having been rendered almost completely useless by scientific standards, social constructionists continue to rely on purely sociocultural arguments for the existence of sex and gender differences. This reliance only gets stronger as we enter the development of gender identity in childhood.

Going back to our fictional baby girl, now that Julia has been exposed to higher levels of androgens, she will likely exhibit male-typical psychological traits and male-typical behavioral preferences despite socialization. However, socialization will still have an effect on her behavior. The question is: how does this socialization arise? And what happens if her parents suppress the effects of her prenatal androgens through gender-based reinforcement and punishment?

In the next chapter, we will explore the development of gender identity, beginning with its theoretical basis as a social construct. Then we will explore how a mix of prenatal androgens and socialization influences Julia's development of her own gender identity. Given that the male and female brains have different organizational structures affected by androgens in the womb, as well as genetic factors which predispose males and females to unique sex-differentiated gene expressions, how does this neural structure affect Julia's behavior and interests? And how do these behaviors and interests interact with the process of socialization?

9

DEVELOPMENT OF GENDER IDENTITY

Social constructionists conceptualize gender identity as an individual's sense of one's own *gender*. Or, one's innermost concept of self as "male, female, a blend of both or neither -- how individuals perceive themselves and what they call themselves. One's gender identity can be the same or different from their sex *assigned* at birth."[238]

From this definition, gender identity is not theorized to be something innate and biological, but rather something which is fluid, transient, and moldable based on feeling and whim, and yet, at the same time, gender identity is also viewed as an immutable trait. And here we find a contradiction in logic: social constructionists argue that gender identity is as flippant as one's personal coffee preferences while simultaneously arguing that gender identity is an objective category which must be protected from discrimination. Sorry, but you cannot have it both ways. Either your gender identity is formed from a mix of biological and social factors linked to your sex, giving it an objective biological reality, or it's something which can be molded and changed by preference and feeling, giving it a subjective reality. Surely those fighting for the rights of transgender and non-binary people can get behind objective biological arguments, right? Well, not so much.

Let's remind ourselves of a quote from sex educators in the UK, who taught that gender is based on fluid feelings and personal preferences:

---

[238] (2019). Sexual Orientation and Gender Identity Definitions. *Human Rights Campaign.*

> "Gender is how you feel on the inside about whether you're a boy or girl, a man or a woman, if you're non-binary, feel like neither, or both. People can also be fluid, feel more like female or more like male based on a different day or time...it's really individual."[239]

If gender identity can be, like they say, based on a different day or time, then this quickly becomes problematic. Like the postmodernists rightfully argue, this subjective feeling of one's own gender proliferates into a near limitless number of potential identities, each one with their own unique feelings and sense of self. From this proliferation we arrive at the logical end: everyone exists as a unique individual with their own desires, interests, and identity. There would be, in a sense, 7 billion unique gender identities, one for every human on the planet.

In this aspect, *gender identity* is not so different from *personality*. Just as there are limitless degrees of difference when it comes to personality traits, social constructionists claim there are limitless degrees of difference when it comes to *gender identity*. Because of this, legislating gender identity into law as if it's a biological reality which must be protected remains a major problem insofar as people's identities are both subject to whim and *feeling* based on a given day or time and infinite in their uniqueness. This proliferation problem makes it easy for laws protecting *gender identity* to be abused and near impossible for them to be enforced; a similar problem would arise if the state forced people to respect other's religious beliefs, which can also be near infinite in number and complexity. *Or, imagine trying to enforce a law protecting each person's unique personality.*

Instead, to protect the rights of all people, use of language should be specific and careful. Like a surgeon's scalpel, laws written to protect people from discrimination should utilize extreme caution and be precise. We do not want to compel people's speech by law, nor do we want the law to be abused by ideologues.

---

[239] Questioning LGBT/CSE Education, (2019). Sexualizing Children with Sex Education. *Twitter.*

Instead of arguing for gender identity as something which is fluid and malleable based on time, day, and feeling, gender identity is better conceptualized as a biologically-based identification (or lack of identification) with one's own birth sex. Gender identity is influenced by a mix of biological and sociocultural factors which become fixed at a very early age. It is not fluid, transient, or malleable after this fixity; it is almost inextricably linked to biological sex and is only rarely at odds with it. Thus, conceptualizing gender identity as biologically-based allows us to form a stronger argument for the rights of all individuals, because something immutable has a much stronger footing than something which can change at any time. Data to support this conceptualization comes from a variety of fields, which I will discuss throughout the chapter.

Despite the potential strength of the biological argument for the source of gender identity, social constructionists reject all of it in favor of the more ephemeral and ever-changing effects of *socialization* and *indeterminate feelings*. Utilizing the Cognitive Social Learning Theory, social constructionists posit that all gender differences arise through reinforcement and punishment of gender-appropriate/inappropriate behavior from infancy onward.[240] Children's sense of gender identity is reinforced through parental encouragement and biased play.[241] Possible deviances from stereotypical gendered norms are punished, as parents either discourage their children from playing with sex-typed toys opposite of their gender or suppress their children's innate tendencies through correction, play manipulation, and even abuse.

Parents influence children's choice of toy through providing only male-typed toys for the boy and female-typed toys for the girl. This continual play with gender-appropriate toys reinforces a binary gender identity in the child and contributes to the child's self-socialization and internalization of expected behaviors.[242] Gender stereotypes are seen as accurate representations of an innate reality, and children align themselves

---

[240] Hyde, J. (2014). 377.

[241] Kane, E. (2006). 'No Way My Boys Are Going to be Like That!' Parents' Responses to Children's Gender Nonconformity. *Gender & Society, 20(2),* 152.

[242] Boe, J., Woods, R. (2017). 3.

with these stereotypes for either fear of social reprisal or fear of cognitive dissonance across their internalized behavioral structures. As mentioned in Chapter 2, because stereotyped gender behaviors often become solidified in childhood, occupational career choice can be significantly affected as men and women sort themselves into careers which mimic the roles they practiced, learned, and enacted as a child.[243] Because of this, a child's gender identity is not allowed to be fully expressed.

And yet, while social constructionists embrace socialization arguments when children *conform* to stereotypical behavioral patterns, the arguments reverse when children are *gender non-conforming*. When a girl plays with male-typical toys and thinks she's a boy, social constructionist ideologues reject socialization arguments in favor of 'innate feelings,' 'authenticity,' and 'essence.' It does not matter if the girl *is* a female or *was* socialized as a girl; all that matters is that she *feels* like a boy. And thus, by saying a gender non-conforming girl *feels* like a boy, and therefore *is* a boy, social constructionists transform gender stereotypes into rigid biological realities. What does it actually *feel* like to be a boy? How can such a question be answered without resorting to stereotypical understandings of what a boy is? Here, the social constructionist faces a major problem: if gender identity is based on a stereotypical understanding of what it feels like to be a boy or a girl, then the social constructionist is no better than the biological essentialist who conflates biological sex with gender stereotypes. If gender identity is a mere social construct based on indeterminate subjective feelings, then what's stopping us from molding children in accordance with our wishes? What if, instead, gender identity is actually a biological reality tied to genetics, hormones, and internal reproductive structures, traits which help determine whether you're a male or female?

Let's think for a moment: what does it mean for something to be *expressed*? Authentic expression requires the trait being expressed to be in some sense *innate*, doesn't it? Expressing oneself means expressing one's uniqueness, one's innate traits which make you who you are. If you are merely the composition of societal conditioning, then *what are you*

---

[243] Kollmayer, M., et al. (2018). 2.

exactly? And what does it mean to express yourself if all you are is a product of society? Identities cannot be at the same time both socially constructed and personally authentic. Either one's gender identity points back to a biological formwork of male or female, a mold from which one's uniqueness develops, or this identity does not exist at all. Once again, it is in the formwork of biology where the strongest arguments reside for the legal protections of individuals, and especially for the protections of biological females.

However, the social constructionist cannot rectify the contradictions between *innate being* and the *socialized self*. When challenged on this, social constructionists will often retreat to sociological arguments by putting all emphasis on the *socialized self*. If societal forces could be removed, they argue, then gender identity would either proliferate into a wide spectrum of identities or cease to exist as a concept. In this utopia free of socialization's effects on gendered behavior, boys and girls would grow up equal, with equal levels of average interests, preferences, and behaviors. Everyone would exist on a spectrum, and everyone could be whoever they wanted to be, regardless of any biological reality. Unfortunately, this is a fantasy world, and I will prove it.

It is an undeniable fact that gender identity is tightly linked to biological sex:

> "The largest psychological sex difference is in gender identity: the overwhelming majority of girls and women are female-identified, and the overwhelming majority of boys and men are male-identified."[244]

However, through socialization and biological abnormalities, gender identity sometimes does not match with sex. For instance, gender identity disorder (mismatch of one's biological sex and perceived gender identity)

---

[244] Berenbaum, S., Beltz, A. (2011). 189.

is prevalent among 2-5% of children, showing that this mismatch is almost as rare as being born intersex.[245]

> "Boys are referred for treatment more than girls, three to six times more in childhood, but only 1.2-1.3 times more in adolescence. This may reflect referral bias due to cultural factors; for example, beliefs that girls may outgrow cross-gender behavior, and less tolerance of cross-gender behavior in boys than in girls."[246]

Among this 2-5% of children with gender dysphoria, most of them grow out of it as they age and recognize that they cannot change their biological sex. Because of this, many psychologists recommend *waiting it out* if your child has a mismatch between their sex and gender identity. Once the child reaches adolescence, the presence of pubertal hormones will often influence the child's gender identity in the direction that is consistent with either androgen or estrogen exposure.[247] This is why, for transgender individuals, hormone treatment should wait until after adolescence to see if the child outgrows their gender dysphoria. Once they realize that just because they enjoy playing with gender-atypical toys does not mean they are actually the opposite sex, they will also accept themselves for who they are, for their biological reality. Thus, attempts to change the child's sex can quickly become not just unnecessary, but incredibly damaging. Without surgery and hormones, healthy development will often proceed unaffected. Even among CAH girls who have been exposed to high levels of prenatal

---

[245] Zucker, K.J., Bradley, S.J. (1995). *Gender Identity Disorder and Psychosexual Problems in Children and Adolescents.* Guilford, New York
[246] Berenbaum, S., Beltz, A. (2011). 189.
[247] Berenbaum, S., Beltz, A. (2011). 190; Note: Androgen exposure leads to a male identity, while estrogen exposure leads to a female identity.

androgens, gender identity is almost always consistent with their sex.[248] Only a small number of CAH girls (less than 5%) exhibit gender dysphoria; masculinized genitals or androgen excess have no effect, showing that regardless of how much these girls defy stereotypical norms, they still understand that they are *females*.[249]

Because of the incredibly rare nature of gender dysphoria, gender identity is not some fluid, transient, and ephemeral preference based on feeling and whim; it is tightly linked to one's biological sex for almost 99% of the population. However, despite the potential pitfalls of placing too much emphasis on biology when it comes to gender identity, there is a much greater danger of viewing gender identity as solely the result of socialization and 'innate feelings' based on gender stereotypes. If differences between boys and girls are all social constructs, then what's stopping physicians, psychiatrists, and parents from molding an individual in ways they see appropriate, justified, necessary, and even *moral*?

GENDER IDENTITY AS A BLANK SLATE

Viewing *gender identity* as a socially constructed category has serious implications for medical science. If it's true that infants are psychosexually neutral at birth and that their gender identity is a learned trait, not a biological reality, then the problem of ambiguous genitalia at birth can be solved through castration and reconstructive surgery. The individual can then be raised in accordance with the appearance of their genitals with no issues. If, however, gender identity is largely a biological reality, then the problem of ambiguous genitalia in intersex infants cannot be solved by simply changing the baby's genitals. Such a surgery, if gender identity is biological, would likely produce gender identity disorder in children who are raised according to the appearance of their genitalia and would

---

[248] Dessens, A.B., Slijper, F.M.E., Drop, S.L.S. (2005). Gender dysphoria and gender change in chromosomal females with congenital adrenal hyperplasia. *Archives of Sexual Behavior, 34*, 389-397.

[249] Berenbaum, S., Bailey, J. (2003). 1102-1106.

therefore be no different from child abuse. Something else must be tied to gender identity if such a biological theory is true.

Social constructionists, however, state that gender identity is culturally determined. Because of this, genes and hormones are theorized to be independent of gender identity. Differences between boys and girls in gender identity, therefore, are produced *only* through socialization. This blank slate view of boys and girls was popularized by Dr. John Money from Johns Hopkins Hospital in the 1950s. We briefly discussed Dr. Money in Chapter Two (Philosophy of Gender), but for this, we will be exploring his views in more detail.

Money was a sexologist interested in studying sexual abnormalities such as intersex infants and so-called *hermaphrodites*. As a major theorist of sex and gender, he was one of the first people to popularize the term *gender identity* by utilizing Simone de Beauvoir's famous separation of sex and gender. And yet, for Money, Beauvoir's view wasn't accurate enough.

After studying abnormal cases where someone's genitals didn't match their chromosomes, Money realized that the concept of *sex* needed to be separated into eight different categories: *chromosomal or genetic sex, gonadal sex, prenatal hormonal sex, internal morphologic sex, external morphologic sex, pubertal hormonal sex, assigned sex,* and *sex of rearing*.[250]

Each of these categories, according to Money, can vary independently from one another, and this is partly true. Abnormalities in sexual development can produce a mismatch of these traits, and yet, for most of the population, these eight categories align. For instance, if someone's chromosomal sex is XY, they will likely have testes, higher levels of prenatal androgen, internal and external male morphology, higher levels of testosterone in puberty, and will likely identify as a man. Money, however, recognized the importance of separating these categories to study and understand sexual abnormalities. *The Genderbread Person* is a famous representation elucidating the variability of these eight categories.[251]

---

[250] Money, J. (1985). The conceptual neutering of gender and the criminalization of sex. *Archives of Sexual Behavior, 14(3),* 280.

[251] See chapter, "Exploring Gender Differences."

9 DEVELOPMENT OF GENDER IDENTITY

The last category, *sex of rearing*, was of special interest to Money. He theorized that infants were psychosexually neutral and undifferentiated at birth. Or, in the words of the social constructionists: "Boys and girls are not born different, but are rather *made* that way."[252] If there was no biological component to the differences between boys and girls, then healthy psychosexual development could be directly linked to the appearance of the genitals. One's concept of gender, therefore, was a direct product of one's genitals. However, this is not to say that the *biology* of the genitals determines gender, but rather the *appearance* of them. In other words, it is the culture's view of the genitals which makes boys and girls different, not chromosomes, genetics, or hormones. *Sex of rearing*, therefore, was *how* someone was raised, and consequently, whether they identify as a man or a woman.

Once John Money began to solidify his *theory of gender neutrality* through studies on intersex infants, the term *sex of rearing* became synonymous with *gender identity*. Psychoanalyst Robert Stoller wrote in 1964 that gender identity is "produced by the infant-parents relationship, by the child's perception of its external genitalia."[253] The idea that gender identity is formed through an infant's relationship to his/her parents and solidified through the perception of his/her external genitalia is one of the core doctrines of social constructionism.

In an infamous case study on a set of twin boys, Money provided the social constructionists with perhaps the greatest evidence they could have ever hoped for, supposedly proving that gender identity was socially constructed. It is difficult to overstate just how ubiquitous this theory has become across our contemporary culture. In the 1970s and beyond, entire books, textbooks, and medical practices were altered in light of Money's findings. Known as the *John/Joan Case*, Money's experiment became the gold standard for social construction arguments which argued gender

---

[252] Thorne, B. (1993). *Gender play: Girls and boys in school.* New Brunswick, NJ: Rutgers University Press. 2.
[253] Stoller, R.J. (1964). A contribution to the study of gender identity. *The International Journal of Psycho-Analysis, 45:* 220-226.

identity was independent of genes and hormones, and therefore a social construct.

It all began in 1965 when a set of normal XY twins were born with healthy chromosomes, hormones, and genetics. For a while, everything seemed perfectly normal, until one of the twins, John (pseudonym), began having difficulty urinating. In a rare case of phimosis, John's foreskin wouldn't retract from the head of the penis, making it difficult to pee. And so, at seven months, John was sent into surgery to fix the tight foreskin through circumcision. Unfortunately, the surgery was botched. John's penis was accidentally burned during the procedure by cautery.[254] The parents and physicians thought, without a normal and functioning penis, could John live a normal and healthy life?

To see what could be done, the parents went to Johns Hopkins Hospital for consultation, where they soon met Dr. John Money. Utilizing the two postulates of his *gender neutrality theory*, that infants are born psychosexually neutral and that the appearance of the genitals is paramount to normal gender identity development, Money recommended that John be raised as a girl. If the differences between boys and girls are mere social constructs, thought Money, then raising John as a girl should present no issues. In fact, it wasn't just seen as the safe thing to do; it was seen as the *moral* thing to do. And so, in consultation with Money and other physicians, the baby boy underwent an orchiectomy and preliminary surgery to construct a vagina.[255] His perfectly healthy and functioning testes were removed, as well as his damaged penis, and the extra skin was utilized to fashion his new vulva and vagina. His parents were told to immediately begin raising him as a girl by giving him female-typed clothes, toys, and behavioral treatment. From now on, he would be known as *Joan*. This feminization was managed, monitored, and reinforced through yearly visits to Dr. Money.

---

[254] Money, J., Ehrhardt, A. (1972). *Man and Woman, Boy and Girl*. Baltimore: John Hopkins University Press.
[255] Diamond, M., Keith, H. (1997). 2.

9 DEVELOPMENT OF GENDER IDENTITY

As John developed and entered adolescence and after frequent visits to Johns Hopkins and his doctors, Money concluded that Joan's life as a girl was not just proceeding as planned, but was flourishing:

> "Although this girl is not yet a woman, her record to date offers convincing evidence that gender identity is open at birth for a normal child no less than for one born with unfinished sex organs or one who was prenatally over- or underexposed to androgen, and that it stays open at least for something over a year after birth. The girl's subsequent history proves how well all three of them [parents and child] succeeded in adjusting to that decision."[256]

Joan's mother echoed this sentiment, saying in the early 1970s that she's never seen a little girl so "neat and tidy" and that she is "so feminine."[257] Around this time, Time magazine famously reported that Joan's successful adoption of a female gender identity proved that psychological sex differences do not exist, that gender identity is fully independent from genes and hormones, and that gender is a social construct:

> "This dramatic case provides strong support that conventional patterns of masculine and feminine behavior can be altered. It also casts doubt on the theory that major sex differences, psychological as well as anatomical, are immutably set by the genes at conception."[258]

Psychology, sociology, and women's studies textbooks were changed in light of the successfulness of the case. Even physicians were not immune

---

[256] Money, J., Tucker, P. (1975). *Sexual Signatures: On Being a Man or Woman.* Boston: Little, Brown, 98.
[257] Money, J., Ehrhardt, A. (1972). 119.
[258] Diamond, M., Keith, H. (1997). 2.

to this change, as pediatricians began adopting the newfound evidence that gender identity could be molded through socialization:

> "The choice of gender should be based on the infant's *anatomy*...Often it is wiser to rear a genetic male as a female. It is relatively easy to create a vagina if one is absent, but it is not possible to create a really satisfactory penis if the phallus is absent or rudimentary. Only those males with a phallus of adequate size which will respond to testosterone at adolescence should be considered for male rearing. Otherwise, the baby should be reared as a female."[259]

After the apparent successfulness of the *John/Joan* case, medical professionals began castrating baby boys who deviated from the subjective criteria of a "satisfactory" penis by constructing a vagina and raising them as females. It is unknown just how many babies were castrated, mutilated, and psychologically damaged from the pervasiveness of Dr. Money's philosophy. Numbers range from the thousands to tens of thousands throughout the 1970s and 80s. If Money's theory was correct, that an infant's gender identity is socially constructed, then such a procedure would have been logical, necessary, and moral.

However, outside the successful reports of Joan's transition, something truly insidious was happening. Despite all the praise, the support, and the claims of incredible success, John never actually accepted his female gender identity. All of Dr. Money's reports, all of the media fawning over the collapse of biological gender, and all the alterations to textbooks and medical practices...all of it was based on *lies*.

Every attempt at socializing John as a girl was an utter failure. From the female-typed clothes and toys to behavioral interventions, John continually rejected the socialization pressures from his physicians,

---

[259] Donahoe, PK., Hendren, WHI. (1976). Evaluation of the newborn with ambiguous genitalia. *Pediatric Clinics of North America, 23,* 361-370.

psychiatrists, and even parents. From the earliest age he could remember, despite his female-looking genitalia, he always felt uncomfortable being a girl.[260] Something outside of John's socialization was making him feel this way. Could it have been the effects of prenatal androgens, sex-specific genes activation, vocal pitch, skeletal structure, bone density, muscle mass and a host of other sex differences which made him, not female, but *male*? Just by observing the girls and boys, John could tell he was not a girl.

Despite the fact that John never felt like a girl, he was still subjected to the annual visits with Dr. Money. Because John had a healthy XY twin, the brothers became important case studies for the social construction of gender identity. During these visits, Money wouldn't just analyze their behavior and ask them questions; he would get them to do things which would shock even the most ardent supporters of social constructionism:

> "Reimer [John] and his twin brother were directed to inspect one another's genitals and engage in behavior resembling sexual intercourse. Reimer claimed that much of Money's treatment involved the forced reenactment of sexual positions and motions with his brother. In some exercises, the brothers rehearsed missionary positions with thrusting motions, which Money justified as the rehearsal of healthy childhood sexual exploration."[261]

This pedophilic exploitation of the twins by Dr. Money did not end there:

> "At least once, Money photographed those exercises. Money also made the brothers inspect one's pubic areas. Reimer stated that Money observed those exercises both alone and with as many as six colleagues. Reimer recounted anger and verbal abuse

---

[260] Gaetano, P. (2017). David Reimer and John Money gender reassignment controversy: The John/Joan Case. *The Embryo Project Encyclopedia*.
[261] Gaetano, P. (2017). 2.

from Money if he or his brother resisted orders, in contrast to the calm and scientific demeanor Money presented to his parents."[262]

Why did Money subject the boys to such psychological and sexual abuse? The answer is in his theory, which provided no room for the effects of biology on the differences between boys and girls. If infants were psychosexually neutral, then they could be molded in preferential ways. Money wanted to prove the accuracy of his theory that children *learn* gender roles through socialization, and that these gender roles are solidified through identification as a male or female. Such an identification was linked to the appearance of the genitals. By having John engage in female-typical sexual behaviors, perhaps Money could socialize the boy to act like a female, thereby supporting his reports. Like a despotic tyrant, Money subjected Reimer and his brother to the some of the worst kinds of psychological abuse for his sick and twisted desire that his theories be proven true. Because of the fabricated reports that John's transition to Joan was a great success, Reimer and his brother became the unknowing case studies which made the mutilation of infants and the psychological abuse of children routine medical practices for the next two decades:

> "According to biographers and the Intersex Society of North America, this data was used to reinforce Money's theories on gender fluidity and provided justification for thousands of sex reassignment surgeries for children with abnormal genitals."[263]

Throughout Reimer's childhood as Joan, he experienced constant and severe gender dysphoria.[264] Everyone told him he was a girl, but for some reason, he never felt comfortable as one. He did not feel this way because of some superficial, transient, and fluid concept of his own gender. Rather, this gender dysphoria was simply the result of *being* a male. John had male-typical sex differences in his body, his brain, and his genetics. This explains

---

[262] Gaetano, P. (2017). 2.
[263] Ibid.
[264] Ibid.

why John would not accept his female gender identity despite all the socialization. Everything from the wearing of dresses and the insistence to participate in female-typed activities to even the injection of estrogen hormones, all of it was useless; John was and always would be a *male*, with a masculinized neural structure, male-specific genetics, hormones, reproductive structure, and body type.

> "Girl's toys, clothes, and activities were repeatedly proffered to Joan and most often rejected. Throughout childhood Joan preferred boy's activities and games to those of girl's; she had little interest in dolls, sewing or girl's activities. Ignoring the toys she was given, she would play with her brother's toys. She preferred to tinker with gadgets and tools and dress up in men's clothing; take things apart to see what makes them tick. She was regarded as a tomboy with an interest in playing soldier. Joan did not shun rough and tumble sports nor avoid fights."[265]

As Reimer began to exhibit more male-typical psychological traits and compared his body to those of other boys and girls, he also began to feel more and more dysphoric. He describes this in detail as an adult:

> "There were little things from early on. I began to see how different I felt and was, from what I was supposed to be. But I didn't know what it meant. I thought I was a freak or something; … I looked at myself and said I don't like this type of clothing, I don't like the types of toys I was always being given, I like hanging around with the guys and climbing trees and stuff like that and girls don't like any of that stuff. I looked in the mirror and saw my shoulders are so wide, I mean there is nothing feminine about me. I'm skinny, but other than that, nothing. But that is how I

---

[265] Diamond, M., Keith, H. (1997). 4.

figured it out. [I figured I was a guy] but I didn't want to admit it, I figured I didn't want to wind up opening a can of worms."[266]

Because of his abnormal behavior as a "girl," many of his classmates bullied him constantly:

> "Joan knew she already had thoughts of suicide brought on by this sort of cognitive dissonance and didn't want additional stress. Joan fought both the boys as well as the girls who were always 'razzing' her about her *boy* looks and her girl clothes. She had no friends; no one would play with her. 'Every day I was picked on, every day I was teased, every day I was threatened. I said enough is enough...' Mother relates that Joan was good looking as a girl. But it was 'when he started moving or talking, that gave him away and the awkwardness and incongruities became apparent.'"

This abuse continued into Reimer's adolescence, as the physicians put him on an estrogen regimen at 12. John rebelled. It was here where a different methodology was used in treating his gender dysphoria: *talk therapy*.

> "They made her 'feel funny' and she didn't want to feminize. She would often dispose of her daily dose. She unhappily developed breasts but wouldn't wear a bra. Things came to a head at the age of 14. In discussing her breast development with her endocrinologist she confessed, 'I suspected I was a boy since the second grade.' The physician, who personally believed Joan should continue to take her estrogens and proceed as a girl, used that opening to explore in a nonjudgmental manner, the possible male or female paths available and what either one would mean. Since the local management team had already noticed Joan's

---

[266] Diamond, M., Keith, H. (1997). 4.

preference for boys' activities and refusal to accept female status, and they had discussed among themselves the possibility of accepting Joan's change back to male, the endocrinologist explored Joan's options with her. Shortly thereafter, at age 14, Joan decided to switch to living as a male."[267]

Using cutting-edge methods such as Cognitive Behavioral Therapy, which involves talk therapy, Reimer was able to fully explore his true feelings. It was this talk therapy that allowed him to transition back to a male. Around the same time, his father finally told him the truth: "All of a sudden, everything clicked," Reimer said, "For the first time things made sense and I understood who and what I was."[268] At 15, he underwent a full mastectomy and phalloplasty to reconstruct his penis, and at 25, Reimer married a woman and adopted her children. In short, he became a gynophilic sexually active male.

Despite his rebound into life as a man, Reimer's childhood trauma continued to haunt him. Dr. Money's sexual exploitation of him and his brother rendered them psychologically damaged beyond repair. At 36 years old, his brother committed suicide by overdose of antidepressants, and two years later, in 2004, after finding out his wife wanted to separate, Reimer drove to a grocery store's parking lot in his hometown of Winnipeg and shot himself in the head with a sawed-off shotgun.[269]

As medical professionals dealt with the fallout of the *John/Joan Case*, Dr. John Money's theories, once viewed as the gold standard for the social construction of gender, were found to be scientifically illiterate and morally bankrupt lies. Reimer's life provided direct evidence that one's gender identity was *not* in fact *independent* from genetics, hormones, body morphology, and internal reproductive structures, and therefore, neither *fluid* nor *moldable*. Reimer didn't accept the sex-typical behaviors pushed onto him. He didn't accept the clothes, the toys, any of it.

---

[267] Diamond, M., Keith, H. (1997). 5.
[268] Ibid.
[269] Rolls, G. (2015). *Classic Case Studies in Psychology*. London: Routledge. 145.

The sex reassignment of Reimer failed exactly where it needed to succeed. This was because 1) *his genes and cells were male,* 2) *his prenatal hormones were male-typical,* 3) *his internal reproductive structures were male (setup to produce male gametes),* 4) *his skeletal structure was male-typical,* 5) *his vocal pitch was male-typical,* 6) *his interests were male-typical,* and 7) *his physiology was male-typical.* In short, he was a biological male.

Just because a boy may be born with ambiguous genitalia does not give anyone the right to castrate him, mutilate him, and damage him psychologically beyond imagining. And yet, this insidious view that gender differences between boys and girls are *only* the result of socialization is the dominant philosophy of our contemporary culture. In an attempt to be inclusive as possible by putting all biological components aside, we have ignored the very traits which are unchanging, innate, and fundamental to a person's unique being: *the genetic, morphological, and physiological characteristics of biological sex.* Because of this, we are in danger, as John Money was, of imposing our subjective societal standards onto unwilling and often times (as in the case of intersex infants) *defenseless* patients. Such a practice would be immoral, and yet, practices such as these are happening more and more as our culture attempts to revitalize the theories of John Money: *that girls and boys are psychologically and biologically the same.*

And yet, such a denial of biological sex is becoming more medically acceptable as doctors surgically and hormonally alter young people's bodies under the guise of 'acceptance' and 'compassion.' Ironically, denying reality does not make it go away; in fact, reality will always come back in a vengeance. Or, in the words of Watson, a young Scottish woman who de-transitioned:

> "I agree with Philip Dick when he says, 'Reality is that which, when you stop believing in it, doesn't go away.' It doesn't matter how many people disregard biological reality; it is what it is and always will be."[270]

---

[270] ImWatson91. (2020). In response to changing one's sex. *Twitter*.

Telling a four-year-old boy he's actually a *girl* because he likes to play with dolls is not progressive; it's not forward-thinking, enlightened, or compassionate. It's *immoral.* It's *damaging.* And it's *child abuse.* Giving a 12-year-old girl puberty blockers because she feels like she's a boy is not progressive. It's immoral. Telling a six-year-old girl she's a male because she loves to play with trains and dresses like a boy is not progressive. It's immoral. And yet such defenses of a child's well-being are met with accusations of *bigotry, sexism, misogyny, homophobia, transphobia,* and other overused words vomited out by vehement activists to shut down disagreement.

While silencing voices who dare counter social constructionist dogma, our society continues to champion the idea that both gender identity and sex are fluid and malleable social constructs.[271] More and more kids, not even young enough to decide what they're going to eat for lunch, are suddenly making life-altering decisions beyond their comprehension with the encouragement of their *enlightened* parents who, in an attempt to be accepting, tell their little girl she's actually a boy simply because she likes male-typical activities. How is this not immoral? A six-year-old can barely decide what he wants to wear on a given day, let alone what gender he is.

How can we not see where we're heading? This is not the recipe for dismantling the oppressive nature of gender stereotypes; this is a recipe for reinforcing and solidifying them! We used to say a girl who plays with male-typical toys was a *tomboy.* Now we say she *is* a boy, that she should *become* a boy, be given puberty blockers, hormones, and change her entire sex. How is this not reinstituting the regressive views social constructionists supposedly want to defeat? How is this not relegating boys and girls to old-fashioned gender stereotypes? How is this actually progressive? And yet, while our society descends back into a confused sexual mess, the idea that gender identity is a socially constructed and ephemeral feeling has been adopted wholesale as an enlightened and moral sign of progress without recognizing the inherent dangers and history of social constructionist practices.

---

[271] See chapter, "The Gaslighting Ideology."

We're all for acceptance, for tolerance, for inclusivity, for compassion, and for people to express themselves how they feel. But pumping small kids full of hormones and reinforcing a strict gender binary on confused children, as Dr. John Money did to those twins? *That's not how the Gender Paradox can be solved.* It's time to get off the postmodern social constructionist train, and it's time to adopt a better model for sex and gender differences!

GENDER IDENTITY AS A BIOLOGICAL REALITY

If infants are not psychosexually neutral at birth, then what are they? Without the doctrine of the blank slate, social constructionism cannot fully explain the origins of gender identity. I propose that infants are *not* psychosexually neutral at birth and instead come into the world with a unique brain structure influenced by their chromosomal and hormonal makeup. Gender differences in behavior and interests, therefore, form from the influence of prenatal androgens, sex-differentiated neural structures, activation of sex-specific genes in the brain, and behavioral reinforcement and punishment. Prenatal androgens and sex-differentiated neural structures occur in the womb, creating a formwork within which gender identity will arise in early childhood and adolescence. Gender differences can be suppressed or exaggerated through socialization, but socialization can only go so far to influence the biological substrates.

Social constructionist theory states that infants are treated in sex-stereotyped ways by their parents through a process of external reinforcement and punishment which is then internalized in the child as innate. This is somewhat true, but it cannot be applied holistically. Gender differences are *partially* reinforced through socialization, but not completely.

For example, in 1991, a meta-analysis of 172 studies explored the effects of socialization on children's sex-typed toy preferences. It was found that parents *do* in fact differentially reinforce sex-typed behaviors, a finding which agrees with social constructionist theory. However, the meta-analysis found that the specific type of behavior being reinforced was most important. Out of many unique behaviors examined, parents were found to encourage sex-specific *toy choices, playmate preferences,* and *play style* in their girls and boys.[272] On the other hand, parents *did not* treat their boys and girls differently in the areas of dependency, warmth or nurturance, encouragement of achievement, and discouragement of aggression.

The meta-analysis also showed that parental approval was sex-typed as well, meaning that parents gave their boys and girls different levels of approval for sex-specific behaviors. For example, parents were shown to give more approval to their girls than their boys for playing with dolls and dressing in a feminine way, a behavioral reinforcement which was directly observed by the researchers.[273] In regards to sex-differentiated verbal patterns, parents also used different types of speech when talking with their girls and boys: "Daughters received more verbal interaction than did sons, and during problem-solving exercises, mothers used more directive and supportive language with daughters than with sons."[274]

While differential treatment from parents exists, it only extends to certain aspects of behavior. Sex-typed treatment does not apply to all of sex-differentiated behaviors in childhood. Some aspects of sex-differentiated behavior are influenced by hormonal and chromosomal influences on the brain. **Here, recent evidence from psychology, behavioral science, and neuroendocrinology provides a solid case for gender identity as a biological reality. This biological reality is not devoid of socialization, but is rather the prerequisite from which sex-differentiated treatment from parents may exaggerate or suppress innate tendencies.**

If a baby's brain is differentiated according to their sex, then there must be some way of measuring these differences. If psychological sex differences such as an interest in things versus people are innate and are

---

[272] Lytton, H., Romney, D. M. (1991). Parents' differential socialization of boys and girls: a meta-analysis. *Psychological Bulletin, 109,* 267-296.

[273] Pasterki, V., et al. (2005). Prenatal hormones and postnatal socialization by parents as determinants of male-typical toy play in girls with congenital adrenal hyperplasia. *Child Development, 76(1),* 265.

[274] Ibid., 266.

not formed from socialization alone, then these neurological differences should be present in infancy. Accordingly, any neurological differences should then impact gender differences in both behavior and toy preferences.

Gendered behavior is influenced by the levels of prenatal androgens a child is exposed to in the womb. Manipulation of androgen levels in rodents and nonhuman primates has provided mammalian-based evidence for the effects of androgens on sex-typed behavior. In the mammalian world, the effects of these androgens has been shown through studies on primates such as rhesus macaques and vervets.[275] Male monkeys, like human boys, showed strong preferences for mechanical objects such as trucks, while female monkeys showed a more varied and diverse preference across both plush and wheeled toys.[276]

Researchers observe sex differences in vervet monkey toy preferences.
Source: Alexander, G. M., & Hines, M. (2002).

---

[275] Alexander, G. M., & Hines, M. (2002). Sex differences in responses to children's toys in a non-human primate (Cercopithecus aethiops sabaeus). *Evolution and human behavior, 23,* 467-479.

[276] Hassett, J. Siebert, E., Wallen, K. (2008). Sex differences in rhesus monkey toy preferences parallel those of children. *Hormones and Behavior, 54,* 359.

Similarly, in human infants from the ages of three to eight months, sex differences have been shown in toy preference. Observational studies utilize eye-tracking technology to measure the level of visual interest in sex-typed toys such as dolls and trucks. Girls tend to show a greater visual preference for a doll over a truck at an effect size of $d > 1.0$ (very large), while boys tend to show a greater visual preference for a truck over a doll at an effect size of $d = 0.78$ (large).[277]

How do prenatal androgens directly affect this toy preference? The answer is in the organizational effects of these hormones on sex-differentiated brain structures.[278] Prenatal androgen exposure increases interest in watching things move in space by altering the development of the visual system.[279] This is why we see boys, more often than girls, exhibit a greater interest in mechanical and moving objects: boys are usually exposed to higher levels of prenatal androgens. This gender difference in toy preferences matches the gender difference in the Things-People dimension in adulthood. In occupational preferences, men tend to be more interested in things and women tend to be more interested in people.[280] Therefore, studies into mammalian species and observational studies into children's toy preferences provide strong evidence that prenatal organizational effects on the brain influence the sex-typed behavior of many mammals.

However, there is more to hormones than just prenatal androgens. In early infancy, a postnatal surge in testosterone occurs, which can be more easily measured through urine and saliva than attempting to measure it during pregnancy. Researchers sampled the urine from infants each month from weeks 4 to 26 to measure testosterone. Gender-typical play was then

---

[277] Alexander, G. M., et al. (2009). Sex differences in infants' visual interest in toys. *Archives of Sexual Behavior, 38,* 427.

[278] Hines, M. (2004). *Brain Gender.* New York: Oxford University Press.

[279] Alexander, G. M. (2003). An evolutionary perspective of sex-typed toy preferences: pink, blue, and the brain. *Archives of Sexual Behavior, 32,* 7-14.

[280] Su, R., & Rounds, J., et al. (2009). Men and things, women and people: a meta-analysis of sex differences in in interests. *Psychological bulletin, 135(6),* 859-884.

measured using a questionnaire to see how testosterone levels related to the infants' toy preferences. The results showed that boys had a significantly larger amount of testosterone during the first 6 months of life compared to girls. Unlike what social constructionists would predict (that testosterone levels have no effect on gender-specific behavior), these higher levels of testosterone positively predicted male-typical play and a preference for trains over dolls.[281] Girls who were exposed to higher levels of testosterone also preferred the train over a baby doll.[282]

Despite such strong predictions, these studies are not without their potential pitfalls, because sociocultural and biological factors interact simultaneously and are often difficult to separate. Unlike the rhesus macaques and vervets who don't have sociocultural structures, humans are heavily cultured and socialized animals. For instance, boys with high levels of androgens will exhibit male-typical behaviors while being simultaneously discouraged from playing with girls' toys. Because of this, research into girls with Congenital Adrenal Hyperplasia provides a unique perspective for the organizational effects of prenatal androgens on the brain and subsequent gender differences in behavior.

Girls with CAH are exposed to higher levels of prenatal androgens than their unaffected sisters. If androgens have a masculinizing effect on the brain structure, then girls with CAH should exhibit more male-typical psychological traits such as an affinity for spatial manipulation, an interest in things, and male-typical toy preferences. This is exactly what researchers have discovered:

> "These behavioral changes may involve a range of sex-typical behaviors, such as sexual orientation, aggression, interest in infants, and specific cognitive abilities (Collaer & Hines, 1995; Hines, 2004; Hines, Fane, Pasterki, Conway, & Brook, 2003). However, the clearest and most consistent influences have been seen for childhood gender role behaviors, including toy choices,

---

[281] Hines, M., et al. (2015). 7.
[282] Ibid.

playmate preferences, and activity preferences. CAH girls were more likely to be labeled by themselves and others as 'tomboys,' to like boys' toys and boys' clothes, and to prefer boys as playmates. In an observational study, Berenbaum and Hines (1992) found that girls with CAH spent more time with boys' toys and less time with girls' toys than did their unaffected female relatives."[283]

The case of CAH girls is problematic for social constructionist arguments. If prenatal androgens have no effect on CAH girls, then socialization should produce female-typical behavior. This is not the case. Male-typical behavior, such as an interest in mechanical things over people, is consistent among CAH girls despite widespread social modeling and labeling of gender-appropriate behavior.[284]

Social constructionists will attempt to argue against prenatal androgen's effects on CAH girls by theorizing that male-typical play in these girls is a result of socialization: differential parental encouragement is theorized to be the likely cause of CAH male-typical play. In other words, parents are said to be biased in interacting with their CAH girls, steering them towards masculine objects and toys.

However, this claim is not just disproved through interviews with the parents of CAH girls; it's also disproved by studies which show that the presence of a parent does *not* affect the child's toy play.[285]

> "In fact, contrary to suggestions that masculine play in girls with CAH is caused by parental encouragement, an observational study of parent-child interactions in a playroom found that

---

[283] Pasterki, V., et al. (2005). 267.

[284] Hines, M. (2016). 2.

[285] Hines, M., Constantinescu, M. (2012). Relating prenatal testosterone exposure to postnatal behavior in typically developing children. *Child Development Perspectives, 6(4)*, 407.

parents encouraged girls with CAH more than their unaffected counterparts to engage in female-typical play."[286]

It makes sense why parents would steer their CAH girls to more female-typical activities. After all, they are still girls! However, even the efforts of parents in manipulating their CAH girls' preferences shows little to no effects on their innate preferences of male-typical toys. In fact, CAH girls are often *more encouraged* than their unaffected sisters to play with female-typical toys, as parents will try and normalize their behavior to fit within gender-appropriate norms.[287] It is in this aspect that children's behavior is not only shaped by parents, but that parents' behavior is often shaped by their children.

These studies suggest that male-typical behavior from CAH girls is the result of prenatal androgens. This is supported by convergent evidence from a variety of studies which show a strong relationship between the amount of testosterone in the mom during pregnancy and her offspring's subsequent development of sex-typed behavior.[288]

By considering the parental treatment of CAH girls' toy preferences compared to their normal brothers and sisters, we can gain a more holistic perspective on the interaction of socialization and prenatal androgens.

> "Thus, our findings overall suggest that, in general, parents attempt to encourage sex-typical play in children and that this encouragement has some effect. At the same time, however, attempts by parents of children who show cross-gender toy choices because of CAH do not appear to be completely successful. One possible explanation is that biological constraints in the form of prenatal exposure to androgen limit the degree to

---

[286] Hines, M., Constantinescu, M. (2012). 407.
[287] Pasterki, V., et al. (2005). 275.
[288] Hines, M., Constantinescu, M. (2012). 408.

which parental encouragement can produce female-typical behavior."[289]

Based on the studies mentioned above, biologically-based toy preferences are likely enhanced or suppressed through the process of socialization.[290] Using the understanding that differential treatment from parents interacts with prenatal androgens, let's pick up where we left off with our fictional baby girl, Julia. She was exposed to higher levels of androgens in the womb compared to her unaffected sisters, creating slightly masculinized structures inside her brain circuitry similar to that of girls with CAH.

From the very beginning, Julia's parents noticed some differences compared to that of her older sister. She exhibited less eye contact than her sister in early infancy and even developed language at a slower pace. Evidence from high prenatal androgen levels has been associated with reduced eye contact and small vocabulary in infancy, confirming her parents' observations.[291] Around the age of two to three months, infants like Julia will show slight sex differences in spatial-visualization ability, responsiveness to faces versus things, and language ability.[292] This responsiveness to faces versus things shows up as a sex difference in looking time at different stimuli. Girls tend to look longer at a face while boys tend to look longer at a moving object.[293] Because of this, responsiveness to faces is higher in female infants.

---

[289] Pasterki, V., et al. (2005). 276.

[290] Hines, M., & Alexander, G. M. (2008). Monkeys, girls, boys and toys: A confirmation [Letter to the editor]. *Hormones and Behavior, 54*, 359-364.

[291] Lutchmaya, S., Baron-Cohen, S., Raggatt, P. (2002). Fetal testosterone and eye contact in 12-month-old human infants. *Infant Behavioral Development, 25*, 327-335; Lutchmaya, S., et al. (2002). Fetal testosterone and vocabulary size in 18- and 24-month-old infants. *Infant Behavioral Development, 24*, 418-424.

[292] Goldman, B. (2017). Two minds: the cognitive differences between men and women. *Stanford Medicine*, 3.

[293] Connellan, J., Baron-Cohen, S. et al. (2000). Sex differences in human neonatal social perception. *Infant Behavior and Development, 23(1)*, 113-118.

Further research into sex differences such as spatial ability continue into adulthood. Men tend to be better at "visualizing what happens when a complicated two- or three-dimensional shape is rotated in space, at correctly determining angles from the horizontal, at tracking moving objects, and at aiming projectiles."[294] In both rats and humans, the females of the species tend to be better at using landmarks while navigating and rely more heavily on spatial location memory than their male counterparts.[295]

However, because Julia was exposed to higher levels of prenatal androgens, she instead exhibits male-typical traits such as an interest in things, a better spatial ability, and less responsiveness to face-based stimuli. Such sex differences in toy preferences and cognitive abilities tend to be influenced by different factors more heavily. For example, because the development of one's gender identity is multidimensional, sociocultural factors may influence certain aspects more than biological factors. At the same time, certain biological factors may influence certain aspects more than society and culture:

> "Gender development is multidimensional, and the combinations of factors that influence the many different dimensions of gender appear to differ. Early hormonal influences appear to play a larger role, for example, in children's toy preferences than they do in cognitive abilities that show sex differences, where social and cultural influences appear to be more important."[296]

The true difference in Julia's behavior began to arise when she started crawling and interacting with her toys. Even though her parents gave her a variety of toys, from male-typed mechanical objects such as wheeled toys to female-typed toys such as plush and dolls, Julia always preferred her mechanical train set over any type of plush or doll. In fact, even while her

---

[294] Goldman, B. (2017). 3.
[295] Ibid.
[296] Hines, M. (2011). Gender development and the human brain. *Annual Review of Neuroscience, 34,* 81.

parents tried to reinforce her play with dolls to construct female-typical behavior, none of their attempts worked. Julia, despite all the encouragement and reinforcement to play with female-typed toys, continued to choose her male-typed and gender-neutral toys over any feminine ones. This is because, in large part, Julia's brain had been organized in a more masculine way through higher levels of prenatal androgens. This masculinization resulted in two structural changes in her brain: higher *suboptimal arousal* and *rightward shift*, as I discussed in Chapter 8.

Her male-typical toy preferences directly relate to the studies on girls with CAH, showing the important role that androgens play in Julia's gender identity development. However, around the age of four, confusion began to arise in Julia's young mind. She observed her sister was obsessed with playing with dolls, kitchen, and plush toys and only sometimes playing with the gender-neutral and male-typical ones such as the train set. When playing with her male friends, however, Julia noticed that she liked playing with the same toys they did. *Perhaps she was really a boy.*

One day, while playing with her favorite train set, tinkering and analyzing the different ways it could be put together, Julia ran over to her mom and told her that she was, in fact, actually a boy. After all, she thought, she doesn't have any girl friends who like to play with trains like the boys do. After hearing this, her mom smiled and knelt down. This was not the time to tell Julia that she was correct, that she was *actually* a boy. It wasn't the time to inform this four-year-old about transgenderism and hormones. And it certainly was not the time to inform Julia that just because she feels like a boy she should become one.

Instead, with grace and acceptance, her mom looked Julia right in the eyes and told her this:

> "Julia, you're a girl. And being a girl, it's okay to play with toys that boys play with. Just because you play with them doesn't mean you're a boy. Some boys like to play with girly toys, and some girls like to play with boy toys. Whatever you want to play with is okay with me."

After recognizing this important distinction that just because she liked boyish toys did not mean she *was* a boy, Julia smiled and ran back off to play with her trains. Because of the way her mom handled the confusion, Julia accepted herself and her body for who she was: a girl. This healthy understanding of herself as a girl, not a *boy trapped in a girl's body*, solidified within the next few years. Never again did she think she was actually a boy. Instead, she learned that playing with boyish toys was okay and that her mom would support her with any choice she made. As Julia went to bed the next coming nights, she went to sleep in peace, for she was proud of being a girl.

Let's imagine the opposite scenario for a moment, where Julia's mom *does not* correct her slight confusion. Instead of telling her that she is a girl, and that some girls like playing with boy toys, Julia's mom tells her that she is a *boy*. Just as David Reimer's parents, physicians, and psychiatrists told him he was a girl despite his chromosomal and hormonal makeup, Julia's mom reinforces that she's actually a boy just because she likes boyish toys. If both social constructionist theory and John Money's experiments were correct, then raising Julia as a boy should not be a problem. However, as both prenatal androgens and socialization effect the development of gendered behavior, it is at this critical moment of development where gender dyshporia can be dealt with correctly or immorally enflamed.

If Julia is told she's actually a boy, she will begin to notice a disconnect between her body and the bodies of boys, especially when she begins puberty. Continual behavioral reinforcement from her mother that she is a boy will create deep gender identity confusion, causing a mismatch between her mental state and her physical state. If behavioral reinforcement is applied for too long, as is the case for David Reimer, then gender dysphoria will be inevitable, and with it, a life of confusion and emotional distress will follow.

Under the guise of acceptance and tolerance to construct her supposed male gender identity, puberty blockers may be given to Julia at the age of ten or eleven to stop the normal development of her female secondary sex characteristics such as breasts. At the same time, hormone treatment with testosterone may begin to further alter her development. Continual use of puberty blockers without caution may cause Julia to be permanently sterile, which would not allow her to bear children. Such a decision should not be possible before the age of consent, as even adolescents do not have the maturity to make such life-altering decisions. And yet, using a social constructionist framework developed by Dr. Money, such decisions are being continually made across our contemporary society.

There are more and more stories of transgender kids and adolescents who are going through hormone treatment, puberty blockers, and even surgery to alter their physiology in line with their gender identity.[297] While some gender dysphoria can be solved through gender reassignment, such a life-changing decision should not be made until after adulthood begins. The data suggests that the majority of children who have gender dysphoria in childhood grow out of it and assume a gender identity which matches their biological sex.[298] Because of this, medical practitioners should err on the side of caution, non-harm, and patience to allow the child's physical and psychosexual development to continue naturally. However, this is not what's happening.

Social constructionists continue to push the idea that gender identity is a fluid and entirely malleable trait which can be changed according to date, time, and feeling. Scientific evidence from the fields of biology, neuroendocrinology, psychology, and behavioral science disproves the theory that infants are psychosexually neutral and that their identities can be molded at whim by parents, physicians, and psychiatrists.

Despite the failures of Dr. Money's attempts at raising a biologically healthy male as a girl, additional evidence can be used to refute strict social constructionist claims. The influence of prenatal androgens, sex-differentiated brain regions, and parental treatment all influence the development of gender identity. It is not a socialized trait, nor is it a purely biological one. **Rather, gender identity's innate structures arise from the formwork of one's sex, hormones, and genetics, within which the forces of socialization are poured and molded.** Such a theory allows for the

---

[297] See chapter, "The Gaslighting Ideology."
[298] Berenbaum, S., Beltz, A. (2011). 189.

critical consideration that a preference for sex-typed toys opposite of gender stereotypes can exist in either sex. Just because a girl enjoys boy-typed toys does not mean she is actually a boy. At the same time, and equally important, just because a boy enjoys girl-typed toys does not mean he is a girl. Therefore, our biopsychosocial model shows that gender identity (identification as a male or female) is a solidified and fixed trait which almost always aligns with one's birth sex, while gender expression (behavior and toy preferences) can be more variable, showing that no child conforms to strict gender stereotypes.

Social constructionists, in an attempt to be tolerant and inclusive, conflate the concepts of gender identity, gender expression, and biological sex. In this view, if Julia is seen playing with male-typed toys, then her gender identity must therefore be male. However, this could not be more wrong. Julia's preference for male-typed toys does not mean she is a male or should identify as one; it means she's a girl who likes male-typed toys, and likewise, it also means that not all girls like playing with female-typed toys! Thus, her gender identity does not need to match stereotypical traits of gender expression, and this is perhaps the most critical point to understand.

Parents, teachers, and peers should allow for variability when it comes to gender expression. *Many boys and girls will show overlap in behavior and toy preferences; it does not mean such preferences show they are members of the opposite sex or have opposite gender identities.* By merging these two concepts of biological sex and gender expression, social constructionists relegate boys and girls to binary gender stereotypes, the very things they say they wish to dismantle. Whether this way of thinking is coming from biological essentialists or highly vocal social constructionists, such examples of regressive thinking must be fought for the psychological and sexual health of our children.

Despite the effectiveness of rhetorical strategies from John Money and aesthetic arguments from social constructionists, biological sex cannot be molded and altered by parents, psychiatrists, or physicians. Attempting to alter 1) *someone's behavior opposite of their innate tendencies* and 2) *someone's physiology opposite of their biological sex* is not just wrong; it's damaging, immoral, and abusive. Perhaps strict social constructionists can finally accept that their theories are scientifically illiterate, philosophically dangerous, and damaging to children. Unfortunately, such a change of perspective in our society's zeitgeist won't be happening any time soon.

10

GENDER STEREOTYPE PATHOLOGY

Before we move further into the gender development of the individual, we must first deal with the critical impediment of healthy development that occurs when gender stereotypes transform from mere generalizations of behavioral patterns into pathological ways of thinking. Conceptualizing gendered behavioral patterns as fixed and binary while judging male and female behavior according to these patterns is known as gender stereotyping. Such stereotyping can quickly turn pathological when applied incorrectly to individuals. For example, we may recognize that more girls than boys have an interest in dolls, and while such an observation is true, its prediction of individual behavior can only extend so far.

Extending to individuals an accurate observation of behavioral differences between groups is stereotyping. Individual behavior, being more variable *within* populations than *between* them, makes stereotyping a slippery slope. Such a practice of applying stereotypes to individuals can quickly devolve into oversimplified judgments which threaten the natural expressions of individual behavior and interests. While more girls than boys enjoy dolls, there are some boys who may have a higher interest in dolls than average boys. Stronger interests in dolls in some boys does not negate the accuracy of the stereotype, but it does limit its application to mere generalizations. Because of this, predicting an individual's behavior and interests based on stereotyped behavioral patterns is not without serious flaws.

Through recognizing the potential danger of relying on stereotypes to judge individuals, we once again show that social constructionism has some merit. Social constructionists rightfully posit the variability of individuals, the arbitrary nature of relying too heavily on strict categories, and the significant effects of socialization forces. And when it comes to gender stereotypes, social constructionists and gender theorists alike rightfully analyze the validity of these stereotypes and question their legitimacy. However, like any philosophical position, its beliefs can be taken too far. Social constructionism, while explaining the importance of individual variability, renders all biological factors for individual differences as constructs of an oppressive power structure, a meta-narrative, which continually reinforces the subordination of women and minorities. Variability between men and women, for instance, is not produced through a complex mixture of biological, psychological, and sociocultural forces, but only through a constructed gender binary which marginalizes non-conformists. Having acknowledged *one* out of the three variables, social constructionists are close to the truth, but not close enough.

ORIGINS OF GENDER STEREOTYPES

Using the biopsychosocial model, I theorize that variability between men and women exists due to an interaction of biological and sociocultural factors, which then produces social systems around these differences. Gender stereotypes, therefore, are generalizations of perceived patterns of behavior, or variability, between men and women. The origins of these stereotypes vary. Some of them originate from perceived patterns which are biological in nature, some originate from socialization factors, and others originate from a mix of both biology and society. For instance, the stereotype that men are more interested in engineering than women can be traced back to first, a biological component, and second, a social one. We'll call this perceived pattern the *male-dominant engineering stereotype*.

The biological component of this stereotype arises *in utero*. An increased exposure to prenatal androgens, usually occurring in males, causes a *rightward shift* in the brain structure of the developing fetus.[299] Rightward shift happens when the brain utilizes the right hemisphere more for perception and tasks and the two hemispheres become less interconnected.[300] Such a rightward lateralization of the brain means that spatial ability and systematizing (organization and arrangement of data in a system) will be emphasized.[301] Such an emphasis in visuospatial cognition also occurs in females who are exposed to male-typical levels of androgens, which influences their toy preferences as children and their occupational interests as adults.[302]

From an early age, our fictional Julia began to experience this as well. Being exposed to higher levels of prenatal androgens than her sister, Julia's brain became more lateralized and compartmentalized within each hemisphere, causing her visuospatial system to strengthen and emphasize moving objects. She quickly began to show an interest in understanding the composition of more mechanically-based toys compared to her feminine sister who enjoyed more people-based role play.

While the right hemisphere is dominant in both sexes for visuospatial cognition (ability to manipulate objects and navigate in three-dimensional space), males are often more lateralized for this function, meaning visuospatial cognition in the right hemisphere is developed and utilized more for males than it is for females, showing less connections between the hemispheres.[303] Because of this *rightward shift* for males, there are more men who are interested in engineering than women, since engineering involves heavy amounts of manipulating and visualizing objects in three-dimensional space. Males' visual and cognitive systems, therefore, are often

---

[299] Ellis, L. (2011). 554.
[300] Khazan, O. (2013).
[301] Ibid.
[302] Berenbaum, S., Beltz, A. (2011). 196.
[303] Pfannkuche, K.A., Bouma, A., Groothuis, T.G. (2009). Does testosterone affect lateralization of brain and behaviour? A meta-analysis in humans and other animal species. *Philosophical Transactions Royal Society London B, Biological Sciences, 364 (1519),* 929–942.

developed towards such tasks. (To be clear, it is not that women cannot accomplish an engineering task just as well as men, but that the average woman is not as interested as the average man in such a task.)[304] While Julia is a genotypic and phenotypic girl, higher androgen levels from her mother most likely predisposed her to have a greater interest in things and objects. Such a preference will likely show up through an interest in engineering-like occupations which deal with manipulating and visualizing objects in three-dimensional space. In this aspect, Julia's psychology is male-typical, and yet, this does not mean she is actually a male. It simply means she exhibits more male-typical interests.

The second component to the male-dominant engineering stereotype arises from society. Socialization pressures from parents, teachers, peers, and the media reinforce and encourage gender-appropriate behavior. For example, a male's predisposition to like occupations such as engineering may be further reinforced through cognitive social learning, feedback loops from parental encouragement, and internalization of expected modes of behavior. Likewise, a female who has a predisposition to also like engineering may be *discouraged* from pursuing such a career due to gender-appropriate behavioral punishment and subsequent reinforcement towards female-typical interests. Our Julia is likely to experience some level of gender stereotyping when pursuing her interests, something which may affect her self-concept and self-efficacy later in life. For instance, if stereotyping becomes particularly pernicious, she may fail to develop a strong self-efficacy (or sense of her own ability) with mathematics and steer towards safer and more gender-typical prospects. The problem, however, is that she may be excellent at math and wish to pursue an engineering path, but strong gender stereotypes may block that path.

Like the forces of socialization, a gender stereotype, while accurate in an observation such as *men are more interested in engineering than women*, can either exaggerate or suppress biological tendencies. Applying the male-dominant engineering stereotype to an already predisposed male will only exaggerate his perceived interest in engineering, while applying it to a non-

---

[304] Su, R., Armstrong, P. (2008). 859-884.

predisposed female will also exaggerate her perceived *non-interest*. On the other hand, applying the male-dominant engineering stereotype to an already predisposed female, like Julia, may suppress her perceived interest in engineering. Telling a girl she's not equipped to be an engineer because such a task is a *man's job* is a form of gender stereotyping, and applying such a stereotype to girls may suppress an innate interest towards engineering. This is where the major problem arises.

If stereotypes are limited to mere generalizations of perceived behavioral patterns, then they can be used accurately and without issue. Such usages are fine for describing general patterns which are often observed, even unconsciously. However, once gender stereotypes leave the area of generalizations and are directed at individuals which have opposite qualities of the applied stereotype, then serious pathologies, or psychological issues, can be created.

A girl who is interested in engineering but is discouraged and shunned for wanting to pursue such a career due to stereotypical gender roles may develop maladaptive strategies for dealing with such behavioral punishment. She may become depressed and anxious while feelings of resentment and anger build within the unconscious for not being able to express her true self. She may then fall into a neurosis and be unable and unwilling to adapt to her environment and life patterns.[305] Such a state is worse than being simply *unhappy*; every level of her being exists in a state of hell, as her internal desires and goals are being suppressed and discouraged through her external reality.

For Julia, however, such a hellish scenario is unlikely. Western society, especially in the past few decades, has continually, and almost obsessively, championed girls' interests in science, technology, engineering, and math (STEM). There is so much institutional and societal pressure to dismantle traditional stereotypes about girls that the push for women in STEM is one of the strongest social forces of our time. However, this is not to say that gender stereotypes about boys and girls, especially girls in STEM, do not

---

[305] Boeree, C. G. (2002). A bio-social theory of neurosis. *Shippensburg University*.

exist; they continue into the present as long as caricatures of the masculine and feminine exist, as I will soon discuss.

Thus, to understand the inherent dangers of using gender stereotypes when judging the behavior of individuals, we must explore a variety of things, from the accuracy of gender stereotypes and what they are to how they are expressed and how they can turn pathological if applied incorrectly. But first, to recognize when gender stereotypes go too far, we have to understand their innate composition when it comes to humans. In other words, what are gender stereotypes made of?

While I have defined the *origins* of gender stereotypes (biology and society), I have not defined the *content* of the stereotypes, the patterns of behavior we perceive between men and women. Such patterns, such as nurturance or aggression, have not just been perceived in humans for tens of thousands of years, but have also existed in mammals for hundreds of millions of years. Sexual reproduction patterns are differentiated between mammalian males and females, with each sharing their own behavioral and perceptual structure formed around the evolutionary need of reproduction.

For example, the male peacock has a glorious set of colorful feathers which spread out while he dances for a hopefully starry-eyed female. The female, on the other hand, is slightly smaller and has no such colorful plumes. Her behavior is differentiated from the male as a sexually selective force which chooses mates she deems attractive. In this aspect, the male and female peacocks have differentiated behavioral structures around sexual reproduction. Dare I say these behavioral structures can perhaps be called *gender roles*? Though such an application may be dubious, I view the term as a perfect description for the male and female peacocks' differentiated behavior. From these behavioral patterns, or gender roles, we can derive concepts of masculinity and femininity for the peacock. Thus, the concept of masculinity for the male centers on his ability to perform an attractive dance for the female, while the concept of femininity for the female centers on her god-like sexually selective force and dull plumage.

When it comes to humans, however, such a sexually dimorphic expression of behavior is much more complicated. In humans, masculinity and femininity are multifaceted traits and patterns of behavior common to males and females which are defined through the interaction of biology and culture. An individual can have any mix of masculine and feminine traits, which produces the variability we see in gender expression between males and females. Masculine and feminine behaviors continually overlap and interact with one another in any given person. Because of this, our gender roles are more interactionist and adaptive towards both masculine and feminine energies compared to other mammals.

For example, imagine we say that male peacocks are stereotyped to have colorful feathers. Such a statement makes no sense, because there is no overlap between male and female peacocks in this regard and no chance for an incorrect application of a stereotype. Yet in humans, gender stereotypes arise because of the existence of overlapping behaviors. There is a relatively high chance that a gender stereotype will not correctly predict an individual's behavior. Since such variability exists, we must realize that our essential natures of the masculine and feminine are built into the physiology of every human not through a binary structure but through average distributions, creating a much more variable system than in the binary peacock. And so if the peacock has its own qualities of masculinity and femininity based in sexual reproduction, what are humanity's varied masculine and feminine qualities? Or, put another way, what are the generalized contents of humanity's male and female behavioral patterns?

THE FEMININE AND MASCULINE ARCHETYPES

To understand what is generally perceived as feminine and masculine behavioral patterns, we have to explore the feminine and masculine archetypes, also known as the *ideals*, the *essences*, or the *embodiments* of feminine and masculine behaviors. We can think of an *archetype* as a representation of behavioral patterns:

> "An archetype is a pattern of behavior grounded in biology; the behavior itself is both the instinct and the manifestation of that instinct, but an archetype is also the *representation* of that pattern."[306]

Thus, an archetype is not solely a pattern of behavior but is also the representation of that pattern through imagery, symbols, and writing. And so we should ask the first critical question in regards to the authenticity of the feminine and masculine archetypes, which is to say, are these archetypes, the patterns of the ideal feminine and masculine, *real*? If they are not real, then such distinctions of feminine and masculine behavior are meaningless. Such a viewpoint is often held by mainstream social constructionists who reject any distinction between the two.

However, opposite of what social constructionism would predict, perhaps the symbols, images, and descriptions of the ideal feminine and the ideal masculine remain similar across time and culture; if they do, then such archetypal structures must be partially biological in nature.

The founder of analytical psychology, Carl Jung, argued the concepts of the feminine and masculine have been built into our physiology through hundreds of thousands of years of human development. He believed that the feminine and masculine archetypes, the patterns we perceive in male and female behavior, have existed throughout all of history and will continue to exist across time and culture. Jung believed this, in large part, through empirical observation of mythological narratives, individual

---

[306] Unbelievable? (2018). Jordan Peterson vs Susan Blackmore: Do we need God to make sense of life? *YouTube*. 10m45s.

dreams, and descriptions from clinical patients. All areas of observation pointed back to a similar structure for the feminine and masculine across time and culture. Thus, if one wanted to challenge the idea that the feminine and the masculine are biological realities, it would have to be shown that such archetypes exhibit no common structures, symbols, images, or descriptions across time and culture. However, this is not what we find: the feminine and masculine archetypes, expressed in art, literature, and religious writing, seem to share a common symbology and imagery across cultures and time periods.

We as humans have ideals in our minds of what is feminine and what is masculine. Part of this perception is certainly influenced through cultural practices and beliefs, but another part of it is deeply embedded in our physiology, our psychology, and even our biology. Because the feminine and masculine archetypes are nested deep into each individual's psyche through symbols, images, and language, such perceptions of the ideal feminine and masculine human are tightly linked to the innate biological structure of our mind and our nature. Such *syzygic* structures can be seen in other natural occurrences such as the split of our brains into two hemispheres, the dimorphism of sexual creatures, and the conjunction of the moon and sun.[307] In religious and philosophical thought, connected opposites are commonplace. The concept of *yin* and *yang* is one example; *chaos* and *order* is another. Just like other occurrences in our natural world, the concept of masculine and feminine is a syzygy, a pair of connected and corresponding things which are dually linked to one another and cannot be separated.

The feminine and the masculine cannot be split into two; if they are, which often occurs in Western culture, then they are unlikely to interact correctly and will instead become mere caricatures of themselves. We can see such caricatures proliferate in the West, which views the feminine and the masculine as extreme opposites rather than corresponding pairs. Such a divide between the two creates a kind of *split-brain* in our society. Men

---

[307] A *syzygy* is a conjunction, or a pair of connected and corresponding things. *Yin* and *yang*, for example, is a syzygy.

are pushed towards the ideal masculine end as their more unconscious feminine instincts become suppressed and discouraged. The reverse is true for women. Thus, rather than integrating one's opposite sex instincts into a complete whole, the masculine and feminine remain separate entities which fail to conjoin.

Splits such as these can also be seen in neuroscience, as separating the brain's right and left hemisphere syzygy can also create pathological and maladaptive behavioral strategies. For instance, in cases of extreme autism, the individual's brain is heavily lateralized to the right, which causes the crucial communication between the hemispheres to be developmentally neglected. Such extreme lateralization causes serious problems for social functioning, as the left hemisphere fails to properly integrate faces, language, and social interaction. Because of this, it is critical that both hemispheres of the brain interact with one another as corresponding pairs and not split opposites. It is therefore not difficult to understand that the hemispheres work as a syzygy; they are interconnected things which are dependent upon the other for proper functioning. Like the hemispheres in the brain, knowing that the feminine and masculine cannot be separated and that they interact with one another, we should be able to better understand the natures, the structures, and the patterns of these feminine and masculine ideals, or archetypes, and how they relate to gender stereotypes.

According to Jung, the feminine archetype, especially when it relates to the *mother*, can be conceptualized as a "maternal [care] and sympathy; the magic authority of the female; the wisdom and spiritual exaltation that transcend reason; any helpful instinct or impulse; all that is benign, all that cherishes and sustains, that fosters growth and fertility. The place of magic transformation and rebirth, together with the underworld and its inhabitants, are presided over by the mother."[308]

---

[308] C.G. Jung. (1990). *The Archetypes and the Collective Unconscious (Collected Works of C.G. Jung Volume 9 Part 1)*. Bollingen Series XX: Princeton University Press, 82.

10 GENDER STEREOTYPE PATHOLOGY

Such archetypal qualities in the feminine, like the masculine, have their own dark sides as well. For the feminine archetype, the negative side may "connote anything secret, hidden, dark; the abyss, the world of the dead, anything that devours, seduces, and poisons, that is terrifying and inescapable like fate."[309]

Knowing that similar descriptions of the feminine *mother* archetype exist across time and culture, it is not difficult to see the image of the ideal feminine in one's own psyche. Here we are dealing with more than mere idiosyncratic quirks or unique cultural practices such as wearing dresses or putting on makeup, but rather overarching abstractions of behavioral patterns that form symbols, images, and descriptions across tens of thousands of years. From Jung's description of the ideal feminine and its negative and positive qualities, we can see that the feminine is something which transforms and renews, something which can tear down decaying structures and replace them with growth and fertility, which sustains life, and which takes life. Our commonly known concept of Mother Nature, the ultimate selective force, is similar to such descriptions in that it too sustains and takes life. And yet this ideal even extends to religions and mythologies. One doesn't have to go far from Christianity to see that the Virgin Mary, through which the rebirth of the world took place, is one such example of the mythological *mother* archetype and the ideal feminine.

The other side of this syzygy is the ideal masculine, or the *father* archetype. Like with the feminine archetype, we are not dealing with transient cultural practices such as men dressing in ties and suits. Rather, when discussing the masculine archetype, we are dealing with abstract concepts which elucidate near metaphysical patterns of behavior. Therefore, if the feminine archetype relates to instinctual nature and material reality, then the masculine archetype must relate to light and spirit, something which transcends nature.[310] A transcending of nature requires that an unformed chaos, or a fragmented structure, be transformed into order. If chaos can be ordered, it must utilize logic and

---

[309] C.G. Jung. (1990). 82.
[310] Edinger, E. (1992). *Ego and Archetype*. C.G. Jung Foundation Book Series (Book 4), Shambhala.

reason, through consciousness, to order it. Therefore, the word *logos*, a Greek word meaning logic, is a perfect description of the masculine archetype, which represents the process of ordering unformed potential into a formal actuality. Such a process is an act of creation and can be described holistically as an exertion of the human mind:

> "The father archetype, in its *logos* function, exerts his influence on the human mind in order to transform undifferentiated--concrete--emotionality bind to body, to less material and abstract form of images and mental representations, but also, in its specific complex-bound function, serves as regulator of boundaries, restrictions, and social values that are imposed as rules and laws."[311]

Such boundaries, restrictions, and social values are formed through the use of consciousness, something which gives order to chaos and transforms unrealized potential into realized actuality. In Christianity, the Holy Spirit is an archetypal representation of *pneuma*, wind, and breath. Its illumination of spirituality in the individual is analogous to the consciousness of the masculine archetype, a light which orders, produces, and creates. And yet, just like the negative feminine, such a power for creation can be destructive. Order, rules, and laws when taken too far, can produce tyrannies in one's personal life and authoritarian regimes across an entire society. Both the Soviet Union and Nazi Germany were exemplars of when the masculine archetype goes too far and rejects the necessity of the feminine, the importance of care for others, renewal, change, and rebirth. It goes without saying that, like the two hemispheres in our brain, both the feminine and the masculine have their positive and negative, healthy and pathological sides, which, when brought into conscious awareness in the individual, can be appropriately integrated.

Of course, both males and females exhibit and use masculine qualities, just as both males and females exhibit and use the previous feminine

---

[311] Solc, V. (2013). Father Archetype. *Therapy Vlado.*

qualities. Feminine qualities in a male are known as the *anima*, whereas masculine qualities in a female are known as the *animus*. The anima and animus are simply labels for patterns of contrasexual behavior between males and females. For a male, integrating one's anima and bringing it into conscious awareness is a critical aspect of becoming a holistic and complete individual. A man who integrates his anima exhibits a plethora of healthy feminine qualities in relationship to the masculine:

> "[Anima integration] gives him a great capacity for friendship, which often creates ties of astonishing tenderness between men and may even rescue friendship between the sexes from the limbo of the impossible. He may have good taste and aesthetic sense which are fostered by the presence of a feminine streak. Then he may be supremely gifted as a teacher because of his almost feminine insight and tact. He is likely to have a feeling for history, and to be conservative in the best sense and cherish the values of the past. Often he is endowed with a wealth of religious feelings, which help to bring the *ecclesia spiritualis* into reality; and a spiritual receptivity which makes him responsive to revelation.
>
> [Anima integration] can appear positively as bold and resolute manliness; ambitious striving after the highest goals; opposition to all stupidity, narrow-mindedness, injustice, and laziness; willingness to make sacrifices for what is regarded as right, sometimes bordering on heroism; perseverance, inflexibility and toughness of will; a curiosity that does not shrink even from the riddles of the universe; and finally, a revolutionary spirit which strives to put a new face upon the world."[312]

Such beautiful integration of the anima in a man is often suppressed and discouraged by our Western culture. Tenderness between men, for instance, is often ostracized as effeminate and gay. A man with good taste,

---

[312] C.G. Jung. (1990). 86-87.

insight, and a strong empathic sense may be rejected by his male peers who have not yet integrated their own anima and instead reject it completely. Failure to allow a man's feminine qualities to be expressed inevitably creates pathological behaviors which are often destructive, prejudicial, and discriminatory.

Likewise, though not as pervasive, women are sometimes discouraged from integrating their animus, the masculine qualities within. If the animus is integrated properly in a woman, then she will exhibit healthy masculine ideals in relationship to her feminine qualities. Such animus qualities are exhibited through the development of a woman's ability to be assertive and aggressive, to speak honestly and truthfully without fear of what others think, to make decisions for one's own sake and not for others, to embrace one's capacities for righteous judgment, and to transcend the natural impulses and beliefs of the instinctual world and utilize the *logos* within to transform the unrealized and fragmented chaos of reality and bring forth an actualized order.[313] Development of the animus is therefore a process through which a woman rediscovers the masculine qualities left behind in childhood and adolescence.

Julia may be especially predisposed to experience the process of animus integration because of her more masculinized brain and psychology. She may already possess an inherent sense of the masculine archetype and therefore develop it more fully compared to other women. She may find that, while society has tried to suppress the masculine within, her masculine tendencies grow into powerful forces that demand the actions necessary from which they will break forth. Her drive to take something unrealized and form it into an actualized reality, such as building and designing something she sees in her mind's eye, is one such aspect of the masculine archetype that she is likely to embody.

Such an integration of anima in a man and animus in a woman does not mean that one's gender identity is rejected or that it *can* be rejected, but rather that integrating one's masculine and feminine qualities is a healthy

---

[313] Peterson, J. (2018). Jordan Peterson explains Jung's animus and anima. *Bite-sized Philosophy, YouTube*.

and critical step to complete individuation, which is to say, to being a fully developed individual. This process can occur once one reaches a mature stage of development, or as Dr. Jordan Peterson explains:

> "Jung thought that when you had established a personality that was sufficiently developed to be acceptable socially, and functional on the individual level, then you could have the opportunity to expand that personality and take into yourself elements of perception, thought, and behavior that you wouldn't have had the sophistication to be able to handle at an earlier stage of development."[314]

Therefore, if one wishes to integrate their anima or animus and develop their contrasexual qualities, an individual must reject the gender stereotypes in their culture, just as Julia must reject hers. Applying gender stereotypes to individuals, similar to cutting the connection to the brain's right and left hemispheres, forces an unbalanced and pathological structure onto an individual's psyche. Such practices are commonplace in a Western culture that views the ideal feminine and the ideal masculine archetypes as categorical opposites rather than an interactive and connective syzygy. Splitting the feminine and the masculine into two separate and distinct categories is analogous to cutting the connection between *yin* and *yang*, *chaos* and *order*, and the *right* and *left* hemispheres. It cannot and should not be done, as separating any of these syzygies ultimately results in pathological consequences to individuals and societies. Thus, any individual who rejects either their anima or animus, the feminine or the masculine, exists in a neurotic imbalance, a split-brain; since the hemispheres are not connected, nothing functions properly.

Now that we have covered the contents of the ideal feminine and the ideal masculine, we will explore how disconnecting this syzygy can quickly turn pathological through the use of gender stereotypes.

---

[314] Peterson, J. (2018). Jordan Peterson explains Jung's animus and anima. *Bite-sized Philosophy, YouTube.*

WHEN STEREOTYPES BECOME PATHOLOGICAL

Gender stereotypes are the oversimplified behavioral patterns and psychological traits of men and women. While the feminine and masculine archetypes may accurately describe patterns of behavior, gender stereotypes occur when these archetypal concepts are applied too narrowly by attaching binary descriptions to individuals. For example, applying the feminine archetype to *only* females transforms the complex archetype from its widely applicable behavioral patterns to a narrow and fixed description of female behavior, which is to say, a stereotype. Here individual variation is not accounted for:

> "When people associate a pattern of behavior with either women or men, they may overlook individual variations and exceptions and come to believe that the behavior is inevitably associated with one gender but not the other."[315]

Narrow stereotypes of what it means to be a man and what it means to be a woman constrain the development and expression of the individual, just as it will constrain Julia's development if the masculine is rejected by her or is prohibited from being identified with. Such is the problem of stereotyping. While understanding the feminine and masculine archetypes is beneficial for integrating one's anima or animus, failing to integrate these contrasexual qualities will result in a fragmented individual who, rather than relying on a deep exploration of the feminine and the masculine, relegates themselves to either/or through gendered stereotypes.

Taken to their extremes, such stereotypes often result in mere caricatures of the feminine and masculine archetypes. Throughout the 19th century, women were stereotyped as pious, pure, submissive, and domestic; they were seen as weak, dependent, and timid; their domain was relegated to the home, and their role was to take care of the children and the husband.[316] However, such stereotypes couldn't be more wrong.

---

[315] Brannon, L. (2016). 160.
[316] Ibid., 161-162.

Nowhere in the feminine archetype exists the word *weak* or *dependent*, *timid* or *submissive*. Rather the feminine is what transforms, renews, and regenerates existing structures; it is the opposite of passive.

Men were stereotyped as active, independent, coarse, and strong; they were prone to sin and seduction; aggressive, tough, and confident; they were daring, violent, and rebellious; the creators and the revolutionaries, the challengers and the changers.[317] And yet these words, too, are incredible oversimplifications of masculine behavior and are mere caricatures of the archetype. Such stereotypes of men and women still exist today in Western culture, which often makes it difficult for individuals to express their innate selves, and to integrate their anima and animus, for fear of challenging existing stereotypes. Failure to integrate the feminine and masculine archetypes into conscious perception, or rejecting the archetypes completely, always results in pathological beliefs.

A rejection of the archetypes occurs in all facets of society and belief systems. One example is in the beliefs of social constructionists who reject the masculine archetype by claiming that masculinity is nothing more than a socially constructed drive for power which marginalizes and oppresses groups of people.[318] In this view, there is no room for the possibility that the masculine archetype has a multiplicity of positive and negative behavioral patterns which represent themselves across thousands of years of human civilization, mythologies, and symbols; rather, the masculine archetype is simply a construct of *power:*

> "Men's lives are structured around relationships of power and men's differential access to power, as well as the differential access to that power of men as a group. Our imperfect analysis of our own situation leads us to believe that we men need more power, rather than leading us to support feminists' efforts to rearrange power relationships along more equitable lines."[319]

---

[317] Brannon, L. (2016). 162.
[318] Kimmel, M. (2000). 217.
[319] Ibid., 218.

These narrow-minded and stereotyped beliefs, that masculinity is structured around power and control, result in sexist and cynical views of anything masculine, thereby becoming pathological in their application; such are the fruits of postmodern thinking. For instance, radical feminists such as Michael Kimmel, who believe such things about masculinity, will not accept anything less than a full endorsement of the political agenda of *equity*, which states that all differences between men and women must be equalized. Because of this view that masculinity is comprised of power and control, radical feminists and gender theorists reject the good and noble qualities of the masculine and replace them with its pathological and tyrannical forms. And yet, such a view disregards the true nature of the masculine just as misogynists disregard the true nature of the feminine. Thus, an individual who perceives the masculine archetype as simply tyrannical and authoritarian has likely not integrated their own masculine qualities and undertaken the process of individuation. Likewise, an individual who perceives the feminine archetype as simply destructive and chaotic lacks a coherent structure of the feminine. Both narrow perspectives of the feminine and masculine result in abhorrent caricatures of their true natures.

Ironically, social constructionists embrace the very stereotypes they claim to want to dismantle; namely, that masculinity is about power and dominance and femininity is about submissiveness and timidity.[320] By conceptualizing the feminine as a submissive behavioral pattern in response to masculinity's domination and power, social constructionists end up reinforcing stereotypical beliefs about men and women, thereby hurting the very cause they are supposedly championing. Outside of social constructionism, however, there are other belief systems which perpetuate gender stereotypes.

One of these belief systems is the stereotypical idea that certain jobs are for men and other jobs are for women because of some innate biological fact. A job as a construction worker may be stereotypically viewed as a

---

[320] Connell, R. W. (1987). *Gender and power: Society, the person and sexual politics.* Cambridge: Polity, 188.

male's job, since more males than females work at construction sites. Likewise, a job as a nurse may be stereotypically viewed as a female's job, since more females than males work as nurses.

While the observation that differences exist in behavioral patterns is accurate, the gender stereotype occurs when it's applied incorrectly. There is no true occupation for only men or only women; rather there are occupations which have more men than women and more women than men because of differences in average interests. Such a statistic doesn't mean that men and women cannot go into gender-atypical jobs, and it certainly cannot accurately predict the behavior of individuals beyond mere educated guesses. And yet, such stereotypical views of the feminine and masculine archetypes proliferate into many aspects of society. For example, interviews of women working in male-dominated industries have elucidated stereotyped treatment, as one incident shows:

> "Several women told us that men questioned their ability to effectively supervise others, and Marie was no exception. One older subcontractor explicitly told her, 'This isn't the job for a woman.' Having just walked through the requirements of some complex paperwork, Marie said, 'I think he just thought I was being a nag and that I didn't know what I was doing.' Later in the interview, she said, 'Just being a female in management is difficult, and guys don't like it--especially guys that work in the field. They think that women should be secretaries.'"[321]

Such beliefs that women should be relegated to specific jobs creates an imbalance in people's perception of the feminine and masculine archetypes. Just because a woman may work in construction doesn't make her any less feminine, nor is the case that a man working in nursing makes him less masculine. There is no room for binary and dichotomous thinking in the archetypes; *one can integrate and develop their feminine side without*

---

[321] McLaughlin, H., et al. (2012). Sexual harassment, workplace authority, and the paradox of power. *American Sociological Review, 77(4),* 635.

suppressing the masculine, and the reverse is also true. Thankfully, the view of stereotyped jobs is fading away in the most gender egalitarian countries. However, there is still much room for societal change, especially when it comes to the development of more accurate and non-stereotypical views of the feminine and masculine archetypes.

Another belief system which stereotypes the archetypal composition of the feminine and the masculine exists in the area of children's toy preferences. Traditional beliefs of gender-typical play have accurately observed that many boys and girls have different interests when it comes to toys. Such a statement is innocuous and accurate on its own. However, the stereotypical belief that boys play with trucks and girls play with dolls can turn pathological when an individual's innate tendencies are suppressed. A girl may have an interest in boy-typical toys, and yet, through the application of rigid stereotypes, such an interest may be discouraged and even punished. Suppressing a child's innate interests because of stereotypical beliefs about boys' and girls' interests is not just wrong; it's likely to cause maladaptive coping strategies in the child which can eventually turn into neuroses and disorders. On the extreme end, such suppression can even result in suicide, as in the case of David Reimer. Dr. John Money's insistence that he be raised as a girl despite being a male was a pathological and deadly application of gender stereotypes.[322]

Knowing this, it is clear that rather than applying gender stereotypes to individuals and suppressing their innate interests, a child should be able to explore what interests them, while also forming a healthy understanding of their biological reality (being a male or female). If they are able to explore their interests through parental encouragement and support, then they are much more likely to have a healthy view of the feminine and masculine archetypes, thereby integrating them into their individualized self as they grow into adolescence and adulthood.

However, not allowing a child to explore their own interests may seriously affect their self-concept (their internal belief about themselves), their self-efficacy (perception of one's own ability), and their self-esteem

---

[322] See chapter, "Development of Gender Identity."

(perception of one's own worth and value). Stereotypes about girls' math ability, for instance, can seriously affect a girl's self-efficacy when it comes to performance on math-based tasks even though boys and girls are equally capable of mathematics. Such a self-efficacy incongruity presents itself, in part, through girls choosing to go into mathematics less than boys.

In adulthood, stereotypes about the masculine and feminine archetypes continue. While such pathological views are easy to dismantle through a simple analysis, stereotypical beliefs are difficult to overcome because of the social and economic structures from which they are reinforced. When people deviate from stereotypical behaviors, backlash, penalties, and rejection can sometimes follow, which makes behaving outside of stereotypes something that takes courage and resilience. Statements such as "walk tall," "be a man," or "don't be a sissy" are phrases which reinforce stereotypical and narrow-minded definitions of the masculine archetype:

> "Men are expected to be successful, powerful, and dominant, show no weaknesses or chinks in the armor, and avoid acting in ways that might be perceived as feminine. In general, masculine stereotypes prescribe men to be 'bad but bold,' demanding that they strive to gain and maintain the respect of others."[323]

Failure to adhere to these stereotypical views of masculinity can result in a variety of consequences for the non-conforming individual, from expressions of discouragement from a peer group to prejudicial hiring practices.

> "Our findings demonstrate that men encounter prejudice when they behave atypically, and raise the possibility that men

---

[323] Moss-Racusin, C. A., et al. (2010). When men break the gender rules: status incongruity and backlash against modest men. *Psychology of Men & Masculinity, 11(2),* 141.

may avoid behaving modestly because they risk backlash for stereotype violation when they do. Perceived violations of both men's proscriptive and prescriptive traits helped to explain backlash against atypical men."[324]

Stereotypical views that men are dominant, powerful, and aggressive and that women are submissive, timid, and agreeable are able to proliferate, in part, through the media and entertainment industries which increase their profits by implementing caricatures of the feminine and masculine archetypes. These caricatures can be seen in things like beer commercials, where the masculine archetype is turned into a cultural stereotype of itself:

> "The men of beer commercials fill their leisure time in two ways: in active pursuits usually conducted in outdoor settings (e.g., car and boat racing, fishing, camping, and sports; often symbolized by the presence of sports stars, especially in Miller Lite ads) and in "hanging out," usually in bars. As it is in work, the key to men's active play is the challenge it provides to physical and emotional strength, endurance, and daring. Some element of danger is usually present in the challenge, for danger magnifies the risks of failure and the significance of success. Movement and speed are often a part of the challenge, not only for the increased risk they pose, but also because they require immediate and decisive action and fine control over one's own responses."[325]

The visual of men filling their leisure time with active and outdoor pursuits reinforces the stereotype that men are the dominant and conquering sex. And this is the core of the commercial's stereotype:

---

[324] Moss-Racusin, C. A., et al. (2010). 148.
[325] Strate, L. (1992). Beer commercials: a manual on masculinity. *Men, Masculinity, and the Media, Sage Publications,* 534-535.

> "The central theme of masculine leisure activity in beer commercials, then, is challenge, risk, and mastery--mastery over nature, over technology, over others in good-natured 'combat,' and over oneself."[326]

While relegating the masculine archetype to an outdoors, activity-based adventurer, the cultural stereotype is reinforced. And yet, unfortunately, the contents of the masculine archetype go well beyond simply liking the outdoors. Such a view is overly simplistic and patronizing. Caricatures such as these are perhaps more ubiquitous when it comes to women in advertisements. Across many ads, from perfume and clothes, women are often depicted in stereotypical ways. One such example is what's known as the *feminine touch*:

> "Women, more than men, are pictured using their fingers and hands to trace the outlines of an object or to cradle it or caress its surface. This ritualistic touching is to be distinguished from the utilitarian kind that grasps, manipulates, or holds. Self-touching can also be involved, readable as conveying a sense of one's body being a delicate and precious thing."[327]

Another example of the stereotypical feminine body language present in the media is the *bashful knee bend*:

> "Women frequently, men very infrequently, are posed in a display of the 'bashful knee bend.' Whatever else, the knee bend can be read as a foregoing of full effort to be prepared and on the ready in the current social situation...Once again, one finds a posture that seems to presuppose the goodwill of anyone in the surroundings who could offer harm."[328]

---

[326] Strate, L. (1992). 535.
[327] McLaughlin, H. (2018). Doing gender. *Sociology of Gender, Lecture at Oklahoma State University.*
[328] McLaughlin, H. (2018).

Caricatures such as the feminine touch and bashful knee bend are stereotypical representations of the feminine archetype and are therefore overly simplistic. Such caricatures make women out to be extra sensitive, timid, and submissive, reinforcing stereotypical behaviors while ignoring the development of the masculine archetype, the *animus*, in women.

If such representations of the feminine and masculine archetypes are mere caricatures of actual reality and if such caricatures about men's and women's gender-appropriate jobs and boys' and girls' gender-appropriate toys can be harmful to individual development, what should we do about gender stereotypes?

As much as social constructionists might want to believe, we cannot say that gender stereotypes are all wrong. Many stereotypes have foundations in reality, as stereotypes are simply generalizations of behavioral patterns. Likewise, rejecting the feminine and masculine archetypes altogether leaves us with no reference point from which to understand our own psychology. Rejecting such archetypes can result in pathological beliefs and behaviors which fail to integrate the good qualities of the feminine and the masculine. However, while we may understand the importance of the archetypes in describing behavioral patterns, we should be cautious in putting too much emphasis on them to the point that we judge an individual's behavior, interests, and life choices in light of them.

Instead we must recognize that the feminine and masculine archetypes are not separate entities but are rather an interconnected and dually linked syzygy which relies on the other to function. Therefore, it is our job as individuals to look within ourselves to see whether we have integrated our more unconscious feminine or masculine qualities, thereby disregarding our culture's idiotic stereotypes as mere caricatures of the truth. If we haven't, we can start now.

What we should not do, however, is reject our contrasexual qualities and identify solely with either a feminine or masculine energy. Such a practice is like cutting off half of one's brain in hopes the existing side will grow and develop on its own. A woman who wants to integrate her animus, therefore, can work to be more assertive and to speak up on her own behalf, to fight for what she believes in, and to not back down from sacrifice and risk. Such is the job of our fictional girl Julia who, upon recognizing and illuminating the masculine archetype within, can begin to develop and strengthen it. Reversely, a man who wants to integrate his anima can work to be more compassionate, caring, and sensitive to others while developing a creative curiosity and drive to uncover the mysteries of life with an empathic sensibility.

Such an integration of the feminine and masculine archetypes and an individuation of the self will allow us to move beyond the stereotypical judgments of ourselves and others. It will break the barriers of narrowmindedness and build a foundational structure out of the ashes. Accomplishing such a courageous and terrifying task won't be easy, but with it comes a knowledge and wisdom about our inner selves that will not just last the test of time but remain strong in the face of our culture's monolithic and pathological gender stereotypes.

11

THE ACTIVATIONAL STAGE

Organizational effects of prenatal androgens produce structural differences between the brains of boys and girls before they're even born; these structural differences influence perception, behavior, interests, and the ultimate solidification of gender identity.[329] From the increase in suboptimal arousal and rightward shift to the development of the visuospatial system and the expression of sex-specific genes, the neuroanatomy and neurocircuitry of boys and girls is significantly altered in the womb through a mix of sex hormone exposure and sex-differentiated gene activation through an X or Y chromosome.

Such claims would not have been possible before the development of technologies which allow neuroendocrinologists to study the effects of sex hormones on the brain. Technologies such as DNA sequencing and RT-PCR are able to further elucidate the complex circuitry and gene activation systems which produce differences in behavior and interests. It turns out that socialization, while certainly a significant force in the development of gender, is not the only factor. Despite all the recent neuroscience evidence which illuminates the complex mechanisms underlying sex and gender differences, social constructionists continue to deny science.

And yet things only get worse for social constructionism as we move further into the gender development of the individual and get closer to the causes for the Gender Equality Paradox. Social constructionists cannot

---

[329] See chapter, "Development of Gender Identity."

explain the most important developmental period which produces the largest sex differences in physical body, brain, personality, and interests through the release of sex hormones. These sex hormones bind to hormone receptors in the brain which then produce specific proteins for the production of morphological differences.[330] The proteins then facilitate the formation of neurostructures which produce differences in behavior, cognition, and interests, which then influence gender differences across society. This final sex-differentiating process between girls and boys is the activational stage of development known as *puberty*.

For those who believe that sex and gender differences are largely the result of socialization forces, puberty provides strong evidence in opposition. This sex maturation stage is the most significant biological process for the production of sex and gender differences in physiology, cognition, behavior, and interests. For this chapter, I will provide strong evidence that sex and gender differences are further amplified through the release of pubertal hormones which activate and develop existing organizational structures in the brain, forming sex-differentiated neuroanatomy and neurocircuitry.

THREE ENDOCRINE EVENTS

Social constructionists do not understand the fact that activational effects of puberty are only possible because of the prenatal organization of the fetal brain.[331] Without the organization of the fetal brain through sex hormones and expression of sex-differentiated genes, hormones would not have their activational ability during puberty. In other words, without a neural structure that is predetermined in the womb, pubertal hormones would have no structure on which to act.

According to neuroendocrinologists, puberty is a "complex set of neuroendocrine processes that occur between child and adulthood to

---

[330] Davey, R., Grossmann, M. (2016). Androgen receptor structure, function, and biology: from bench to bedside. *Clinical Biochemist Reviews, 37(1)*, 3-15.
[331] See chapter, "Conception to Birth."

produce internal and external physical changes to primary and secondary sexual characteristics allowing for sexual reproduction."[332] This hormone activation process can be separated into three unique endocrine events known as *adrenarche, gonadarche,* and *activation of the growth axis.* These events directly produce sex differences in physiology, neuroanatomy, neurocircuitry, behavior, and interests.

Adrenarche is the first event of puberty which activates the hypothalamic-pituitary-adrenal (HPA) axis between the ages of six and nine in females and seven and ten in males.[333] This axis comprises a complex set of interactions between the hypothalamus, pituitary gland, and the adrenal glands which control reactions to stress and regulate digestion, immune system, mood and emotions, and sexuality.[334] Activation of this axis produces an increase in adrenal androgens (weaker hormones than gonadal testosterone) which eventually help the development of underarm and pubic hair, sweat glands, and body odor.[335] Lastly, adrenarche may give rise to maturational effects prior to puberty, though these processes are not fully understood.[336]

Gonadarche is the second event of puberty which activates the hypothalamic-pituitary-gonadal (HPG) axis between the ages of eight and fourteen in females and nine and fifteen in males. This axis is conceptually similar to the HPA axis in adrenarche. However, instead of interacting with the adrenal glands, the hypothalamus and pituitary glands interact with the gonads (ovaries or testes). The process begins when the hypothalamus releases the *gonadotropin-releasing hormone* (GnRH), which then

---

[332] Herting, M., Sowell, E. (2017). Puberty and structural brain development in humans. *Frontiers in Neuroendocrinology, 44,* 122-137; 123.

[333] Blakemore, S.J., Burnett, S., & Dahl, R. (2010). The role of puberty in the developing adolescent brain. *Human Brain Mapping, 31,* 927.

[334] Malenka, R.C., Nestler, E.J., Hyman, S.E. (2009). "Chapter 10: Neural and Neuroendocrine Control of the Internal Milieu." In Sydor A, Brown RY (ed.). *Molecular Neuropharmacology: A Foundation for Clinical Neuroscience (2nd ed.).* New York: McGraw-Hill Medical, 246, 248–259.

[335] Herting, M., Sowell, E. (2017). 123; Blakemore, S.J., Burnett, S., & Dahl, R. (2010). 927.

[336] Blakemore, S.J., Burnett, S., & Dahl, R. (2010). 927.

stimulates pituitary production of *luteinizing hormone* (LH) and *follicle-stimulating hormone* (FSH). These two hormones, LH and FSH, activate sexual changes in the gonads so that gametes (sperm or eggs) can be produced.[337] As the ovaries and testes become exposed to LH and FSH and become more mature, they release increasing amounts of estrogen and testosterone, respectively.[338] The increase in estrogen and testosterone produces maturational effects in the reproductive organs and helps form secondary sex characteristics.[339] Thus, gonadarche is not just the process which helps develop secondary sex characteristics such as breasts or facial hair; it's the event which gives adolescents fully functional reproductive organs.

The activation of the growth axis is the third and final hormonal event during puberty which differentiates the sexes through changes in growth patterns. This results in a linear growth spurt for girls around the age of 12 and for boys around the age of 14. Once adrenarche, gonadarche, and the activation of the growth axis are complete, puberty is finished.

As she grows up into adolescence, our fictional girl experiences these processes as well. She develops into a healthy young woman with average levels of sex hormones and a functional reproductive system, but it is in the brain where interesting differences from normal women arise. Since Julia was exposed to high levels of prenatal androgens in the womb, her brain has already been organized in a male-typical way through heavier lateralization, a process where the brain's connections form more heavily *within* the hemispheres than *between* them. This causes male-typical traits such as an interest in visuospatial tasks to become more developed. Julia's psychology, therefore, is already organized for the activational effects of puberty which will bring about more structural changes in the body and brain. Such additional effects of sex hormones and sex-differentiated genes will further amplify her masculine tendencies and interests as well as develop her mature female form and brain functions.

---

[337] Blakemore, S.J., Burnett, S., & Dahl, R. (2010). 927.
[338] Ibid.
[339] Blakemore, S.J., Burnett, S., & Dahl, R. (2010). 927.

PUBERTY INITIATES MORPHOLOGICAL CHANGES

The largest sex differences in males and females arise through the activational stage of puberty, which activates the development of the gonads into mature gamete-producing organs. Higher concentrations of hormones produced through the gonads bind to sex hormone receptors in the brain to produce structural changes through protein synthesis. The largest sex differences produced through this process lie not in the brain itself but through how the brain influences structural changes in the physical body's morphology.

As the hypothalamic-pituitary-gonadal (HPG) axis matures through the process of *gonadarche*, gonadal hormones (estrogen and testosterone) influence the appearance of the male and female body. For males, as testosterone increases by substantial amounts, the testicles and penis will enlarge and become more mature while hair will begin to grow under the arms, around the groin, and on the face. For females, as the ovaries mature and produce more estrogen, breasts will develop and menstrual periods will begin. Through the production of testosterone in the adrenal glands, pubic hair will grow very similar to that of boys. But, since higher levels of testosterone are not present, facial hair will be zero to minimal in females.

Outside of sexual reproduction, sex differences begin to form in the morphology of our bodies. Through the exposure to higher levels of testosterone through the testes, males develop stronger and denser bones which have around 50% more bone mass than the average woman.[340] Men tend to develop larger heads and longer arms and legs, while women's elbows and shoulders tend to be bent a little further from their bodies and are more mobile at both joints.[341] When it comes to the hands, the length of the ring finger is often longer than the index finger in men, while the reverse is true for women; such a sex difference is likely influenced by prenatal androgen exposure. The higher the androgen exposure, the longer

---

[340] F.P.F.W. (2017). Biological sex differences: bones & muscles. *Fair Play for Women.*
[341] Ibid.

the ring finger is than the index.³⁴² Oddly enough, Julia's ring finger is measured to be much longer than her index finger, providing indirect evidence for her higher levels of testosterone exposure in the womb.

Other than the finger ratio, the average woman tends to have a longer torso and shorter legs than the average man so that the internal reproductive organs such as the uterus and ovaries can be accommodated. Because of the need for childbearing, the pelvis is larger in females than it is for males: "It's wider, longer, and held together by ligaments that soften during pregnancy, allowing the two halves to slide apart because of [the] narrow pelvis. Women's slanted thigh bones put extra pressure on the knee joints, which have to rotate while men's do not."³⁴³

Women tend to have thicker skulls by about 10% while men tend to have a "bony ridge on the brow line and a heavier jaw." This sex difference in facial structure correlates to sex hormone exposure. Estrogen relates to "higher eyebrows, fuller lips, and a rounder jawline with a more pointed chin," while testosterone relates to "a square chin and sharply angled mandible, a more prominent Adam's Apple, more facial hair and male-pattern baldness with age."³⁴⁴ All of these average differences in the physical body result in a unique bone structure for males and females, a structure so distinct that the sex of an adult skeleton can be determined with "95% accuracy by measuring the hip bones alone, 83% accuracy by the skull, and 80% accuracy by the long bones (femur and tibia)."³⁴⁵

In terms of the senses, women tend to be better at seeing nuances in color, while males tend to be better at tracking fast-moving objects and discerning detail at a distance.³⁴⁶ Women tend to be better at hearing high frequencies while their ability to hear low frequencies decreases with age.³⁴⁷

---

[342] Cohen-Bendahan, C., Beek, C., Berenbaum, S. (2005). 373; See chapter, "Conception to Birth."
[343] F.P.F.W. (2017).
[344] F.P.F.W. (2017).
[345] Ibid.
[346] Owen, J. (2012). Men and women really do see things differently. *National Geographic*.
[347] F.P.F.W. (2017).

Men's voices are almost always deeper than women's due to the sex-differentiated growth of the vocal tract and vocal folds during puberty.[348]

In terms of muscle, 30-35% of the average woman's weight is comprised of muscle; for a man, it is 40-50%. Muscle can be built through physical exercise and strength-building, but there is no way for women to match men's muscle composition other than to utilize androgen steroids by a significant amount.[349] This sex difference in muscle mass is exemplified through *grip strength*: "The strongest 10% of females can only beat the bottom of 10% of men."[350] Men tend to have a much greater upper body strength than women, whereas women tend to do better with lower-body strength than men. "Females generally have 40% less skeletal muscle than males on the top half; 33% less below the waist. Men have 66% more upper-body muscle than women, and 50% more lower-body muscle."

In terms of how the muscles are put together, men's muscles contain more proteins:

> "Men's muscles are more solid, due to a higher proportion of Type 2 fast-twitch fibers. This type of muscle fiber contains a lot of protein but not much blood. It can expand/contract rapidly with great force and generate its own energy, but it draws on the rest of the body for oxygen. Because of this, Type 2 muscle fibers get tired more quickly. Although it varies by body type & genetics, men's muscles are, proportionally, about 50% more Type 2 than women's."[351]

Although men tend to have higher amounts of Type 2 muscle fibers, women's muscles tend to be more impacted by the Type 1 muscle fiber, which are more long-lasting:

---

[348] Ibid; there is almost no overlap in male and female vocal pitch.
[349] F.P.F.W. (2017).
[350] Ibid.
[351] F.P.F.W. (2017).

> "Women's Type 1 muscle fibers protect health: Type 1 (slow-twitch) fibers are more loosely packed and have their own capillaries. This means they can keep going for a long time. They interact with other metabolic processes, which helps to protect against insulin resistance and heart disease, supports the immune system and promotes hormonal functions. Olympic athletes aside, both female & male bodies use Type 1 fibers first, then shift to Type 2 when those get fatigued, but women have been found to switch between the two throughout exertion. This is the secret of [women's] famed endurance: not the slow-twitch fibers themselves, but neuromuscular activity which uses all the fibers."[352]

Throughout all the sex differences we see in muscle mass, bone density, and physical morphology produced through pubertal hormones, it is clear that men and women outperform each other in different aspects rather than one sex being better than the other. It would be a dumb generalization to say that men are the stronger sex and women are the weaker sex, and yet, such generalizations continue throughout a culture which fails to recognize that the strengths of men and women represent themselves in different, specialized areas. For bone and muscle performance, men tend to have the advantage, while for reproductive fitness, specialized strength, and long-term stamina, women tend to have their own advantage. Evolutionary biologists would say that these sex differences evolved due to different capacities and functions between male and female reproduction; for instance, many of these morphological sex differences may have originated from natural selection processes selecting and developing males for more physical combat and aggression. However, regardless of where these sex differences came from, it is clear that males and females are further differentiated during puberty through the exposure to sex hormones and that these differences remain relatively stable throughout adulthood.

---

[352] F.P.F.W. (2017).

Table: Sex differences which suggest male design for combat in humans.

Male humans tend to have:	Reference
Greater upper body strength	Lassek and Gaulin 2009
Taller bodies	Alexander et al. 1979
Heavier bodies	Loomba-Albrecht and Styne
Higher basal metabolic rates	Garn and Clark 1953
Faster reaction times	Der and Deary 2006
Thicker bones in the jaw	Humphrey et al. 1999
Faster mental rotation and spatial ability	Voyer et al. 1995
More accurate throwing	Jardine and Martin 1983
More accurate blocking of thrown objects	Watson and Kimura 1989
More interest in the practice of combat skills	Gibbons et al. 1997
Stronger bones	Schoenau et al. 2001
Greater bone density, specifically in the arms	Wells 2007
Easier heat dissipation	Burse 1979
More hemoglobin in the blood	Waalen and Beutler 2001
Higher muscle-to-fat ratio	Loomba-Albrecht and Styne
Larger hearts	Tanner 1970
Higher systolic blood pressure	Tanner 1970
Broader shoulders for efficient weapon use	Brues 1959; Tanner 1989
Larger sweat capacity	Burse 1979
Larger circulating blood volume	Burse 1979
Greater resistance to dehydration	Burse 1979
Tolerance for risk and dangerous activities	Wilson et al. 2009
Faster sensory frame shifting	Cadieux et al. 2010
Thicker skin	Shuster et al. 1975
Larger lung capacity	Gursoy 2010

All of these sex differences in the physical body which arise during puberty also happen to Julia, as she too experiences the development of her ovaries and the secretion of more estrogen and progesterone. Just because she was exposed to high prenatal androgens does not mean she will develop into a male at puberty. Rather, while her brain will be predisposed to male-typical traits, her physiology will develop into a healthy adult female once puberty is complete. Some physical traits may tilt towards the masculine end, but most of her physical body will resemble that of a typical woman because of her mature female gonads.

Knowing that sex differences grow larger during puberty through exposure to sex hormones, social constructionists cannot continue to believe that all differences between men and women are the result of socialization forces. While there are instances of gender maltreatment which result in an altered development of either sex, such instances do not contribute to the large sex differences in physical attributes. For instance, there is evidence from epigenetics that parental touch during infancy causes direct structural changes in DNA functionality. During the Great Chinese Famine, maltreatment of girls through lack of touch and a preponderance of abuse resulted in an epidemic of disability and illiteracy. Their brothers, however, whose needs had been prioritized during the famine, didn't have such problems.[353]

Aside from extreme instances such as that one, differences in physical attributes between girls and boys actually grow *larger* in more gender-equal countries, as each individual is able to achieve the full extent of their physical and cognitive development without impediment from sociocultural forces.[354] However, because of the insistence that all gender differences are the result of society, social constructionists fail to

---

[353] Cortes, L, Cisternas, C., Forger, N. (2019). Does gender leave an epigenetic imprint on the brain? *Frontiers in Neuroscience, 13(173)*, 3.

[354] Lippa, R., Collaer, M., Peters, M. (2010). Sex differences in mental rotation and line angle judgments are positively associated with gender equality and economic development across 53 nations. *Archives of Sexual Behavior, 39*, 990-997.

accommodate all of the physical and morphological differences which are tightly linked to the influence of hormones on the anatomy of the brain.

The effects of puberty on physical sex differences can also be seen in the animal kingdom, where researchers have tested the direct effects of sex hormones on the activation of sex-differentiated structures. When androgen receptors in the brain are deleted in male mice, for example, masculinized morphological differences fail to develop.[355] Erasing the androgen receptors leaves bones which fail to strengthen, sexual and aggressive behaviors which fail to develop; a reduced heart size and impaired contraction; increased subcutaneous and visceral fat mass, decreased voluntary activity, and late-onset obesity; decreased skeletal muscle and perineal muscle mass, decreased strength; and no prostate development.[356] Without these androgen receptors, the male mice failed to develop masculinized morphology and neuroanatomy, showing that the binding of testosterone to androgen receptors in the brain is a crucial mechanism of pubertal development.

ACTIVATIONAL EFFECTS ON THE BRAIN

Because sex hormones travel along the hypothalamic-pituitary-gonadal (HPG) axis, they are therefore integrated throughout the nervous system, including the brain. It is in the brain where sex hormones bind to sex hormone receptors which then help encode proteins, or molecular machines, that carry out specific jobs designed to create and maintain sexually dimorphic traits. The brain, as we have covered in previous chapters, is organized prenatally through the absence or presence of androgens.[357] What this means is that both the neuroanatomy and neurocircuitry of the fetal brain is organized differently in males and females depending on the level of prenatal androgen exposure. Throughout a narrow window of development in the womb, high levels of prenatal

---

[355] Davey, R. & Grossmann, M. (2016). 10.
[356] Ibid.
[357] See chapter, "Conception to Birth."

androgens masculinize the fetal brain's neural circuitry, which increases suboptimal arousal and causes heavy lateralization of the hemispheres.[358] Again, such is the case of our girl, Julia, who has been exposed to higher levels of prenatal androgens than the average female. Reversely, the absence of these high levels of androgens results in a default female neural phenotype that maintains an interconnectivity between the hemispheres.

While these major structural changes of the brain occur in the womb, a second major change occurs during the activational stage of puberty once higher levels of sex hormones bind to sex hormone receptors in the brain, which then help initiate further sexual development. This activational stage affects both the neurocircuitry and neuroanatomy of men's and women's brains, from which gender differences in behaviors and interests arise. To understand how this process works and what behaviors it affects, we will first explore the development of sex-differentiated neurocircuitry.

Neurocircuitry refers to the control mechanisms which regulate neuronal activity in the brain. Understanding such circuitry allows us to see how different brain regions are connected, how they interact, and what regions are activated during a specific task. In humans, this neurocircuitry is complex beyond imagination:

> "The human brain contains around 86 billion neurons networked through 100 trillion synapses, and imaging a single cubic millimeter of tissue can generate more than 1,000 terabytes of data."[359]

Understanding how this complex neurocircuitry can be altered through exposure to sex hormones and activation of sex-specific genes can help us better understand how sex differences are produced in male-typical and female-typical brains. Because the brain is, what we call, *plastic*, its

---

[358] Suboptimal arousal refers to the likelihood a person will seek new environments and experiences and get bored of old ones, even in the face of risk. The higher the suboptimal arousal, the more prenatal and pubertal androgen exposure.

[359] Rainmaker1973. (2019). Google researchers create AI which maps brain's neurons. *Twitter*.

neurocircuitry can change through exposure to hormones, genes, and environmental inputs. Such is the case when the brain is exposed to sex hormones such as testosterone, which initiates changes in neurocircuitry that produce changes in behavior.

These hormone-initiated behavioral changes result from the brain's inherent structural plasticity--the ability for the brain to restructure itself in light of new environmental inputs. Plasticity can be seen in people who become impaired later in life. Their brains often reorganize themselves through changing the connectivity of certain areas, thereby creating new areas of specialization. This is why people who have become completely blind will often develop an acute sense of hearing, and the same concept of plasticity applies to the effects of sex hormones on the brain which modulate male and female neurocircuitry.

Changing levels of sex hormones throughout life cause continual changes in the connections of the brain. During puberty, and specifically *gonadarche*, gonadal sex hormones such as estrogen and testosterone act on dormant neural circuits to elicit adult social and reproductive behaviors.[360] These dormant neural circuits are organized in the womb but remain inactive due to the threshold of sex hormones necessary for mature sexual development.

Knowing that the male and female brain is differentiated prenatally, neuroendocrinologists call puberty the *activational* period: sex hormones *activate* the existing sex-differentiated organizational structures of the dormant neurocircuitry. This claim, that pubertal hormones trigger a second period of structural organization through plasticity, is backed up by decades of research on nonhuman mammals.[361]

The first major effect on male and female neurocircuitry can be found through the hypothalamus which, upon exposure to specific sex hormones, facilitates direct reproductive behaviors.[362] The second effect of *gonadarche*

---

[360] Blakemore, S.J., Burnett, S., & Dahl, R. (2010). 927.
[361] Sick, C.L. & Foster, D.L. (2004). The neural basis of puberty and adolescence. *Natural Neuroscience, 7,* 1040-1047.
[362] Blakemore, S.J., Burnett, S., & Dahl, R. (2010). 927.

on neurocircuitry reorganizes the sensory and association regions of the brain, including the visual cortex, amygdala, and hippocampus, resulting in "altered sensory associations, e.g. to the smell or sight of potential sexual partners or competitors which may facilitate some attentional and motivational changes at puberty."[363] The second effect can be conceptualized as a reorganization of sense perception for emphasis on sexual reproduction, a reorganization which affects the wiring of the systems which control navigation through the world in the context of a motivational structure. Thus, during puberty, our brains are reorganized to prioritize a new set of goals around sexual reproduction, and our perceptions of the world change in light of this. The third effect of puberty on neurocircuitry causes changes in reward-related brain structures such as the nucleus accumbens and dopaminergic pathways to the prefrontal cortex.[364] This final effect, therefore, applies gratification and reward chemicals to the changes in perception so that sexual behavior is repeated and solidified as a pleasurable action.

Knowing these three effects during puberty, it becomes clear that our neurocircuitry is rewired so that we become more motivated to seek out sexual experiences, become aroused when seeing sexual stimuli, and reorient our perceptions for sexual gratification and reproduction. Such claims are backed up by additional evidence from the animal kingdom.[365]

Outside of neurocircuitry, the second major brain change during puberty occurs on the neuroanatomy, which deals with the morphology of the brain and its regions rather than neuronal connections. Sex differences in neuroanatomy between males and females arise and become larger during puberty. Specifically, changes in gray matter and white matter produce sex differences in line with *gonadarche*, the exact time that the gonads begin producing higher levels of sex hormones through activation of the HPG axis.[366] Gray matter consists of neuronal cell bodies known as

---

[363] Blakemore, S.J., Burnett, S., & Dahl, R. (2010). 927.
[364] Ibid.
[365] Sato, S.M., Schulz, K.M., Sisk, C.L., Wood, R.I. (2008). Adolescents and androgens, receptors and rewards. *Hormone Behavior, 53,* 647-658.
[366] Blakemore, S.J., Burnett, S., & Dahl, R. (2010). 929.

soma which house the neuron's nucleus, whereas white matter consists of axon structures which relay information from the soma and form connections between brain cells. Gray matter controls muscular and sensory activity areas such as the cerebral cortex and cerebellum, areas which are crucial to attention, memory, coordination, and motor control.[367] White matter joins these complex areas of gray matter together so that their functional capacities can work with one another. Therefore, gray matter can be conceptualized as the action-based parts of the brain whose functionality is enabled through the connecting axons of white matter.

Sex differences in these gray and white matter volumes arise during puberty, as activational hormones produce additional organizational effects on the neuroanatomy of male and female brains. Peak gray matter volume in girls and boys is reached around the same time at the age of 12, corresponding to *gonadarche*, but it is in the *densities* of gray matter where sex differences arise. Global gray matter density has been positively correlated with testosterone levels in males, whereas estrogen levels in females has been negatively correlated with global gray matter density.[368] For white matter, global volume increases between childhood and adolescence and finally stabilizes in adulthood. Males tend to show higher rates of increases in white matter volume than females, since variances in white matter development relate to specific gene-expression levels which encode androgen receptors. For instance, a shorter androgen receptor genotype usually results in higher volumes of white matter, whereas longer androgen receptor genotypes result in lower volumes of white matter.[369] Therefore, white matter volume may be related to sex-specific gene expressions which are activated through testosterone exposure. Sex differences in gray matter and white matter volume and density do not prove anything about intelligence, ability, or interest, but are rather acute

---

[367] MacKenzie, R. (2019). Gray Matter vs White Matter. *Technology Networks*.

[368] Peper, J.S., et al. (2009a). Sex steroids and brain structure in pubertal boys and girls. *Psychoneuroendocrinology, 34,* 332-342; Blakemore, S.J., Burnett, S., & Dahl, R. (2010). 929.

[369] Herting, M., Sowell, E. (2017). 126.

structural differences that may affect sex differences in cognition and behavior.

Knowing that differences in gray and white matter arise through puberty, why is such a fact important? One reason for knowing sex differences in gray and white matter is if you're a neuroscientist studying the makeup of male-typical and female-typical brains. Knowing these structural differences lets you more effectively treat people for neurological diseases; it lets you find specific areas of the brain which have sex-specific hormone receptors which can activate or suppress certain genes; and it allows you to more fully understand how the overall makeup of male and female brains differ. These sex differences are also likely to impact differences in behavior, as certain brain regions will function differently than others. Such knowledge can help us further understand how sex affects the brain.

There are also sex differences in the neuroanatomy and neurocircuitry of the amygdala and hippocampus, two parts of the brain which have high concentrations of sex hormone receptors.[370] The amygdala plays a critical role in sex drive, fear, and the archiving of intense emotions, whereas the hippocampus forms, organizes, and stores new memories and has a role in the process of learning.[371] In males, the amygdala tends to be larger in gray matter density and volume because of exposure to both prenatal and pubertal hormones.[372] Males who have low testosterone or who have been castrated tend to have smaller amygdalae and lower sex drives.[373] Knowing this, it makes sense why males tend to have incredibly intense sex drives compared to women. The binding of testosterone to androgen receptors and the production of specific proteins seems to create a more intense motivational and pleasure-seeking structure in the male by increasing the

---

[370] Goldman, B. (2017). 6.

[371] Williams, J. (2018). The amygdala: definition, role & function. *Study.com*; Cherry, K. (2019). Hippocampus role in the limbic system. *Verywellmind.com*.

[372] Uematsu, A., et al. (2012). Developmental trajectories of amygdala and hippocampus from infancy to early adulthood in healthy individuals. *PLoS ONE, 7(10)*, 6.

[373] Williams, J. (2018).

intensity of the amygdala. And yet, such a structure is not solely confined to male genotypes; it can also be developed in girls who have been exposed to abnormally high levels of testosterone. For instance, CAH girls, who I have discussed previously, show male-typical enhancements in amygdala activity when shown fearful or angry faces, compared to their control sisters.[374] This evidence from CAH girls, among other research studies conducted on nonhuman animals and human males with low testosterone or missing testes, suggests that testosterone has at least some mediating effect on the neurocircuitry of the amygdala. Julia, having been exposed to higher levels of androgens in the womb compared to her unaffected sisters, may share such a similarity in amygdala activation with her male counterparts.

While such a structure is usually denser in males, the other brain structure with high concentrations of sex hormones is the hippocampus, which tends to be denser in females. The hippocampus, as mentioned previously, plays a critical role in learning, memorization, and object location memory, something which females tend to perform slightly better on than males.[375] Neurons in the hippocampus show large sex-differentiation which grow or diminish based on exposure to estrogen. In nonhuman mammals, the dendritic spine density in parts of this brain region fluctuate over the estrus cycle (meaning these neurons respond to estrogen), while estrogen treatment can enhance hippocampal-dependent memory in both rodents and humans.[376] Such evidence shows that estrogen can affect the neurocircuitry of the hippocampus in similar ways that testosterone can affect the neurocircuitry of the amygdala.

Further research from the frontiers of neuroendocrinology has elucidated a plethora of additional structural and anatomical differences between male-typical and female-typical brains in specific regions, but such an exploration is beyond the scope of this book. These changes in neuroanatomy and neurocircuitry in males and females enhance or diminish certain cognitive functions through the binding of sex hormones

---

[374] Blakemore, S.J., Burnett, S., & Dahl, R. (2010). 931.
[375] Berenbaum, S., Beltz, A. (2011). 192.
[376] Marrocco, J., McEwen, B. (2016). 375.

on sex-specific receptors and through the activation of sex-differentiated genes.

One of the most important cognitive changes which occurs prenatally and is further amplified during puberty is known as *lateralization*, which alters the neuroanatomy and neurocircuitry of the brain so that cognitive functions are more localized to specific hemispheres. Lateralization is a common occurrence in people who have been exposed to high levels of androgens either in the womb or during puberty (namely, males, CAH girls, and our fictional girl, Julia). These androgens bind to androgen receptors in the brain, which then help produce proteins that build masculinized neural structures. This process of lateralization can be initiated through gonadal hormones, activation of genes encoded on sex chromosomes, or both.

Because heavy lateralization is largely the result of high androgen exposure, males tend to be more lateralized, and therefore will exhibit less interconnectivity between the right and left hemispheres. Reversely, female brains tend to be more *equally* lateralized, or balanced, so that their neurocircuitry is more interconnected *between* the left and right hemispheres. This interconnectivity is the default state of neural development and will stay this way unless subjected to high levels of androgens, or testosterone. Such lateralization, or lack thereof, creates significant cognitive differences between males and females in perception and behavior, a topic I will now discuss.

Lateralization is one of the largest sex differences in male-typical and female-typical brains and cannot be accounted for through socialization alone. Such a sex difference represents itself through the way the brain is connected between the hemispheres, which then affects specific cognitive functions. Heavy connections *between* the hemispheres are much more common in females than in males, whereas connections *within* the hemispheres, not between, are much more common in males than in females. Feminine lateralization, therefore, can be represented as high interconnectivity between the right and left hemispheres, whereas masculine lateralization can be represented as a low interconnectivity between the right and left hemispheres and a high connectivity *within* the hemispheres. (I use the terms *feminine* and *masculine*, rather than female and male, so that it is clear some males can have more feminized brains and some females can have more masculinized brains. There is no such thing as a strictly 'female' brain or strictly 'male' brain.)

The degree of lateralization is directly affected by a mix of prenatal androgen exposure, sex-differentiated gene expression, and pubertal testosterone, from which androgen receptors are activated and facilitate protein synthesis. Testosterone binds to androgen receptors and activates male-chromosome-dependent genes, which then produce proteins to build masculinized structures in the brain. Since females are not usually exposed to high levels of testosterone prenatally or during puberty, and since they do not have male-chromosome-dependent genes, the effects of testosterone on brain lateralization are minimal. Because of this lack of lateralization, females tend to use both hemispheres for information processing, which means they tend to integrate linguistic, analytical, and emotional processing into a complete package.[377] In specific regions, women tend to have more symmetrical neurostructures, especially in areas rich with estrogen receptors. The hippocampus, for example, tends to have more balance in the left and right volumes as females grow into adolescence, whereas men tend to have less of a balance and more lateralization.[378] Overall, the two hemispheres of a woman's brain interact with each other more than the average man's.[379]

Less lateralization allows the brain to connect between the hemispheres in a greater degree, which then allows cognitive functions to be more generalized and less domain-specific. Since the left hemisphere is able to interconnect with the right, a natural emphasis on language, verbal learning, spatial memory, vocabulary, reading comprehension, and verbal fluency is produced as a byproduct.[380] Since the right hemisphere is less specialized, there is less of an emphasis on organizing systems and data and on visuospatial cognition (a function which heavily relies on front-to-back

---

[377] Uematsu, A., et al. (2012). 7.
[378] Ibid.
[379] Goldman, B. (2017). 5.
[380] Berenbaum, S., Beltz, A. (2011). 192.

lateralization). This is partly why females tend to be better at verbal tasks and skills, because the right and left hemispheres show less lateralization and more interconnectivity. Because of this interconnectivity, females tend to be good at both math and verbal skills, whereas men are more specialized in specific domains in the right hemisphere. Thus, having been more heavily lateralized, Julia may be more specialized in visuospatial tasks like her male counterparts.

On the masculine end, more testosterone causes the neurocircuitry to become more lateralized and less interconnected, which is to say, to grow more connections *within* each hemisphere than between the two.[381] Males show right asymmetry in multiple regions of the brain, but especially the amygdala, where androgen exposure contributes to the asymmetry in the volumes of the gray and white matter.[382] This higher density of amygdala volume also correlates to a higher sex drive: Men tend to have a higher interest in casual sex, in multiple sex partners, and in visual-sexual stimuli such as pornography.[383]

This right lateralization produces a more masculinized brain with a de-emphasis on language, spatial location memory, and verbal learning, while instead *emphasizing* systematizing and organizing systems and visuospatial cognition with manipulating objects and tracking moving projectiles.[384] Because of this, men tend to perform better on spatial tasks and mathematical problem solving due to the production of highly specialized brain regions through lateralization. And yet, males also show more behavioral aggression than females through foul language, imitation of aggressive models, and violence and physical aggression.[385]

---

[381] Tomasi, D., Volkow, N. (2011). Laterality patterns of brain functional connectivity: Gender effects. *Cerebral Cortext, 22,* 1461.
[382] Uematsu, A., et al. (2012). 8.
[383] Ngun, T., et al. (2010). 229.
[384] Ibid., 231.
[385] Ibid.

II THE ACTIVATIONAL STAGE

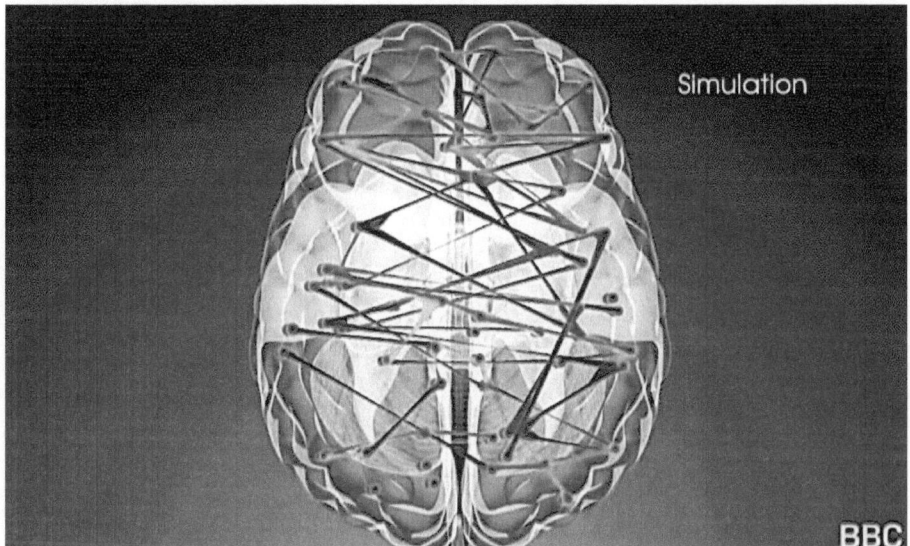

(2014). Female-typical brains show more interconnectivity between the hemispheres. *"Is Your Brain Male or Female?"* BBC.

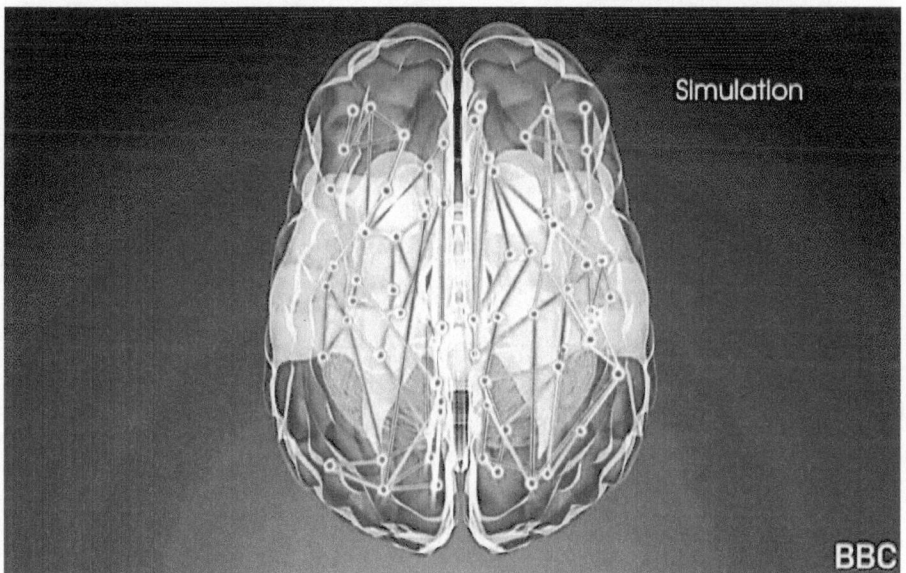

(2014). Male-typical brains show less interconnectivity between the hemispheres and more connections within the hemispheres, connections which move front-to-back. *"Is Your Brain Male or Female?"* BBC.

Evidence from neuroendocrinology suggests that this tendency for aggression and risk-taking behavior is related to the serotonin transporter gene, which regulates perception of one's dominance in a given hierarchy.[386] The binding of testosterone to androgen receptors may have a role to play in this relationship of serotonin with impulsivity, aggression, and the larger size of the amygdala in highly masculinized brains. Overall, therefore, men's brains tend to be more lateralized, which is to say, specialized in specific cognitive tasks such as spatial ability and manipulation of objects. And CAH girls and others exposed to high levels of androgens exhibit a similar pattern in performance on such tasks.

These differences in brain lateralization between males and females, especially when analyzed on the extreme ends of the distribution, most likely influence gender differences in behavior, personality, and interests as boys and girls grow into adulthood. For example, a girl like Julia who has been exposed to high levels of testosterone may develop a more lateralized brain, and therefore will be more interested in occupations which deal with complex visuospatial tasks.

Such claims of lateralization have been reproduced in nonhuman mammals, showing that socialization alone cannot account for structural differences in male-typical and female-typical brains. Altering sex hormone exposure during puberty can result in changes in neuroanatomy and neurocircuitry in a variety of mammals. For example, male and female rats show anatomical sex differences in the hippocampus, the brain region related to object location memory and learning. When females are treated neonatally with testosterone, however, this results in the growth of a more masculinized hippocampus and better performances on spatial tasks compared to untreated females.[387] Furthermore, when males' androgen

---

[386] Manuck, S.B., et al. (2000). A regulatory polymorphism of the monoamine oxidase-A gene may be associated with variability in aggression, impulsivity, and central nervous system serotonergic responsivity. *Psychiatry Research, 95*, 9-23.

[387] Roof, R.L., Havens, M.D. (1992). Testosterone improves maze performance and induces development of a male hippocampus in females. *Brain Research, 572(1-2)*, 310-313; Little, A. (2013).

receptors in the amygdala are blocked, lower sexual arousal results. Such tests have been demonstrated in fish, where injecting the male fish with testosterone increases approach responses towards females.[388] Rhesus macaques also show sex differences in preferences for face-based visual stimuli. Female rhesus macaques, during the peri-ovulatory phase when estrogen is in heavy circulation and pregnancy is possible, looked longer at the faces of male conspecifics (member of the same species) than female faces.[389]

Similar effects can be seen in human females: administration of testosterone can improve visuospatial and mental rotation abilities in women, providing additional evidence that sex differences such as visuospatial ability are not influenced by socialization alone but are affected by a mix of gonadal hormones, sex-differentiated gene expression, and social forces.[390] **In fact, the effects of sex hormones on brain lateralization are so overwhelming that preventing the production of gonadal hormones pre-puberty prevents such sex differences in brain structure from arising at all.**[391] Concentrations of sex hormones during puberty may therefore heavily influence behavioral differences between men and women by causing the structural changes in neuroanatomy and neurocircuitry that we explored above.

Such evidence, combined with the nonhuman animal studies, is incredibly difficult to refute. Understanding this mountain of evidence of hormonal effects on the pubertal brain means that social constructionists have no good research to defend their univariate position that all gender differences are the result of sociocultural forces. Social constructionists must therefore reevaluate their ideological presuppositions and drop rhetorical and aesthetic arguments for gender differences if they wish to be

---

[388] Lord, L.D., Bond, J., et al. (2009). Rapid steroid influences on visually guided sexual behavior in male goldfish. *Hormones and Behavior, 56(5)*, 519-526; Little, A. (2013). 324.
[389] Lacruese, A., King, H.M., et al. (2007). Effects of the menstrual cycle on looking preferences for faces in female rhesus monkeys. *Animal Cognition, 10(2)*, 105-115; Little, A. (2013). 324.
[390] Little, A. (2013). 323.
[391] Ibid., 324.

taken seriously by the scientific community and the public. Ironically, those who argue that gender differences are merely the result of sociocultural forces ignore the critical neuroscientific evidence of male-typical and female-typical brain differences which can increase the treatment efficacy for sex-specific neurological diseases and pathological neuroses.

Organizational and activational effects produce sex differences in neurological diseases which can be more effectively treated by understanding how sex hormones and sex-differentiated genes affect male and female psychopathology. It has been clearly demonstrated that males and females suffer in differing amounts from specific neurological diseases and neuroses, and that these diseases often become amplified during puberty. Much of the sex differences in neurological diseases do not result from mere socialization experiences like bullying or peer pressure, but rather result from structural differences in the brain which can be further enhanced or suppressed through social forces. Therefore, the effects of pubertal hormones can predispose males and females to different diseases which can be activated through environmental stimuli such as a major negative emotional event. These sex differences should not be dismissed by social constructionists if we wish to apply more effective treatments.

> "Sex differences are seen in mental health problems that arise during [puberty], with disproportionate increases in anxiety and depression seen in girls (Angold et al., 1998, 1999) and an increased prevalence of substance abuse and externalizing disorders in boys (American's Children: Key National Indicators of Well-Being, 2009). In addition, both adaptive and maladaptive changes in risk-taking, impulsivity, and reward processing are also seen during adolescence. Thus, pubertal-related timing of region specific brain changes may contribute to sex-specific differences in the rapid and disproportionate increases in rates of psychopathology seen between girls and boys."[392]

---

[392] Herting, M., Sowell, E. (2017). 134.

Knowing how puberty can influence sex differences in neurological diseases can therefore help medical scientists produce more effective treatments. For instance, anxiety disorders and depression are more common in females and have been positively correlated with levels of estrogen and progesterone, female sex hormones.[393] Testosterone, however, has been negatively correlated with depression and anxiety. In fact, low testosterone levels in young men predict an increased risk for depression and anxiety disorders.[394] Therefore, exposure to male-typical or female-typical levels of these sex hormones predisposes boys and girls to sex-specific neurological diseases. For boys, higher levels of testosterone increase the chances of autism, heightened physical aggression, and ADHD, whereas for girls, higher levels of estrogen and lower levels of testosterone increase the chances of depression, anxiety, and eating disorders.[395] Hormone treatment in specific cases can solve many of these neurological disorders. For example, treating men with testosterone who have hypogonadal syndrome (characterized by excessively low androgens), greatly improves mood, reduces anxiety, and mitigates symptoms of depression.[396]

Psychoneuroendocrinologists have further elucidated the mechanisms which produce these sex differences in neurological diseases through studying specific receptors in the brain. Such receptors respond to specific hormones such as estrogen or testosterone and produce structural differences in perception. For instance, 5HTTLPR is a region of the gene that encodes for serotonin, the neurochemical associated with the regulation of feelings of well-being and happiness. Through studies on neuropsychiatric diseases such as anxiety disorders, the 5HTTLPR region

---

[393] Ngun, T., et al. (2010). 232.
[394] Ibid.
[395] Berenbaum, S., Beltz, A. (2011). 193.
[396] Wang, C., et al. (1996). Testosterone replacement therapy improves mood in hypogonadal men. *Journal of Clinical Endocrinology and Metabolism, 81,* 3578-3583; McHenry, J., et al. (2013). Sex differences in anxiety and depression: Role of testosterone. *Frontiers in Neuroendocrinology, 35,* 42-57.

has been associated with anxiety, insomnia, and negative emotionality in humans.[397] Testosterone or estrogen may then bind onto this region to produce differences in serotonin regulation. Therefore, genetic vulnerability combined with sex hormone exposure may lead to the development of sex-specific psychiatric diseases.[398]

Exploring these sex differences in neurological diseases and studying the mechanisms which produce them will allow researchers and medical scientists to develop more effective treatments and safer pharmaceutical drugs for males and females alike. *Social constructionists ignore this evidence at their peril.*

IDENTIFYING SEX THROUGH THE BRAIN

A critical first step for understanding sex-specific neurological diseases is to identify whether the person being studied has a male-typical or female-typical brain. Such knowledge can produce better treatments tailored to the specific neuroanatomy and neurocircuitry of the individual being treated. For instance, as a neuroscientist, knowing that you're dealing with an anxiety disorder patient with a female-typical brain can help you identify the specific region which has become pathologized and needs to be treated; or it can help a psychologist focus on maladaptive behaviors which are associated with the person's specific neuropsychology and personality. Such specificity allows treatments to be exquisitely tailored to the individual's unique neuroanatomy and neurocircuitry, and knowing these sex-differentiated structures of the brain can increase the effectiveness of such treatments. And yet, through ignoring the evidence of sex differences in neuroendocrinology, the social constructionist will object to such a claim that male-typical and female-typical brains truly exist.

Social constructionists will claim that every individual's brain is simply a concoction of masculine and feminine structures distributed along a spectrum of variability, meaning that each individual has a mosaic-like structure of the feminine and the masculine. However, such a belief is

---

[397] Ngun, T., et al. (2010). 232.
[398] Berenbaum, S., Beltz, A. (2011). 194.

fundamentally flawed and clichéd. While it is true we all have a mix of feminine and masculine traits, even in our brains, this does not mean that this mixture exists on a flat spectrum. Just like any other sex or gender difference, trait differences in male-typical and female-typical brains exist on a bimodal distribution, with a female average and a male average. For example, most females have a neural structure which is more feminized, and therefore bilateral, whereas most males have a neural structure which is more masculinized, and therefore *lateral*. Of course, some people do not fit the mold, but, like other sex and gender differences, exceptions to the pattern do not disprove the existence of the pattern. In fact, utilizing technologies which map the neuroanatomy and neurocircuitry of individual brains, neuroscientists can accurately predict someone's biological sex without knowing anything about chromosomes, genitals, or hormone levels. And it is here, in the field of neuroimaging, where social constructionism suffers one of its greatest defeats.

Contrary to social constructionist claims that an individual's sex cannot be reliably determined through the brain, six unique neuroimaging techniques from contemporary neuroscience are able to predict someone's sex with 80-95% accuracy. The first mapping technique used resting state functional magnetic resonance imaging (rfMRI) to measure *resting brain connectivity*, which represents the inherent functional networks of the brain, the networks which are operating together during a resting state. Functional connectivity (FC) is highly predictive of sex, as it describes how the networks of the brain are connected and how the different regions interact (a concept similar to lateralization). Females and males tend to have unique neural structures which form sex-specific functional connectivity features such as heavy lateralization in men and low lateralization in women. Analyzing this connectivity using rfMRI data from 820 individuals, the neuroscientists achieved a sex prediction accuracy of 87%.[399]

---

[399] Zhang, C., et al. (2017). Functional connectivity predicts gender: evidence for gender differences in resting brain connectivity. *Human Brain Mapping, 39,* 1765.

The second technique used electroencephalograms (EEGs) to record brain rhythms (or waves). Such rhythms are electrophysiological signatures of brain function and contain sex-specific differences, specifically differences in the beta frequency range which relate to cognition and emotion. Using these EEG measures, sex was predicted with more than 80% accuracy, a surprising level of accuracy for simple brain waves.[400] And yet, as the techniques become more focused on neuroanatomy and neurocircuitry, the predictions become even more impressive.

The third technique for predicting sex through the brain used structural MRI scans from 1,500 individuals. Sex was predicted with a 93% accuracy.[401] The fourth technique utilized a more physical approach by looking at neuroanatomical features across 1,000 people, specifically differences in cortical thickness, occipital lobe volumes, and overall brain morphology. Sex was predicted with an 83% accuracy.[402] The fifth technique measured gray matter volume and density with MRI scans from 1,300 individuals. Certain regions showed higher density of gray matter in females, while other regions of the brain showed higher density of gray matter in males. Sex was subsequently predicted with 93% accuracy.[403]

Instead of measuring gray matter or functional connectivity, the sixth and final technique analyzed cortical three-dimensional morphology using a method known as Hierarchical Sparse Representation Classifier (HSRC). Cortical three-dimensional morphology relates to the shape and form of the portion of the brain known as the cortex. Differences between males and females can often be found in the cortex, and neuroscientists found

---

[400] Van Putten, M., et al. (2018). Predicting sex from brain rhythms with deep learning. *Scientific Reports*, 8, 1.

[401] Chekroud, A., et al. (2016). Patterns in the human brain mosaic discriminate males from females. *PNAS, 113(14)*, 1.

[402] Sepehrband, F., et al. (2018). Neuroanatomical morphometric characterization of sex differences in youth using statistical learning. *NeuroImage, 172*, 217-227.

[403] Anderson, N., et al. (2018). Machine learning of brain gray matter differentiates sex in a large forensic sample. *Human Brain Mapping, 40,* 1496-1506.

that the frontal lobe had the most important features for gender differences in human brain morphology. After analyzing the cortex morphology of over 1,100 individuals, sex was predicted with 97% accuracy.[404]

Knowing that these six neuroimaging techniques for predicting biological sex are incredibly accurate and utilize the most contemporary technology for analyzing neuroanatomy and neurocircuitry, it is now rather difficult to claim that the brain is essentially sex-indistinguishable. Such conclusions and their implications are further described by one of the neuroscientists:

> "These findings demonstrate that the brains of males and females are highly distinguishable. Understanding sex differences in the brain has implications for elucidating variability in the incidence and progression of disease, psychopathology, and differences in psychological traits and behavior. The reliability of these differences confirms the importance of sex as a moderator of individual differences in brain structure and suggests future research should consider sex-specific models."[405]

With the realization that male-typical and female-typical brains are highly distinguishable after puberty (up to 97% accuracy), social constructionists must seriously reconsider their assumptions that all gender differences are the result of sociocultural forces. Here the social constructionists and gender theorists might claim that these differences in neuroanatomy and neurocircuitry are caused by cognitive social learning, that through reinforcement and punishment, certain behaviors are

---

[404] Luo, Z., et al. (2019). Gender identification of human cortical 3-D morphology using hierarchical sparsity. *Frontiers in Human Neuroscience, 13(29)*.

[405] Anderson, N., et al. (2018). 1496; For a deeper understanding of how sex hormones and sex-differentiated genes produce sex differences in neuroanatomy and neurocircuitry throughout development, read "Representing Sex in the Brain: One Module at a Time" and other articles in *Frontiers in Neuroendocrinology*.

solidified and embodied. This embodiment of sex-typed behaviors would then produce structural changes in neuroanatomy and neurocircuitry which would construct the illusion that gender differences come from biological sources.

In Julia's case, we could imagine that, as a girl, behavioral reinforcement and punishment would steer her towards behavior and interests which comprise stereotypical gender norms. As these female-typed behaviors are reinforced, her brain goes through structural changes which alter her neuroanatomy and neurocircuitry to behave in feminine ways. The problem, however, is that like other girls who have been exposed to higher levels of prenatal androgens, Julia does not exhibit female-typical behaviors and interests. Precisely the opposite: our girl exhibits *male-typical* behaviors and interests. She's interested in mechanical objects, in systems, and in taking material and forming it into an ordered form. This masculine behavior persists *despite* constant socialization pressures to act like the stereotypical feminine girl. Since Julia's brain has already been wired in male-typical ways in the womb, and since male-typical behaviors could be measured during infancy (such as an interest in mechanical and moving visual stimuli), claims that structural brain differences in Julia are merely the result of socialization do not hold up to any serious scientific inquiry.

Without scientific evidence, such a cognitive social learning theory for brain differences seems plausible, yet it only remains plausible in the absence of evidence from neuroendocrinology, the animal kingdom, and studies into girls with CAH. While it is true that behavioral differences can be produced through socialization and epigenetic modification of DNA, this theory of gender differences fails to accommodate not just the overwhelming evidence of sex hormones' effects throughout the animal kingdom but the organizational and activational effects of these sex hormones in humans from conception through adolescence.

Despite the presence of socialization forces, evidence from girls exposed to higher levels of prenatal androgens than their unaffected sisters and males exposed to lower testosterone than their unaffected brothers, among many other case studies which elucidate more than 50 sex-differentiated genes in the brain, provides a multiplicity of reasons to believe that both sex hormones and genetics directly affect the neuroanatomy and neurocircuitry of men and women from conception through adolescence.[406]

In other words, for the social constructionists, there is no more escaping the reality of biological sex, hormones, and puberty. The walls have closed in and the game is up; your theory has been hit by the equivalent of a nuclear blast. Everything is now in rubble, and you are left with two options: *either you adapt your model to the new evidence and deal with reality, or you move further into your ideological beliefs.* Unfortunately, social constructionists will likely choose the latter so that their influence, power, and control over our social institutions can continue; such is the pattern of postmodernist thinking.[407]

So far, we have covered how biological sex is determined through activation of the SRY gene; how exposure to sex hormones in the womb produces differences in fetal brain structure; how gender identity develops in accordance with one's genetics, brain structure, and hormones despite socialization forces; how gender stereotypes can turn pathological and construct caricatures of the masculine and feminine archetypes; and how puberty further differentiates the sexes through 1) physiological changes which produce maturational effects on the gonads (ovaries and testes), 2) physical changes which produce secondary sex characteristics such as breasts and facial hair, 3) hormonal changes which produce structural differences in male and female bones, muscles, voices, and height, and 4) additional hormonal changes which produce activational effects on dormant neural circuits, which then produce differences in the neuroanatomy and neurocircuitry of men's and women's brains. Lastly, we have explored how sex can be identified with up to 97% accuracy by analyzing the morphology of the brain.

---

[406] Marrocco, J., McEwen, B. (2016). 372.
[407] See chapter, "The Gaslighting Ideology."

Integrating all this evidence for sex differences in the brain, we discovered that the brain of each individual is best conceptualized as a complex interaction of modules. Each module consists of a neural or genetic pathway in charge of specific behaviors which respond to genetic and hormonal signals.[408] Some of these modules can be masculinized through exposure to high levels of prenatal androgens and pubertal testosterone or feminized through exposure to high levels of estrogen and progesterone. Many males will have a more masculinized brain (heavy rightward lateralization) while many females will have a more feminized brain (heavy hemispheric interconnectivity). Other people will fall outside the average: some males will have more feminized brains and some females will have more masculinized brains. Thus, this mix of sex-differentiated modules in the brain gives you your overall degree of masculine and feminine traits.

As we move closer to the precise causes for the Gender Equality Paradox (*that gender differences grow larger as a country becomes more gender-equal*), we will explore how sex differences in the brain result in gender differences and similarities in men's and women's *personalities*. While affected by some aspects of society and culture, personality is also heavily influenced by one's genetics, hormones, brain structure, and life experiences.

In the next chapter, I will lay out the literature on personality differences and similarities among men and women, and with it, I will further dismantle the rhetorical arguments of social constructionists who claim that all gender differences are the result of sociocultural forces. From conception to adolescence, genetic and hormonal evidence for the causes of sex differences provides some of the greatest arguments against social constructionism. And yet it is in the personality psychology literature where the unique fields of biology, neuroendocrinology, and psychology form the second-to-last piece of evidence against social constructionist dogma.

---

[408] Goldman, B. (2017).

12

PERSONALITY AND THE BIG FIVE

Imagine two miniature Australian Shepherds both bred from similar parents. Both dogs are females with some different patterning. One has a thick black coat accented by white fur across her paws and chest, while the other, slightly smaller in stature and build, has a tricolor coat of browns, reds, and whites. Each dog, while they have different coats, is raised in the same home with the same human owners, and they constantly play together. However, despite their upbringing being exactly the same, they exhibit completely different personalities. The black and white one, Dakotah, is more timid and shows fear more easily, while the smaller tricolor one, Izzy, is more rambunctious and hyperactive and shows more confidence and curiosity.

Being close to sisters nonetheless, Dakotah and Izzy share completely unique personalities which express themselves as different patterns of behavior. In fact, Izzy is so rambunctious that she'll often instigate playtime with Dakotah by jumping on her, even if Dakotah's simply wanting to chill out and relax. In a phrase, Izzy's the *red-hot-rabble-rouser* of the pack. She sees the world through different eyes than Dakotah, and even though they share many similarities across traits, neurological differences rise from the commonplace and into unique perceptual structures. These perceptual structures, patterns of behavior, and ways of seeing the world can be conceptualized with a single word: *personality*.

Insofar as social constructionists believe individual differences arise from socialization, personality is just as much a social construct as gender. And yet, can the social constructionists argue that personality differences in dogs are the result of sociocultural forces? Such a view would be absurd, as 1) dogs do not have cultures or societies, and 2) each dog, just like any other biological being, has a set of innate traits formed from a mix of genetics, hormones, and the environment. Each of their personalities are tailored to their genetics, which form their unique perceptual structures. So too is the case with humans who are as part of the animal kingdom as any other species. While it is certainly true that we as humans have societies and cultures which shape us, this does not mean that genetic influences are negated.

The same is true for gender, which utilizes many unique perceptual structures, or personalities, to express itself within the framework of sex.[409] While gender is constructed around the critical component of biological sex, the systems around gender can be further differentiated, suppressed, or exaggerated through societal forces. For instance, the natural differences between boys' and girls' average interests may be exaggerated through gender stereotypes, but this does not mean that the average interests do not exist or are socially constructed. Rather, because behavior is influenced through hormones and genetics, behavioral differences in gender expression such as the average boy's greater interest in moving objects are the result of perceptual structures built into our physiology and neurology *in utero* and during puberty. These perceptual structures can be altered, but only so much before you reach a breaking point, as was the case with David Reimer (the healthy boy who was raised as a girl, leaving him with gender identity disorder).[410]

Like the mix of personality traits in dogs, humans have their own mixture of personalities. Each of us sees the world through a unique framework and perceptual structure which informs us how we should act. Taken to its extreme, a personality can be represented as an archetype, which is the ultimate abstraction of an ideal. Just as there are the masculine and feminine archetypes, there are also personality archetypes, or patterns of behavior, which are highly variable among individuals in a similar way

---

[409] The term *gender* is used here to mean "how individuals express their sex through behavior, interests, actions, and preferences."

[410] See chapter, "Development of Gender Identity."

that they are variable among dogs. However, unlike dogs, humans have a much more complex psychology which allows for more profound personalities that mix with biological and sociocultural forces.

So if we can agree that a certain amount of variance in personality can be accounted for through biological factors, then perhaps we can also agree that gender differences in personality traits may partially arise from genetics and hormones. For instance, we may see behaviors and traits that are more common in girls than boys and vice-versa; such differences must be the result of an interaction of biological and sociocultural forces. Therefore, the question is *where do these behavioral differences originate?* The answer is in the details. Many differences in personality may be, in part, the result of structural differences in the human brain which become differentiated through exposure to sex hormones and activation of sex-specific genes.[411] In other words, personality may be affected by our brain structure, and our brain structure may be affected by our sex hormones and genetics.

Such a claim that personality is influenced by hormones and genetics is so astonishingly apparent that it is likely true, but before we explore the biological origins of gender differences in personality, we must accurately define personality, what it's comprised of, and how researchers measure it.

---

[411] For evidence for the average structural differences between male and female brains, see chapter, "The Activational Stage."

WHAT IS PERSONALITY?

In the words of psychologist Dr. Jordan Peterson, an expert in personality psychology and individual differences, *personality* can be defined this way:

> "A personality is an evolved solution to the problem of an overly complex world. Personality functions by influencing our perceptions, motivations, emotions, and actions such that people of different personalities actually experience the world differently."[412]

Personality is not some theoretical or mystical term used to describe the qualities of individuals like a poet would describe the beauty of nature. Rather, personality is a way of seeing the world that can be empirically and scientifically categorized, observed, and measured. It is not some fluid and transient or ephemeral thing; it's something which lays deep at the heart of our neurology and physiology. It's something we use to orient ourselves in the world and act upon it with a set of goals and values.

Personality psychologists study personality not because they want to find the *one* way of seeing the world but because there is, in fact, no correct or incorrect way of perceiving the world in a given scenario. By this I do not mean that there is no truth to reality or objective knowledge from which to ascertain truth, but that every perceptual structure has an evolved a set of pros and cons to best meet a given scenario. The correct perceptual structure, therefore, is not one that can be used across all possible scenarios, but one that can be best used for a specific context. This is why psychologists are interested in personality: so that the large diversity of perceptual structures can be fully understood.

For instance, there are times when a highly anxious way of seeing the world is more valuable than one characterized by low negative emotion: sometimes the environment is dangerous and deadly, and you should keep an eye out! And yet, sometimes the highly anxious perceptual structure is

---

[412] Peterson, J. (2019). Introduction to Personality Psychology. *Discovering Personality*.

maladaptive to the situation and useless: perhaps everything's going right for you and your career is on the right track, but you feel highly anxious about it anyways. An anxious perceptual structure may not be the best personality for that situation. The context-dependent nature of these perceptual structures means that a personality trait for one scenario may not correctly apply to a different scenario. Because of this need for highly-contextualized perception, personality has proliferated among humans into many unique characteristics, traits, and behaviors.

Personalities are so diverse, in fact, that personality comprises the largest difference among individuals. Even compared to race, sex, gender, and sexual orientation, personality is by far the most important category for variability and diversity among individuals. This high degree of variability works to bring about the most effective way of solving a given problem. Such diverse perceptual structures can change depending on the need that arises. Even our most basic needs, such as hunger, thirst, and sexual desire, can be conceptualized as separate and unique personalities each with their own goals and ways of seeing the world: a person who is hungry will view the world much differently than someone who has just had something to eat! And this is perhaps the most important concept to understand about personality.

We may think that when we perceive the world, we are perceiving a set of objective facts which lay themselves out to us and others in the same way. This could not be more wrong. We do not see the mere facts of the world; we see the world of facts through our goals and values. Because of this, the world of facts presents itself differently to you, differently to me, and differently to your best friend all based on what perceptual structure we are using. It is not just that we see different facts; it's that the entire world presents itself to us differently through our individual perceptual structures and goal-directed value systems.[413] This means that we should view our perception as something based on action towards specific goals rather than an objective assessment of observable facts. This can be

---

[413] Peterson, J. (2019). Introduction to Personality Psychology. 25m30s.

encapsulated in a single phrase from Dr. Peterson: "*What you see is a reflection of your value structure.*"[414] Having different value structures in a group means that each individual will see a given situation differently, and perhaps because of this, a more effective solution can be reached compared to if each person did not differ.

Such intense variability among individuals seems to be the ultimate diversity social constructionists are truly searching for, and yet, instead, gender theorists have placed diversity of sex, gender, and race as the highest of values. Ironically, such categories provide some of the *least* amount of diversity among individuals. Personality, rather than sex, gender, and race, is the real game in town!

Knowing that personality is the most diverse category comprising individual behavior, where do such unique traits come from? Are these different personality traits something we're born with or something that we learn? For the social constructionists who view human behavior as learned rather than innate, the large variability in personality is difficult to explain through socialization. In a highly egalitarian society, a nation who has produced the most equality of opportunity for every individual, personality differences among individuals should decrease and minimize. And yet, social scientists have found the exact opposite phenomenon: personality differences in the freest societies are some of the most diverse and variable traits among individuals in the world, and the differences only grow larger as societies become freer.[415] This makes intuitive sense if one believes that biology and genetics form one's perceptual structure: *less societal forces allows biological variability to express itself.* This is precisely the view of many personality psychologists who study individual differences across cultures, and it seems to be overwhelmingly true.

From as early as infancy, personality differences can be measured among babies. Like other animals, we seem to be born with a proto-personality, a set of basic perceptual structures given to us by genetic selection which we use to orient ourselves in the world and act upon it

---

[414] Peterson, J. (2019). Introduction to Personality Psychology. 25m30s.
[415] Such a paradoxical claim also applies to gender differences in the most gender-equal societies. See this chapter for more, as well as the chapter, "Solving the Gender Paradox."

using basic goals. This set of behavioral patterns may be exhibited through different interests in toys, different responses to external stimuli, and different actions and behaviors towards others. These behavioral patterns seem to be heavily influenced by a mix of genes and hormones which create unique neurological structures for motivation, perception, and action. We certainly learn, adapt, and grow as we develop from infancy, but such learning and growth can only take place in the context of a pre-existing perceptual structure, or a proto-personality.

For instance, a baby may be more naturally anxious and fearful as an infant, crying more often to visually distressing stimuli, but, as they grow older, such a person may become more emotionally stable and learn to manage their fear and anxiety. After adapting to the environment, the proto-personality from infancy transforms from its basic structure and into something more mature. *But to get to a more mature structure, you have to have the foundation.* Therefore, the proto-personality formed *in utero* through genes and hormones is the foundation from which personality can mature into a fully developed perceptual structure. This biological basis for individual differences in personality has been accounted for throughout the scientific literature. A variety of studies have shown that genetic influence on the most common model for personality accounts for about 40-60% of the variance, which is to say: *biology has a large effect on the diversity of personality.*[416]

Social constructionists may argue that this variance is likely to be inconsistent across different societies, as cultural forces will likely be applied differently. However, biological variance for personality differences is consistent across cultures and actually *grows larger as countries become freer and more prosperous.* Gender differences in personality follow a similar model, as I will soon show. First, however, we must understand how personality is measured.

---

[416] Jang, K.L., Livesley, W. J., Vernon, P.A. (1996). Heritability of the Big Five personality dimensions and their facets: A twin study. *Journal of Personality, 64,* 577-591; Riemann, R., Angleitner, A., Strelau, J. (1997). Genetic and environmental influences on personality: a study of twins reared together using the self- and peer report NEO-FFI scales. *Journal of Personality, 65,* 449-475.

HOW DO WE MEASURE PERSONALITY?

To measure behavioral patterns in personality, there is no better tool at our disposal than language. Language allows us to describe patterns of behavior using adjectives, phrases, and qualifiers. For example, a pattern of altruistic and self-sacrificing behavior may be described as compassionate, kind, and empathic. We can then lump these descriptors together into one word, such as *agreeableness*. The idea is that, throughout the thousands of years of linguistic development, all important differences between individuals will have been described through language, and by decoding these descriptors of individual behavior, we can illuminate the dimensions of personality.[417]

So, how do you go about measuring personality through language? You begin by coming up with hundreds of questions which ask individuals about their behavior. Utilizing a wide range of behavioral descriptions, you create an immense questionnaire evaluating whether people agree or disagree with the statements being presented to them. For example, to evaluate how people feel during social situations, you might present the statement "I find social situations anxiety-inducing" and let them pick five different choices ranging from *highly disagree* to *highly agree*. Now imagine creating over 400 of these statements, each one with a unique description of perception and behavior. You find as many people as you can who are willing to take your test, and you evaluate the results.

While evaluating the scores, you're not just looking for what people said on a given question; you're looking for what questions are answered in the same way by the same individuals. For instance, people might consistently answer questions 2, 4, 10, and 20 with *highly agree* and another set of people might consistently answer those four questions with *highly disagree*. This consistency among the answers to questions 2, 4, 10, and 20 shows that there is some common behavior that these four questions are addressing. In other words, perhaps 2, 4, 10, and 20 are describing the same trait, and therefore, perhaps they can be condensed into a *factor*, or an

---

[417] McCrae, R., Oliver, J. (1992). An introduction to the five-factor model and its applications. *Journal of Personality, 60(2),* 184.

overarching description of similar behavioral patterns. For example, 2, 4, 10, and 20 may be asking people about their negative feelings during social situations. If the answers are consistent, then perhaps these four questions can be consolidated into a single factor called "neuroticism," or the tendency to experience negative emotions. This process of consolidating questions into single factors is known as *factor analysis,* which is to say, to analyze which questions group together to form a common description of a single trait. In our example, questions 2, 4, 10, and 20 group together to describe the trait of neuroticism.

Factor analysis extends past the single trait and into other traits. We may find that another set of questions shares a common pattern of answers, forming a second factor. For instance, questions 1, 3, 5, and 9 may each present slightly different scenarios about how people prefer to interact with others at a party. After these questions show similar answers across individuals, they could be grouped together into a factor known as "extraversion," and this is where factor analysis gets really cool. After doing more analysis of the factors, we may find that *neuroticism* and *extraversion* share a relationship with one another. Perhaps it's a *negative* relationship, where the higher someone scores on the *extraversion* factor, the more likely they are to score low on the *neuroticism* factor. Of course there will be outliers and exceptions, but factor analysis allows personality researchers to see patterns and relationships between descriptions of behavior.

Outsiders unfamiliar with this process may object to its validity by claiming that it is near impossible to agree on how to group behavioral patterns into specific traits. And yet, we *can* agree on how the behavioral patterns should be grouped and labeled, and it's because of the process of factor analysis, which finds how descriptions of individual behaviors relate to one another across hundreds or even thousands of unique descriptors. While it is true that there are a wide variety of ways to describe behavior, these diverse descriptions of behavior will group together to form specific traits. Critics of factor analysis would be correct if the answer to each question showed no relationship to other answers; we'd be left with so many personality traits that they'd be impossible to analyze. And ultimately, it would show that there are no behavioral patterns among individuals and only individualistic attributes that share no common structure. Yet this is not what happens.

Using meta-analyses to study how language describes behavior across societies and cultures, researchers found that the immense variation of behavioral descriptions can be consolidated into just *five* personality factors.[418] It turns out that descriptions of behavior, through specific adjectives and phrases, group together into five major clusters. For example, the words *appreciative, forgiving, generous, kind, sympathetic, and trusting* can be grouped together into a single trait psychologists label as *agreeableness*. Individuals who score high on forgiveness and generosity will most likely score just as high on appreciativeness, kindness, and sympathy. Because of this interdependency of descriptors, the trait *agreeableness* can be conceptualized as a single personality factor which is comprised of an entire set of interrelated behavioral patterns. In terms of gender, psychologists have found important gender differences and similarities in the five major personality factors.

After decades of factor analysis and consolidation of behavioral descriptors through questionnaires, psychologists developed the Five-Factor Model (FFM, or Big Five) to measure individual differences and similarities in personality. The relationship of these five factors to their descriptors is so tightly linked that it exceeds .90 for all five dimensions, which is incredibly accurate.[419] However, before we delve into the discoveries, let's lay out what the factors are and how they can be conceptualized. Utilizing factor analysis, behavioral patterns group together as these five factors:

1. **Extraversion**
2. **Agreeableness**
3. **Conscientiousness**
4. **Neuroticism**
5. **Openness to Experience**

---

[418] McCrae, R., Oliver, J. (1992). 180.
[419] Ibid., 181.

12 PERSONALITY AND THE BIG FIVE

Each of the five categories has two sub-factors which describe, in more detail, individual behavioral traits. However, for the scope of this book, we will only delve into these sub-factors when necessary. Some gender differences in personality, for example, are further differentiated through the sub-factors, but I will continue by describing the five factors as a whole.[420]

The first factor of the Big Five, **Extraversion**, can be conceptualized as the *tendency to experience positive emotions.*

> "In terms of behavior, people who are high in extraversion are sociable, fun-loving, talkative, and spontaneous. People who are low in extraversion--in other words, people who are introverted--are reserved, inhibited, quiet, and aloof. Because the trait is normally distributed, most people fall somewhere in between these two extremes."[421]

The second factor, **Agreeableness**, can be thought of as the tendency to exhibit maternal-like behavior, further described below by psychologist Dr. Jordan Peterson:

> "Agreeable people can be described as kind, warm, polite, and accommodating, while disagreeable people can be described as selfish, ruthless, and vengeful. There is an obvious language bias that makes it sound like being agreeable is better, but there are pronounced advantages and disadvantages at any point on the spectrum. Agreeableness is composed of two aspects: compassion and politeness."[422]

---

[420] For example: while the third FFM factor, Conscientiousness, has practically no gender difference, it also has two sub-categories of Orderliness and Industriousness which show slight gender differences in each one. More factors means more resolution.

[421] Peterson, J. (2019). Extraversion: Enthusiasm and Assertiveness. *Discovering Personality.*

[422] Peterson, J. (2019). Agreeableness: Compassion and Politeness. *Discovering Personality.*

The third factor, **Conscientiousness,** can be thought of as the tendency to be responsible, organized, and goal-directed:[423]

> "Conscientious people are careful, reliable, organized, self-disciplined, and persevering. Unconscientious people are carefree, laid-back, happy-go-lucky, messy, and inattentive. Conscientiousness is composed of two aspects: industriousness and orderliness. Industriousness has to do with commitment to diligent work, while orderliness is about organizational skills. Conscientiousness is also related to disgust sensitivity, and this relationship has important social and political ramifications."[424]

The fourth factor, **Neuroticism,** can be conceptualized as the tendency to experience negative emotion:

> "People with neurotic dispositions are more prone to mood disorders, loneliness, self-consciousness, and hypochondria, to name just a few related experiences. In reality, neuroticism is rarely fun for anyone, though some research has shown that neuroticism can predict student success and may be correlated with certain reproductive benefits. Neuroticism may have provided evolutionary advantages, as well—paying more attention to negative outcomes or risks could have helped certain early humans survive."[425]

The fifth and final factor, **Openness to Experience,** can be thought of as the tendency to be open to new experiences, ideas, and stimuli. Openness is the trait that is most closely linked to IQ:

---

[423] What Is Conscientiousness? *Psychology Today.*
[424] Peterson, J. (2019). Conscientiousness: Industriousness and Orderliness. *Discovering Personality.*
[425] What Is Neuroticism? *Psychology Today.*

> "Open people are creative, intellectual, and curious. Openness to Experience is composed of two aspects: openness and intellect. Openness is linked to creativity and interest in aesthetic experience, while intellect is linked to IQ and interest in ideas."[426]

Each of these Big Five factors originate from the highly reliable linguistic factor analysis which analyzes how certain descriptions of behavior form specific clusters that can be represented through a single trait. Each of the factors is normally distributed, where many people fall on the average and many others fall on the extremes. Or, in more technical terms:

> "Many psychological phenomena, including Big Five personality traits, are normally distributed. This means that 68% of people score within one standard deviation on either side of the mean (average), and 65% of people fall within two standard deviations. In other words, the vast majority of people score somewhere near the middle."[427]

The Big Five Model is now the gold-standard for personality models and is used across many disciplines and fields to assess individual behavior. It has three major aspects which make it the superior model compared to other personality frameworks:

> "The appeal of the model is threefold: It integrates a wide array of personality constructs, thus facilitating communication among researchers of many different orientations; it is comprehensive, giving a basis for systematic exploration of the

---

[426] Peterson, J. (2019). Openness to Experience: Intellect and Openness. *Discovering Personality*.
[427] Peterson, J. (2019). Men & Women: Personality Differences, Lecture Notes 6. *Discovering Personality*.

relations between personality and other phenomena; and it is efficient, providing at least a global description of personality with as few as five scores."[428]

Using the Big Five Model allows psychologists to assess the differences and similarities in behavioral patterns among individuals. We can evaluate whether these behavioral patterns extend to the categories of sex and gender by asking whether or not there are personality differences between men and women. If there are, we can then see whether these personality differences are inaccurate stereotypes based in sociocultural constructs, innate traits based in biological structures, or a mix of the two.

GENDER DIFFERENCES IN THE BIG FIVE

Before we delve into the Big Five model for gender differences in personality, we need to ask the critical question of *why?* Why study such a topic? Can studying personality differences reinforce and perpetuate gender stereotypes? I'd argue this fear is unfounded, as understanding the complexity of gender differences in personality can actually help dismantle stereotypical thinking by showing people just how much similarity men and women have across a given trait. Here in personality research, there is no room for broad generalizations or stereotyping; there is only precise statistical data. Personality psychologists have described the goal of understanding gender differences in personality, from which behavioral patterns between men and women can be better understood.

> "The goal of investigating gender differences in personality, therefore, is to elucidate the differences among general patterns of behavior in men and women on average, with the understanding that both men and women can experience states across the full range of most traits. Gender differences in terms of mean differences do not imply that men and women only

---

[428] McCrae, R., Oliver, J. (1992). 206.

experience states on opposing ends of the trait spectrum; on the contrary, significant differences can exist along with a high degree of overlap between the distributions of men and women."[429]

Now that we can understand the purpose of gender-based personality research, we can ask two critical questions about gender differences in personality. First, *are there Big Five personality differences between men and women?* And second, *are these differences real?* Before we can answer the first question, we must define a word that most people might simply pass over as something already understood, and that is the word *real*. It might seem redundant to define such a word, but knowing the difference between what is real and what is not is critical to understanding individual differences in personality.

In personality psychology, what would be an indicator that a given trait is *real*? One way is to see whether the trait changes across time and culture or whether it stays relatively consistent across time and culture. If it changes significantly and shows no cross-cultural universality, then we could safely say that such differences in trait expression are not *real* but are rather subjective sociocultural constructs. On the other hand, if the differences in a given trait do not show much change across cultures, then such differences in trait expression are most likely *biological* and therefore *real*.

The first question *(are there personality differences between men and women?)* can be answered by looking at meta-analyses of Big Five personality scores across millions of men and women. A meta-analysis, as discussed previously, combines individual studies together and consolidates their scores to produce one immense review utilizing hundreds of thousands or even millions of people across hundreds of cultures. Such meta-analyses have been conducted for gender differences in personality. For brevity, I'll cover one of them here and list the others to

---

[429] Weisberg, Y. (2011). Gender Differences in Personality across the Ten Aspects of the Big Five. *Frontiers in Psychology, 2(178)*.

show that such a study is backed up by a plethora of additional meta-analyses which confirm its findings.

Conducted in 2008, one study illuminates the incredible diversity in personality among individuals. Psychologist David Schmitt worked with over 100 social, behavioral, and biological scientists to conduct an analysis on gender differences in personality across 55 nations and 17,000 participants. Each of the participants from the 55 nations received the Big Five personality questionnaire, comprised of 44 items, and filled it out through a scale of 1 (disagree strongly) to 5 (agree strongly) for each of the 44 items.[430] The results showed moderate to large differences between men and women in some factors, and small to moderate differences in others. The psychologists used the common measurement known as *effect size* to quantify the gender differences in the five factors, with a negative number meaning the trait has more men than women and a positive number meaning the trait has more women than men.[431] After calculating the effect sizes for each factor across every nation, the psychologists then combined these nation-level effect sizes to form the global effect sizes. What they found was that gender differences in personality existed, not just in Western societies, but across all the nations studied. However, it is in the details where the true enlightening knowledge resides.

On a global aggregate level, the gender differences across the Big Five tend to be large to moderate to small, with Neuroticism (tendency to experience negative emotion) having the largest difference favoring women. For **Extraversion**, the effect size was 0.10, which means this trait slightly favors women across cultures. **Agreeableness** had an effect size of 0.15, also slightly favoring women. **Conscientiousness** had an effect size of 0.12, **Neuroticism** an effect size of 0.40 (a moderate difference), and

---

[430] Schmitt, D., et al. (2008). 170.

[431] Cohen's *d* effect size is usually calculated so that positive values indicate men being higher than women in a given trait. However, many personality psychologists invert the positive and negative values so that positive indicates a larger difference for women, whereas negative indicates a larger difference for men.

Openness to Experience a size of -0.05 (a negligible difference).[432] Put in more adjective-based descriptors, women scored higher than men on "depression, self-consciousness, impulsiveness, vulnerability, warmth, gregariousness, positive emotions, openness to aesthetics, openness to feelings, straightforwardness, altruism, compliance, and modesty, and men scored higher than women on excitement-seeking and openness to ideas."[433] These differences grow larger in magnitude in certain cultures.[434]

The five factors can be further subdivided into their two sub-factors, from which we can get a higher resolution into the nuances of gender differences in personality. Dr. Peterson lays out a few of these sub-factor differences:

> "The highest difference between men and women [in personality] is in compassion, with women scoring higher than men. In order of highest differences, women also score higher than men in withdrawal, politeness, volatility, openness, orderliness, and enthusiasm. Men score more highly in assertiveness than women."[435]

And so, knowing that women show slightly and moderately higher levels than men across 4 of the 5 major factors across almost all cultures, it seems safe to say that there are gender differences in personality. However, to see whether these differences are truly *real* and not mere social constructs, we have to dig a little deeper into the patterns across different cultures, as well as look at the theories for the origins of personality, from social constructionism to the biological hypothesis. Before we do that

---

[432] Schmitt, D., et al. (2008). 173.

[433] Mathews, G., et al. (2009). Personality and congenital adrenal hyperplasia: Possible effects of prenatal androgen exposure. *Hormones and Behavior, 55,* 285; Costa, P., et al. (2001). Gender differences in personality traits across cultures: Robust and surprising findings. *Journal of Personality and Social Psychology, 81(2).*

[434] As you will see, gender differences in personality grow larger in certain regions of the world, but exactly where in the world was not what anyone expected.

[435] Peterson, J. (2019). Men & Women: Personality Differences, Lecture Notes 6. *Discovering Personality.* 2.

though, we must learn how these differences in personality distribute themselves.

To understand the effect sizes of gender differences in personality, imagine that you take random people out of the population (half are men and half are women), and you were instructed to guess whether a given man or woman in the group was more agreeable. If you guessed the woman was more agreeable, you'd be right about 60% of the time, and if you guessed it was the man, you'd be right about 40% of the time.[436] *That's not that much!* Agreeableness, therefore, has a 60-40 female to male ratio. This shows that for many personality traits such as Agreeableness (as well as other gender differences) large amounts of overlap exists between men and women. This also means that you cannot accurately predict an individual's personality traits merely based on their sex. *You can make an estimated guess, but that's about it.* In graph-based terminology, this is what such a gender difference looks like:

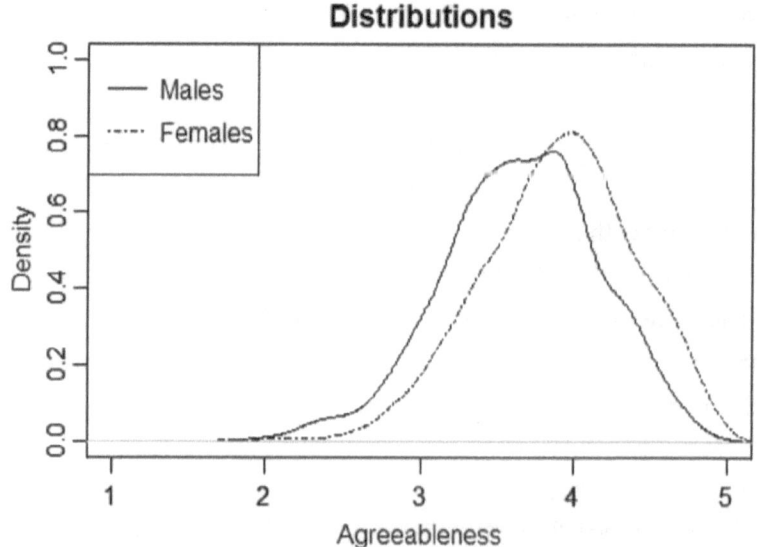

Women show slightly higher average levels of agreeableness. But there is still a lot of overlap. Source: Weisberg, J., et al. (2011). 9.

---

[436] Peterson, J. (2017). Jordan Peterson on Diversity. *Bite-sized Philosophy, YouTube.*

You can see by looking at the graph that the two bell curves overlap significantly. While women and men do not share the exact same average level of Agreeableness, they do share significant overlap. The significant differences arise, however, in the extremes, or the tail ends of the bimodal distributions. Notice the far left end of the graph. There is significantly more men than women that score exceedingly *low* in trait Agreeableness, showing that the *most agreeable people* tend to be women and the *most disagreeable people* tend to be men. *Low* agreeableness partially predicts one's predisposition to be imprisoned at some point life: men make up around 93% of prison inmates.[437] However, these extremes do not negate the importance of the overlap and averages. Understanding such a graph, once again, is the key to understanding how gender differences exist on bimodal distributions and not spectrums.

To further understand this complexity, let's break down the trait of Openness to Experience, which has two sub-factors of *Openness* and *Intellect*. Openness to Experience relates to one's aesthetic sensibilities, enthusiasm for new experiences, and interest in ideas and theories. Out of the other Big Five factors, Openness to Experience is the closest factor to IQ, which measures one's overall intelligence. Someone who has high Openness to Experience will be highly interested in art, knowledge, and learning, and will likely excel in highly intellectual fields. Therefore, understanding the gender differences in this trait can be incredibly important for knowing how men and women might differ in aesthetic and intellectual sub-factors. And yet, unlike Agreeableness which shows a moderately large gender difference, Openness to Experience shows almost no gender difference as a whole.

Women and men share almost equal levels of aesthetic sensibilities, enthusiasm for new experiences, and interest in ideas and theories. However, it is in the sub-factors where some differences arise. The sub-factor *Openness*, associated with creativity and an interest in aesthetics and art, shows a slight difference favoring women, whereas *Intellect*, associated with an interest in ideas, shows a slight difference favoring men. This

---

[437] Carson, A. (2013). Prisoners in 2013. *Department of Justice, Bureau of Justice Statistics.*

makes sense that we may find slightly more women who are interested in artistic aesthetic experiences compared to men who, slightly more than women, are interested in talking about abstract ideas and theories. Although most men and women share fairly equal levels of these traits, this slight gender difference in sub-factor *Intellect* may partially explain the tendency for a small minority of men to become more interested in highly complex and abstract mathematics. This is not to say women don't have the ability to perform complex and abstract mathematics, but that more men on the extreme end of the distribution show greater *interest*.

And finally, to wrap up the example of Openness to Experience, personality researchers have found that, despite the claims of sexist researchers from the past, there is actually no gender difference in average IQ across multiple measures of personality models and IQ tests. Such a fact should give a breath of fresh air to strong social constructionists who dismiss research into sex differences as inherently sexist. It is not sexist to research the complexity of behavioral patterns in men and women using statistical methods, factor analysis, and average distributions. What is sexist, however, is applying inaccurate stereotypes to individuals based on their unchanging characteristics. Both gender ideologues and regressive conservatives tend to do this, as I discussed in a previous chapter.[438]

Unexpectedly, however, while there is no average difference in IQ, there is a slight gender difference in the *distribution* of the IQ scores. Whereas most men and women have similar IQs, the bell curve distribution for men tends to be slightly more *flat*. What this means is that there are more men than women on both extremes of IQ. In other words, there tends to be both more male geniuses and more male idiots. This is known as the *greater male variability hypothesis,* which "argues that the greater variability in traits seen in men is a consequence of the fact that men are more disposable as they are less likely to reproduce successfully."[439]

---

[438] See chapter, "Gender Stereotype Pathology."
[439] Peterson, J. (2019). Men & Women: Personality Differences, Lecture Notes 6. *Discovering Personality*. 2.

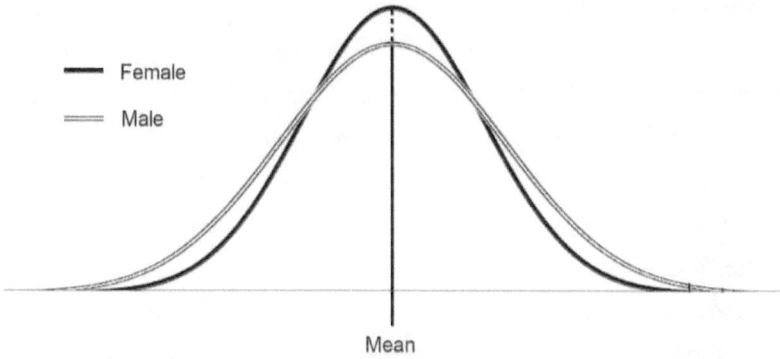

Bimodal distribution of men's and women's IQ scores. While males and females have the same average IQ, males tend to be slightly more variable.

This hypothesis is found in the so-called *Darwin Awards*, which are awards given to those who have discovered the most idiotic way of eliminating themselves from the human gene pool. Taking unnecessary risks which get you killed, such as stepping in front of a moving train, tends to be more of a male behavior, and therefore, the Darwin Awards tend to be bestowed upon men much more often than they are given to women. In other words, there seems to be more male idiots than female idiots, an interesting and hilarious hypothesis indeed.

Knowing that there are gender differences in personality across cultures, this does not answer the second question: *are gender differences in personality real?* Again, for something to be real, it has to be relatively consistent across cultures. Social constructionists would argue 1) *that gender differences in personality are wildly different across cultures* and 2) *that these differences minimize as countries become more gender-equal.* Because of these two claims, social constructionists would view gender differences in personality as sociocultural constructs and therefore not *real*. Is this what the evidence shows?

Here the social constructionists are yet again half correct. The first claim (*that gender differences in personality are wildly different across cultures*) is actually true! Personality traits between men and women show differences in the effect sizes across different nations, proving that the *magnitude* of these gender differences depends on the sociocultural context. For example, the effect size for Agreeableness in South Africa was 0.00, showing no difference between men and women, whereas the difference in Agreeableness in the Netherlands was 0.41, showing a moderate difference. This variability shows that in different world regions, different magnitudes of gender differences in personality arise.

However, it is not that the gender differences completely swap across cultures (as in men being higher than women on Agreeableness) but rather that the *magnitude* of gender differences in personality grows or diminishes based on the culture. For instance, women across almost all nations show higher levels of trait Agreeableness than men, but this difference grows larger depending on the world region being studied. So, in some sense, the social constructionist claim that gender differences change across cultures is partially true; however, a more accurate statement would be that gender differences change in *magnitude* rather than favorability.

So, to answer the second question (are gender differences in personality real?), we can see that gender differences in personality across the Big Five stay the same in favorability but not in magnitude. **This means that women across almost all cultures score slightly to moderately higher than men on Neuroticism, Extraversion, Agreeableness, and Conscientiousness, whereas men across almost all cultures tend to exhibit slightly less negative emotion, more introversion, less agreeableness, and less conscientiousness.**

If the social constructionist hypothesis was correct, we'd expect to see a trait such as Neuroticism score higher in men across many other cultures. But this is not the case. Out of the 55 nations studied, only 2 countries showed a difference in Neuroticism favoring men, and the effect sizes were small to negligible; in all the other 53 nations, women scored higher in trait Neuroticism.[440] So, in other words, while the *magnitude* of the gender differences changes, the *favorability* towards women or men across a given trait stays the same. Therefore, the answer to this critical question is...*yes:* **gender differences in personality are consistent across cultures in**

---

[440] Schmitt, D., et al. (2008). 173.

favorability towards women or men and only change in *magnitude*, showing that these differences are *real*. The social constructionists are right in the fact that gender differences in personality change in magnitude, but ironically, this fact doesn't help their theory much; instead, such a claim actually puts the final nail in the coffin for social constructionism. The second claim (*that gender differences in personality minimize as countries become more gender-equal*) is not just incorrect, it's exactly the opposite of what the Big Five scores show.

Gender differences in personality do not minimize as countries become more prosperous, freer, and more gender-equal; they do not go away, they do not diminish, and they do not subside and fade. **Instead, gender differences in personality** *grow larger in magnitude* **as a country becomes more prosperous, freer, and more gender-equal**:

> "With greater human development and with greater opportunities for gender equality, the personalities of men and women do not become more similar (see also Costa et al., 2001; McCrae, 2002; McCrae et al., 2005). To the contrary, in more prosperous and egalitarian societies, the personality profiles of men and women become decidedly *less* similar. In more traditional and less developed cultures a man is, indeed, more like a woman, at least in terms of self-reported personality traits."[441]

How could such a paradox exist? Could it be that when you let people express themselves freely so that they participate in the economy and society as equal citizens, you allow innate tendencies to rise from the depths? Such a thing seems to be happening across most Western countries, including Scandinavia, which have done the most for gender equality according to the UN (adopting policies which encouraged and facilitated women's participation in the economy and society).[442] These are the most progressive societies in the world for women, and yet, they have

---

[441] Schmitt, D., et al. (2008). 178.
[442] Gender Inequality Index (GII).

the largest gender differences in Big Five personality. So, to put it nicely, *what the hell is going on?!*

You really have to wonder, from the social constructionist perspective, how could such findings be explained? Gender theorists in Scandinavia would respond to this problem of greater gender differences by saying that their countries, the ones doing the most for gender equality, are still the *most* backwards! *Perhaps we still socialize boys and girls differently,* they say.[443] And yet, such a claim of continued differential treatment seems to be unfounded. Scandinavian countries such as Norway have the strongest policies for women's equality (paid maternity leave, economic programs, and social incentives). These policies do not just encourage women to participate in society and the economy but tend to *favor* them. Every business person or politician worth their salt should want to increase their share of women workers or leaders with the large amount of social incentives for women's equality. Almost every incentive structure in our contemporary societies push for more women representation across many hierarchies, including STEM fields.

Compared to children in gender-unequal countries, children in the most gender egalitarian countries have the most economic and social opportunities to explore their innate interests. And boys and girls growing up into adulthood in these cultures are encouraged to go into all types of fields, even fields which are gender-atypical. For instance, girls seem to be continually encouraged in Scandinavia to go into science, technology, engineering, and math, and yet, average differences in occupational preferences between girls and boys continue in large magnitudes.[444] Such a disparity is rather non-existent in gender-unequal countries such as India or Iran, where women make up much more STEM graduates and workers across science and math-related fields.[445] Intuitively, this is incredibly

---

[443] Hjernevask (Brainwash), "The Gender Equality Paradox." *YouTube.*
[444] Lippa, R. (2010). Sex differences in personality traits and gender-related occupational preferences across 53 nations: Testing evolutionary and social-environmental theories. *Archives of Sexual Behavior, 39,* 634.
[445] Hjernevask (Brainwash), "The Gender Equality Paradox." *YouTube.*

strange and unexpected. Why is there a more equal representation of girls in STEM in Iran than in Scandinavia?

Knowing this paradoxical data, that gender differences in personality increase as gender equality increases (measured with eight independent equality models), it is rather difficult to argue for the validity of purely social constructionist theories which state that men and women will exhibit more differences in more traditional and patriarchal societies:

> "These findings may seem paradoxical because in traditional and economically deprived countries the division of labor between the sexes appears more disparate than it does in wealthy and egalitarian societies ... An accumulating body of evidence, including the current data, provides reason to question the social role explanations of gender and personality development (Baron-Cohen, 2003; Campbell, Shirley, & Candy, 2004; Geary, 1998; Lytton & Romney, 1991; Maccoby, 2000; Mealey, 2000; Spiro, 1996; Tiger & Shepher, 1975). In this study, a collection of eight different gender equality indicators provided a comprehensive set of measures that assess disparity between male and female roles in society. In every case, *significant findings suggest that greater nation-level gender equality leads to psychological dissimilarity in men's and women's personality traits.*
>
> If differences in personality traits are controlled by the drastically different social roles that men and women play in the society then in cultures in which women earn considerably less than men, in which they have limited access to education, and in which only few of them become professionals, women's personality profiles should be very different from men's. In reality, these women's personality profiles are more similar to those of men. Whatever the source of men's and women's

personality trait differences across cultures, differences in social roles appear unlikely to play a significant causal role."[446]

One explanation for the minimal gender differences in personality reporting in gender-unequal countries is that in more traditional societies, perceived differences are more likely to be associated with required roles rather than innate personality traits:

> "In traditional cultures, perceived differences between men and women in general might be attributed to role requirements rather than intrinsic differences in personality traits."[447]

Because of this, it is likely that men and women in gender-unequal countries, when reporting on personality, see gender differences as merely differentiated sex roles rather than different personalities, and therefore, do not self-report through stereotyped perceptions of personality. So if gender differences in personality across cultures do not correlate to traditional social roles (such that gender-equal countries show *more* gender differences in personality), then what if these differences could be explained through self-report bias, or men and women reporting internalized stereotypes? Perhaps stereotypical ideas of what men should be like and what women should be like informs the gender differences in self-report personality scores, especially in Western cultures.

Psychologist David Schmitt and the researchers considered this bias, which they refer to as *method artifacts* (bias in the results). They found that, after controlling for the level of human development across societies (known as the HDI), there was a correlation between internal consistency of responses to positive and negative items in the questionnaire and the level of economic and social prosperity of the respondent's nation. In other words, the more egalitarian and free the society, the more likely people will answer with balanced responses:

---

[446] Schmitt, D., et al. (2008). 178.
[447] Costa, P., et al. (2001). 330.

"It can be difficult to determine whether self-reports reflect role requirements (stereotyped beliefs), intrinsic differences in personality traits, or some interactive combination of both. It is, however, much easier to observe simpler biases such as dissimilar responding to positively and negatively worded items or overall inconsistency in responses to personality items. Schmitt and Allik (2005) observed that in countries with lower differences between positively and negatively worded items and higher internal consistency of responses, people tend to live longer; be economically more prosperous; and support individualistic values and equality in rights, wealth, and power. In this regard, this study provided no strong evidence for the artifact explanation."[448]

This shows that external factors on evaluation of one's own personality, such as countries which are less free, affect one's responses. And the more free the country, the more consistent the responses are between positive and negative items. In other words, in the countries with less economic and social freedom, people tend to report more strongly with negative items in the questionnaire compared to positive ones.[449] This is known as a *negative item bias,* where respondents put more emphasis on negative items than their positive counterparts. Countries like Scandinavia, people are more likely to answer more consistently across the positive and negative items than people in less egalitarian societies like Iran, who will tend to be biased towards negative self-reporting.

And so, there is definitely a bias in self-reporting, just not in the way we would expect; bias was shown towards reporting of negative emotionality, but not towards gender differences.

---

[448] Schmitt, D., et al. (2008). 178.

[449] Schmitt, D., Allik, J. (2005). Simultaneous administration of the Rosenberg Self-Esteem Scale in 53 nations: Exploring the universal and culture-specific features of global self-esteem. *Journal of Personality and Social Psychology, 89(4),* 637.

BIOPSYCHOSOCIAL MODEL OF PERSONALITY

If gender differences in personality are consistent across cultures and only grow larger in the most gender-equal societies, then where do such unique personalities come from? Where can the origins of these perceptual structures be found? The first theory, social constructionism, states that differences in personality between men and women arise due to "differences in the way men and women are socialized."[450] And the second theory, the biological hypothesis, states that differences in personality between men and women are linked to "biological differences such as genetic differences."[451]

In contemporary personality research, few theorists endorse either the purely social constructionist theory or the purely biological theory. Instead, personality differences in men and women seem to come from a mix of biology, psychology, and society:

> "Sex differences in personality have been suggested to result from both inborn factors and socialization (Costa et al., 2001). For instance, the differing adaptive problems faced by the two sexes during evolution might be expected to produce inborn sex differences in personality traits (Buss, 1995). In contrast, the social role model (Eagly, 1987) posits that gender differences in behavior and personality arise from sex differences in social roles."[452]

Both evolutionary theory and social role models, when integrated together, can more accurately explain gender differences in personality than if either theory was ignored. On the biological level, gender differences in personality are partially the result of sexual selection pressures which

---

[450] Peterson, J. (2019). Men & Women: Personality Differences, Lecture Notes 6. *Discovering Personality*.
[451] Ibid.
[452] Mathews, G., et al. (2009). 286.

have "caused men to be more prone than women to take risks and seek social dominance, whereas women are thought to have been selected to be more nurturing and cautious (Buss, 1997; MacDonald, 1995)."[453]

This is not to say that men and women are wildly different, or that there is no overlap between these sexual roles. However, evolutionary sexual selection seems to have had somewhat of an effect on the psychology of men and women and their specialized perceptual structures. The reason for these slight differences in personality seems to come from a mix of biological and social effects dating back hundreds of thousands of years:

> "In the ancestral past, as hunter-gatherers, men and women naturally developed sexually selected differences in personality traits such that men were more risk-taking and dominance-seeking and women were more nurturing and cautious."[454]

Though such a hypothesis is speculative, we can currently observe how prenatal androgens and pubertal hormones affect neuroanatomy and neurocircuitry, producing gender differences in how the average man and average woman perceives.[455] Of course, while men and women share most traits in common, there are specific differences in certain personality structures.

Take, for example, trait Agreeableness. There is a reason why women tend to score higher on average. Agreeableness can be broken down into Compassion and Politeness, through which maternal-like behavior is exhibited. Males and females both exhibit levels of agreeableness, but women tend to show higher levels. This may be due to the fact that, historically, women spent the most time caring for infants. They don't just have to bear the child for nine months; they have to breastfeed it, touch it, hold it, and clean it. Males can do the last three, but not the first two.

Because of this difference in sex roles, it not a stretch to say that

---

[453] Schmitt, D., et al. (2008). 178.
[454] Ibid., 179.
[455] See chapters, "Conception to Birth," "Development of Gender Identity," and "The Activational Stage."

women's nervous systems have been more tuned to the caring of infants throughout the millions of years of human development. Caring of infants requires high levels of compassion, kindness, and sensitivity to negative emotion. Throwing the baby out the window because it's crying and screaming is not the best solution for the baby or the species. Even during the most distressing and tiring episodes of infancy, the baby's needs must be constantly attended to without fail. Such a relationship to the infant requires that your perceptual structure be tuned to its needs. Therefore, higher levels of Agreeableness and Neuroticism in women tend to be shifted towards this exact evolutionary need. And such a hypothesis seems to reasonably explain the gender differences that we see in these two traits across the most gender egalitarian societies.

Such a hypothesis also integrates the role that the environment may play in gene expression, through which gender differences in perceptual structures like personality may be passed down to offspring. This biopsychosocial approach integrates epigenetics, which partially explains the gender differences in personality in gender-equal countries. Gender differences, while originating in biological sex, may grow larger as sex roles become more divided between caring of infants and finding food, thereby producing differences in gene expression through environmental influences.

However, as men's and women's sex roles become more united (as in the case of gender egalitarian countries), these gender differences may eventually grow smaller through epigenetic effects as well. As of now though, such changes in gene expression of personality differences are unlikely to show up. *We would need more time to see if personality differences between men and women in gender egalitarian nations become more attenuated.* As of current research, however, gender differences in personality are still the largest in the most gender-equal societies, whereas gender differences are the smallest in the most gender-unequal societies.

Another piece of evidence showing that some gender differences in personality are the result of biological factors can be found through the effects of sex hormones. Sex hormones such as androgens, as I have shown throughout the previous chapters, play a critical role in the organization of the fetal brain by altering neuroanatomy and neurocircuitry. And these hormonal effects are a critical part of mammalian sexual differentiation which cannot be ignored:

> "One of the most powerful proximal determinants of sex differences in non-human mammals is the early hormone environment (De Vries and Simerly, 2002; Goy and McEwen, 1980; Hines, 2004). Experimental studies in species ranging from rodents to non-human primates support the view that prenatal or neonatal levels of gonadal hormones (especially the androgen, testosterone) permanently influence behaviors that show sex differences. Gonadal hormones also appear to influence the development of at least some human behaviors (Collaer and Hines, 1995; Hines, 2004). This conclusion is based largely on studies of individuals who have experienced anomalies in hormones during fetal life."[456]

The direct effects of testosterone on gender differences in personality can be elucidated through the common case studies of girls exposed to high levels of prenatal androgens. These girls have the condition of Congenital Adrenal Hyperplasia (CAH), where the adrenal glands, or the mother's blood, overproduce androgens in the womb. Julia, our fictional girl, also exhibits clues that she was exposed to high levels of prenatal androgens through her male-typical psychology and interest in boy-typed toys despite socialization as a normal and healthy female. Such is the case of CAH girls who exhibit male-typical toy, playmate, and activity preferences in comparison to unaffected female relatives and controls.[457] Socialization, postnatal hormone treatment, and surgical feminization at infancy do nothing to change the girls' interest in male-typical toys, playmate

---

[456] Mathews, G., et al. (2009). 286.
[457] Ibid.; Ehrhardt et al., 1968; Ehrhardt and Baker, 1974; Slijper, 1984; Dittmann et al., 1990; Berenbaum and Hines, 1992; Zucker et al., 1996; Nordenstrom et al., 2002, Pasterski et al., 2005).

preferences, and activity preferences; nor do these things change their *disinterest* in female-typical toys such as dolls.[458]

Because of these male-typical behaviors in CAH girls, personality differences should show similar results. Across multiple studies (specifically Mathews et al., 2009), the average man scored higher than the average woman on *dominance* and *physical aggression* (low trait Agreeableness) whereas the average woman scored higher than the average man on *tender-mindedness* and *interest in infants* (high trait Agreeableness). Using the average scores of men and women, boys and girls alike, researchers found that females with CAH differed from the average girl in three of four personality measures.[459] Specifically, females with CAH were less *tender-minded* and *less interested in infants,* and "gave more physically aggressive responses than unaffected female relatives."[460] Likewise, our fictional girl will likely exhibit these more masculine tendencies due to the organizational effects of high prenatal androgen exposure on her brain structure. As she grows into adolescence, she may exhibit slightly more physically aggressive tendencies compared to her unaffected sister, and she may also be less tender-minded and less agreeable. Thus, from these results and additional studies on prenatal androgen's effects on personality, it seems clear that the traits of *tender-mindedness, physical aggression,* and *interest in infants* relate to the levels of androgens a fetus is exposed to in the womb, regardless of sex. Such a personality difference in CAH girls compared to their unaffected sisters cannot be explained away through social construction theory.

Social constructionists will posit that because the parents know their CAH girl's genitals have a more masculine appearance, they treat their girl more like a boy, thus causing them to exhibit male-typical behavior. However, such a claim is not just unfounded; it's been continually disproven by a variety of studies. For instance, parents who have a girl with CAH have self-reported that they treat their daughter the same as their

---

[458] Mathews, G., et al. (2009). 286.
[459] Ibid., 289.
[460] Mathews, G., et al. (2009). 289.

unaffected daughters in encouragement to "act as a girl should."[461] Such self-report can be biased, so researchers have also conducted observational studies on play behavior in CAH girls. These studies show that parents actually give *more* encouragement of female-typical behavior to their daughters with CAH than to unaffected sisters:

> "A study that observed parents interacting with their children as they played with sex-typical and sex-atypical toys found that parents gave more, rather than less, encouragement of female-typical behavior to daughters with CAH than to daughters without CAH (Pasterski et al., 2005)."[462]

With these studies, it has become clear that parents do not just treat their daughters with CAH the same but actually encourage more female-typical behavior than they otherwise would while knowing that their girls have more masculinized genitals. Despite this socialization towards female-typical behavior, girls with CAH continually reject female-typical toys. Prenatal androgens, therefore, seem to play a large role in toy preferences, behavior, interests, and personality factors. Or, described a more direct way by sex hormone researchers:

> "Levels of maternal testosterone during pregnancy correlate with the amount of male-typical behavior in female offspring (Hines et al., 2002), and testosterone measured prenatally in amniotic fluid positively predicts male-typical behavior in both girls and in boys (Auyeung et al., 2009) despite none of the offspring exhibiting genital abnormality."[463]

In the face of this hormonal evidence, social constructionists cannot continue to deny the effects of prenatal androgens on a child's behavior,

---

[461] Berenbaum and Hines. (1992).
[462] Mathews, G., et al. (2009). 289.
[463] Mathews, G., et al. (2009). 290.

specifically personality factors.[464] If traits such as Agreeableness can be heavily influenced by exposure to androgens and grow larger as countries become more gender-equal, then what about Neuroticism (another trait which shows one of the largest gender differences)? Neuroticism, or the tendency to experience negative emotion, shows a large gender difference favoring women which only grows larger the more gender-equal the country becomes.

Neuroendocrinologists theorize this difference in Neuroticism levels can be critically affected by levels of testosterone. Studies on mice show that, when male and female mice are administered testosterone, their anxiety levels decrease in a given maze.[465] Repeated doses of testosterone in rats show a similar effect on anxiety during burying behavior.[466] In rhesus monkeys, pharmacological castration resulted in higher anxiety levels, which then were normalized after administration of testosterone.[467] In humans, administering testosterone to both men and women resulted in reduced fear; however, the effectiveness of testosterone decreases once the levels reach a certain threshold.[468] Other studies throughout the decades have shown relationships between testosterone levels and behavior across both sexes.[469] Higher levels in men correlate to lower levels of anxiety.[470] Higher testosterone also correlates to decreased levels of depression in both humans and animals through numerous methodologies and studies.[471] It seems as though testosterone can modulate the neurotransmitter serotonin, which plays a critical role in the development of depression in both sexes.[472] Past Neuroticism, testosterone levels also correlate positively

---

[464] See chapters, "Conception to Birth" and "Development of Gender Identity."
[465] Aikey et al. (2002).
[466] Fernandez-Guasti and Martinez-Mota. (2005).
[467] Suarez-Jimenez et al. (2013).
[468] Celec, P., et al. (2015). On the effects of testosterone on brain behavioral functions. *Frontiers in Neuroscience*, 2.
[469] Edinger and Frye. (2004); Frye et al. (2010); Khakpai. (2014); Walf and Frye. (2012).
[470] Celec, P., et al. (2015). 2.
[471] Ibid.
[472] Jovanovic et al. (2014).

to increased spatial ability and visuospatial memory.[473] Knowing these highly correlative effects of testosterone across age groups and populations on the tendency to experience negative emotion, it is highly likely that gender differences in Neuroticism may be influenced by exposure to testosterone.

From the evidence for the effects of androgens on the male and female brain to the evidence of larger gender differences in personality as countries become gender egalitarian, it can no longer be postulated that gender differences in personality are mere social constructs. Each of us has a unique perceptual structure formed during the womb through sex-specific gene expression and exposure to hormones. As infants and young children, we operate through the world using our pre-determined perceptual structure, our proto-personality, which then develops into a more mature personality as we grow up. Therefore, gender differences in personality result, in part, from biological factors such as genes and hormones; sociological factors such as behavioral reinforcement and punishment of appropriate social roles; and epigenetic factors such as expression of genes to specific environmental stimuli.

The claims of Cognitive Social Learning Theory, which state that behavioral differences between people are reinforced or punished through social learning, cannot fully explain the gender differences in personality. Opposite of what the social constructionists would predict, gender differences in Extraversion, Agreeableness, and Neuroticism grow larger in magnitude as countries become more egalitarian according to eight unique gender equality indices.[474] Specifically, the Big Five personality model and its meta-analyses across 55 nations has shown that personality differences between men and women *do* exist across cultures and are *real*, and that these differences only increase in magnitude as a country becomes freer, more prosperous, and more gender-equal in economic and social conditions. It turns out that when you flatten out the state so that people

---

[473] Celec, P., et al. (2015). 5-9.
[474] Schmitt, D., et al. (2008). 178.

are given the freedom to follow their interests, you simultaneously allow for biological variation to fully express itself.

Such a view makes sense in light of the biopsychosocial model, which integrates the three factors of biology, psychology, and society into a complete framework. If people, like animals, are born with innate behavioral structures that can adapt and evolve to their environment, then such a view that gender differences in personality are partially biological in nature becomes almost common sense. Throughout hundreds of thousands of years of human development, males and females shared unique reproductive functions which produced slight psychological differences in their behaviors and personalities. As societies and cultures grew, innate personalities could also develop into more advanced perceptual structures. Socialization can exaggerate or suppress innate personalities, behaviors, and interests, and yet these factors can only do so much to alter individual behavior.

The enlightening work of the Big Five Model shows us that individual differences in personality abound and are highly nuanced, forming the most diverse category of individual differences, much more diverse than race, sex, or gender. Personality differences set us apart from one another by bestowing upon us unique ways of seeing the world through our own customized perceptual structure. It allows us to solve complex problems by integrating viewpoints different from our own, and once we have understood how unique personalities perceive the world, it gives us the much-needed ability to show more empathy, compassion, and kindness to those who see things differently from us. Such an understanding of personality can give each of us the power to communicate more effectively, solve problems with greater skill and finesse, and lead the world into a better future where we accept each other as we are.

Finally, gender differences in personality show us that, while men and women are much more similar than different across all the traits, there are some traits which show more difference than others, especially on the extremes. The most agreeable people in the world tend to be women, whereas the most disagreeable people tend to be men (as witnessed in the prison population). Understanding these bimodal distributions, that men and women share slightly different averages of a given trait and that the differences are more pronounced on the extremes, allows us to better understand the disparities in occupations across liberal democracies.

Opposite of what social constructionists would posit, perhaps gender differences in average interests lead people to choose different career paths. Maybe occupational disparities between men and women, rather than resulting from discrimination, injustice, and patriarchal oppression, are actually the result of the free choices of individuals, choices which are modulated through genetics, hormones, and the environment? To complete our understanding of how the factors of biology, psychology, and society affect sex and gender differences, we will explore what men and women become interested in as they grow into adulthood and enter the workforce.

According to social constructionists, the workforce has many glass ceilings where women are still not equally represented. And yet, if we allow individuals to choose their own path in life, perhaps such equal gender representation across all fields won't happen the way we think. Such a view is in direct opposition to the postmodern social constructionists who believe that any disparity is the result of injustice and that equity across all hierarchies must be achieved. However, maintaining liberty requires we allow people to make their own life choices, help those in need when they need it, and let people's interests develop on their own. It's a view which continually seems to be revolutionary in our postmodern era, and giving people this freedom to choose will not result in equal outcomes.

Rather, if we choose equality of opportunity over equality of outcome, we will see a world that is highly variable, highly diverse, and highly unique, just like personality. However, if we, like Dr. John Money, wish to mess with the free expression of individual behavior, interests, and personality, we do so at our peril, for what we do against others may be done to us. Because of this, we should strive for a world where any individual, regardless of their sex or gender, can express themselves how they see fit, as long as this expression does not infringe on the rights of others. Such a world may seem like it is impossible to achieve. But understanding the complexity of sex and gender differences and their origins will allow us to more fully accept others as they are and not try to mold them in ways we see fit.

For personality psychology and the Big Five, the true diversity seen across traits shows us that we must embrace the idea that individuals are different; doing so will improve our societies, our relationships, and even our inner life as we accept ourselves for who we are, through our biology, our genetics, our hormones, and the unique characteristics which form our bodies and brains. Embracing the idea that each person is genetically formed into a unique individual is not sexist, racist, homophobic, or transphobic. It is the ultimate idea of Western society, an idea that has produced some of the greatest prosperity in world history, and it must be protected at all costs.

Illiberal forces, which dare to tear down the individual and divide us across categories of group identity, push against the foundations of our liberal democracies, our academic institutions, and our scientific communities. Unlike more conventional means, these illiberal forces can only be fought through language, the strongest tool at our disposal when backed up by reason, science, and evidence. Like the postmodernists have come to realize, it is only through language where such a battle can be won. Knowing this, do not let your words be taken over by ideologues who wish for nothing more than power and control over what you say, do, and believe. *Whatever you do, do not give an inch of ground to them!*

Instead, bring forth your best self; judge people for their values and ideas, not their identities; speak the truth about what you believe with grace and wisdom; and most importantly, champion the sovereignty and diversity of the individual, from which true equality can be forthrightly embodied and victoriously achieved.

13

INTERESTS AND INEQUALITY

After decades of social pressures for girls to express their innate selves without the barriers of discrimination and gender stereotyping, postmodern social constructionists have still not achieved the equity they so desire. Men and women, to the dismay of gender theorists, are still not equally represented across the major occupations of our liberal democracies. It's not that men and women are just slightly underrepresented in gender-atypical occupations; they are still highly underrepresented even in the most prosperous societies. From engineering and computer science to nursing and elementary school teachers, society's most distinguished careers have still not reached parity between men and women, which is to say, fifty-fifty representation across all hierarchies and occupations.

For the social constructionist, patriarchy is to blame for such inequalities: men still hold power over women, even in the most gender egalitarian countries, they say. *Such power maintains and perpetuates both gendered behavioral reinforcement and gender stereotypes which sublimate women's roles into submissive domestic caregivers*, or so they claim. It is not even that every man consciously thinks that he's helping to maintain the patriarchal power structure across society, but that everyone, men and women, are unconsciously complicit in the construction of this unequal hierarchy by continuing to treat boys and girls differently.[475]

---

[475] Johnson, A. (1997). Patriarchy, The System. *The Gender Knot: Unraveling our Patriarchal Legacy*, 96.

While women are underrepresented in male-dominated occupations such as engineering thanks to patriarchal structures, the forces of injustice are even more insidious: when women do enter such fierce and competitive arenas, they so commonly reach the illusive glass ceiling, the unbreakable barrier to their ascension in the upper echelon of occupations. Such forces are ruthless and destructive, halting women's progress and equality. Even with plenty of laws barring discrimination, instituting quota-based hiring practices which push for more women representation, and increasing social pressures on girls to go into male-dominated STEM fields, women's underrepresentation in both male-typical occupations and the upper echelons of society continue to plague social constructionists.

And yet, such a dramatic lack of progress is even worse than at first glance, as feminists have painfully noticed, paradoxically, that other countries with more traditional and patriarchal gender roles have much higher percentages of women in STEM.[476] For example, females make up 67% of science graduates in Iran, an incredible number compared to the West's much lower percentages. Social constructionists feel rather good about such statistics which show that Asia is leading Western societies in the road to gender equality, however, such a statistic does not paint the entire picture. Iran, for example, still *stones* women to death, not just for committing adultery, but for being *raped*.[477] So much for leading the way in women's equality!

Because of this insane paradox (that women are highly represented in Middle Eastern and Asian science fields), things look rather bleak for the social constructionist who, upon any investigation into the inequalities across egalitarian societies, concludes that we still seem to be living in the regressive 1950s, where many women were truly discriminated, suppressed, and halted by patriarchal structures from expressing their true career interests and desires. In the eyes of the radical feminists, we still seem to be living in an oppressive patriarchy. Such terrifying circumstances must

---

[476] A Complex Formula: Girls and Women in Science, Technology, Engineering, and Mathematics in Asia. *United Nations Educational, Scientific, and Cultural Organization (UNESCO).* 38.

[477] Aslan, R. (2017). Stoning a woman in Iran for adultery. *The Daily Beast.*

be overcome: we must double down, treat boys and girls even more the same, and fight even harder for our utopian goal of equity.

As social constructionists do such things, they look out across the most prosperous and privileged societies in history and, rather than professing gratitude for all the material wealth and equality brought forth from the courage of their ancestors, decry the wonders laid out before them as an oppressive patriarchy which must be dismantled. They decry Western patriarchy without understanding the irony that this so-called *patriarchy* has given them the freedom to say such things in the first place. Failing to recognize the unspoken freedoms the West has secured, it goes without saying that a social constructionist feminist sent to live in Iran, for example, would quickly realize where the real patriarchal devils reside.

However, the claims of the social constructionists, those who believe that the West is oppressive, are partially true. There is no discounting the struggles women have faced and continue to face in our Western societies. Discrimination and injustice were common things for women who wished to overcome the shackles of the domestic life and make something of themselves in a career, especially in careers dominated by men. Such extreme circumstances of the 1950s, however, have changed drastically. Contemporary liberal democracies are some of the most egalitarian societies in the world. Institutional barriers towards women's equality, such as discriminatory hiring practices and unfair wages, have mostly been removed not just through the passing of laws but through entire shifts in how society views and treats women.

According to the UN's own indices for gender inequality, the liberal democracies of the West are some of the least sexist, least patriarchal places on earth which have done the most, in legal and social terms, for women's equality. Such institutional measures cannot be denied or handwoven as ineffective against the perpetuation of discrimination. But at the same time, we should not ignore evidence of injustice committed against women, let alone any other person. And this is where the ideology of social constructionism begins to go off-the-rails, in the area of injustice. Any reasonable person would acknowledge true discrimination and injustice, but just what exactly is discrimination and injustice?

Can such inequities be defined as women's underrepresentation in male-dominated jobs, as women's lower average income, or as women's lack of representation in the upper echelons of the Fortune 500 companies? Are these examples of discrimination and injustice? Or are they examples of free and rational choices made by individuals, each with their own interests and desires? In Iran and many other places across the Middle East, women are highly represented in science fields, and yet, this does not negate the fact that they are still living in oppressive patriarchies which stone them to death for being *rape victims*. Therefore, we cannot make broad-sweeping claims about institutional discrimination from mere occupational disparities; we have to instead look deeper to find the root causes for women's underrepresentation in Western STEM fields. Rather than examples of inequity, could these disparities be examples of average differences in men's and women's *interests*--interests which naturally produce some degrees of inequality in occupational representation?

Such a claim, that occupational disparities between men and women may be due to differences in average interests, was the same argument made by James Damore, the software engineer at Google who was promptly fired for daring to claim that men and women might have different interests when it comes to career choices. Using Damore's claim that not all inequality is necessarily indicative of injustice, I will show that the largest sex difference between men and women can be found in the category of *interests*. In doing so, I will argue that occupational disparities between men and women in the most gender-equal societies can be attributed, in large measure, to average differences in interests. Such a claim is the final step towards solving the Gender Equality Paradox.

THE THINGS-PEOPLE DIMENSION

Social constructionists claim that differences in interests between men and women are the result of socialization, that due to differential treatment of boys and girls, children grow up and enter gender-typical and stereotypical careers. Combined with discrimination and injustice, this unequal socialization further divides men's and women's interests. If socialization forces could be removed, then men and women would go into all careers in equal proportions. We would see a decline in the disparity in STEM, women would be equally represented, and inequality would be eliminated. However, such claims can only be true if socialization is the only force affecting men's and women's interests and choices. Instead, I propose the final aspect of the biopsychosocial model: *gender differences in occupational preferences are largely the result of biological, psychological, and social factors which interact to form an individual's interest in a specific profession.*

Gender differences in occupational preferences do not arise from socialization alone but follow a chain of events which originate at conception. Sex chromosomes at conception largely determine the developmental pathway the fetus will go down. For instance, a fetus with an XY chromosome will have the SRY gene, which initiates male sexual development and differentiation from the default female path.[478] Once testes form and begin producing spikes in prenatal androgens, these sex hormones alter the neuroanatomy and neurocircuitry of the fetal brain, causing structural changes such as increased suboptimal arousal and rightward shift.[479] These changes tilt males towards tasks which involve visuospatial cognition and the manipulation of objects. Heavy rightward lateralization causes language to be de-emphasized in exchange for a specialized visuospatial tasks.[480] A girl exposed to high levels of androgens

---

[478] See chapter, "Conception to Birth."
[479] Ibid.
[480] See chapters, "Conception to Birth," "Development of Gender Identity," and "The Activational Stage."

can also exhibit such psychological and neurological traits, as studies on girls with CAH have repeatedly shown.[481] Julia, our fictional girl, also exhibits increased suboptimal arousal and rightward shift. Since she was exposed to high levels of prenatal androgens, she shows more masculine psychological and neurological traits compared to her unaffected sisters: she exhibits more of an interest in mechanical and moving objects and more of an affinity for visuospatial tasks such as manipulating objects in three-dimensions.

Once the neuroanatomy and neurocircuitry has been adjusted through exposure to sex hormones and activation of specific genes, a unique proto-personality develops, through which the infant perceives the world.[482] This perceptual structure grows and develops as the person becomes more mature, and eventually, this structure forms a personality, which can be differentiated into the Big Five dimensions.[483] Each individual's personality, based on genetics, hormones, and the environment, becomes tailored to specific tasks, interests, and desires. On the extreme ends, men and women tend to have very unique personalities which develop into gender-specific interests. We see these interests develop in childhood, as many boys and girls tend to like different activities and toys, while others show gender-atypical preferences. Socialization can dampen or exaggerate innate preferences, but such attempts can only affect a child so far before their innate interests will inevitably rise up from within.[484]

From all the cascades of development (conception to adulthood), such gender differences in preferences then form one of the largest sex differences social scientists know of. This sex difference we call the *Things-People* dimension, which is measured by an individual's interest in *things*-oriented careers versus *people*-oriented careers. Things-oriented careers

---

[481] See chapters, "Conception to Birth," "Development of Gender Identity," and "The Activational Stage."
[482] See chapter, "Personality and the Big Five."
[483] Ibid.
[484] See chapter, "Development of Gender Identity" and the John/Joan Case.

involve scientific, mechanical, and technical activities compared to people-oriented careers which involve social and artistic activities.[485]

In 2008, psychologists at the University of Illinois at Urbana–Champaign and Iowa State University conducted the first ever meta-analysis on sex differences in vocational interests, with over 500,000 participants, an unheard of number in psychological studies.[486] Their goal was to examine sex differences in interests on a broad scale. What they found may not be shocking to the average reader, but for the social constructionists who believe men and women share no differences in average interests, such results were staggering. The researchers found, after aggregating the responses from more than 500,000 people, the *Things-People* dimension had an effect size of about $d = 0.93$, one of the largest effect sizes in all of the social sciences.

Let's pause for a moment to make sure we understand just how profound such an effect size is. This means that men and women differ by almost a full standard deviation in their interest in people versus things, indicating that "only 46.9% of the male and female distributions of interest on the Things-People dimension overlaps."[487] It means that around 82% of males have a stronger interest in things-oriented careers than the average female and that only 18% of females have a stronger interest in things-oriented careers than the average male.[488]

These two dimensions (Things-People) were subdivided into additional factors of interests, which were *Realistic and Investigative (Things) and Social* and *Artistic (People).* These factors, when analyzed separately, illuminate the beautiful complexity of the bimodal distributions between men and women in regards to vocational interests:

> "Men have stronger Realistic, Investigative, and STEM interests, and women have stronger Artistic and Social interests that parallel the Things-People sex difference. These differences

---

[485] Su, R., & Rounds, J., et al. (2009). 861.
[486] Elliott, Z. (2017). *Sex Differences: A Land of Confusion.* Lulu Press. 39.
[487] Su, R., & Rounds, J., et al. (2009). 873.
[488] Ibid.

were large, with the mean effect size of 0.84 for Realistic interests and 1.11 for engineering interests, equal to a 50.9% and 40.7% overlap of male and female distributions, respectively. The mean effect size for Social interest ($d = -0.68$) was moderate, equal to a 58.4% overlap of distributions. In other words, only 13.3% of female respondents were more interested in engineering than an average man, whereas 74.9% of female respondents showed stronger Social interests than an average man. These findings echo Thorndike's (1911) statement that the greatest differences between men and women are in the relative strength of the interest in working with things (stronger in men) and the interest in working with people (stronger in women)."[489]

Such strong differences in interests in *things* versus *people* cannot be ignored. Effect sizes this large show us that many common sense observations in behavioral patterns often have an underlying truth to them: that men tend to be more interested in mechanical and technical activities whereas women tend to be more interested in artistic and social-based activities. Such an observation should not be surprising. However, this is not to say that there are no exceptions.

Of course there are women who are more interested in engineering than the average man, just as there are men who are more interested in nursing than the average woman. For example, girls with Congenital Adrenal Hyperplasia (the condition which exposes XX fetuses to high levels of prenatal androgens) exhibit not just male-typical toy preferences in childhood but male-typical interests in things-oriented careers over people-oriented careers as they grow into adulthood. In most people, the effects of prenatal androgen levels versus gender socialization are often difficult to separate, but for girls with CAH, the effects of androgens are more apparent. Aligning with the research which shows that androgen levels have direct effects on the organization of the fetal brain and contribute to masculinized neural structures, the evidence of girls with

---

[489] Su, R., & Rounds, J., et al. (2009). 873.

CAH provides strong support for the idea that prenatal androgens influence a child's predisposition towards liking *things*-oriented activities over *people*-oriented activities.[490] Despite continual socialization pressures to act like normal girls, women with CAH nevertheless continue to pursue male-typical careers.

Likewise, Julia was exposed to higher levels of prenatal androgens than her unaffected sisters. These androgens formed structural differences in her brain, specifically causing heavy rightward lateralization, which then led to a specialization of the visuospatial system and an interest in mechanical and moving objects. Like normal males, our fictional girl exhibits an interest in more things-oriented activities. Her interest in trains and mechanical objects as a child eventually leads to her a design and engineering profession where she can utilize her skills and satisfy her innate interests. Despite the proliferation of gender stereotypes which may have relegated her to people-oriented careers, Julia's parents encouraged her to pursue her own interests, regardless of what society, peers, or the media told her. Because her innate interests were accepted, not suppressed, she was able to pursue a profession she truly loved.

While such exceptions between girls and boys do exist, they do not negate the profundity of the immense Things-People dimension--a dimension which can help explain many of the occupational disparities we see across Western societies.

Data from the United States Department of Labor illustrates the importance of the Things-People dimension, from which career choices are affected. Dental hygienists, for example, are almost exclusively women (98%), while brick and cement masons are *exclusively* male (near 0% women). For the people-oriented careers, women constitute large majorities as therapists, licensed and vocational nurses, registered nurses, elementary school teachers, and mental health counselors. And for the things-oriented careers, men constitute large majorities as carpenters, mechanics, truck drivers, software developers, and sales representatives.

---

[490] Beltz, A., Swanson, J., Berenbaum, S. (2011). Gendered occupational interests: Prenatal androgen effects on psychological orientation to Things versus People. *Hormones and Behavior, 60,* 313.

The distribution of men and women is more equal in professions which have more of a balance of things-oriented and people-oriented activities, such as physician assistants, accountants, marketing specialists, medical scientists, postsecondary teachers, database administrators, and physicians.

Opposite of what social constructionism would predict, this sex difference in the Things-People dimension only becomes more apparent as a society becomes more egalitarian: *the more gender-equal the society, the greater the gender differences in occupational preferences.*[491] How can such a paradoxical claim be true? Cognitive Social Learning Theory, which states that gender differences are constructed and learned, cannot fully explain this phenomenon. If socialization was the major force on gender differences, even gender differences in occupational preferences, then why do gender differences in personality and interests grow larger in the most gender-equal societies? And why are there far more women in STEM fields in Iran than in Norway? Shouldn't gender differences diminish and minimize as we treat boys and girls more equally?

The answer is this: *gender differences will minimize only if the differences between boys and girls are socially constructed.* If gender differences are not mere constructs, then if we treat boys and girls more equally, we will not get less differences between them; we will see *more* differences between them as the absence of social forces will allow the full expression of innate interests and desires.

Refusing to accept the fact that gender differences may partially originate from biological factors, social constructionists in Scandinavia have doubled down on their ideology by claiming that their society, the most progressive in history, still treats boys and girls differently; *we need to treat them even more equally, they say.* And yet, ironically, they fail to see that the more they treat their boys and girls the same, the more different their boys and girls will be in personality and interests.[492]

---

[491] Schmitt, D. (2017).

[492] Peterson, J. (2019). Men & Women: Personality Differences, *Discovering Personality*.

For people who champion the diversity of individuals, such differences should be welcomed and cheered. However, due to an incredible lack of understanding of any scientific research into sex and gender, strong social constructionists view such differences between men and women, boys and girls, as oppressive constructs which must be eliminated rather than accepted. Such a constricting view on gender differences has profound implications for society, which I will discuss in a coming chapter.[493]

Rather than originating from socialization alone, gender differences in personality and interests are the result of biology, psychology, and society. Biology impacts the Things-People dimension through genetics and hormones, which then affects the psychology of every individual, man and woman. To see just how much biological sex relates to the Things-People dimension, researchers found that biological sex "accounted for 33% of the variance in occupational interests."[494] This means that sex accounts for a third of the variables affecting vocational interests, which is quite large for social science standards. Therefore, it is clear that one's sex has a considerable impact on how one's interests will be oriented.

Whether it's an interest in more things-oriented careers, more people-oriented careers, or a mix of both, it can no longer be denied that there are large gender differences in vocational interests. These gender differences in *interest* then produce occupational disparities across the workforce as men and women choose largely different careers.[495]

---

[493] See chapter, "The Gaslighting Ideology."

[494] Lippa, R. (2010). Sex differences in personality traits and gender-related occupational preferences across 53 nations: Testing evolutionary and social-environmental theories. *Archives of Sexual Behavior, 39*, 619–636.

[495] Peterson, J. (2019). Men & Women: Personality Differences. *Discovering Personality, Lecture 6 Notes*, 4.

WOMEN IN STEM

The Things-People dimension partially explains why women may not be going into STEM as much as men in the West, as it shows that many women are not as interested in things-oriented careers compared to their male counterparts. However, such a sex difference does not explain why women are going into STEM in droves in the Middle East and Asia, especially in countries which still treats them as second-class citizens. Many of these Middle Eastern and Asian societies have the highest scores on the Gender Inequality Index, making them some of the worst places for women in the world. Countries like Malaysia, Iran, Indonesia, and India have horrible ratings for women's equality, with discriminatory policies and practices against females and LGBT people, and yet, these countries have some of the highest rates of female science graduates. From a sociological perspective, such a strange phenomenon can be explained economically, and the answer is actually quite simple.

Though the Things-People dimension still applies to men and women in these countries, showing that men and women still share relatively different interests, there is much less economic opportunity and income mobility in these societies. And so, with few career prospects at their disposal, many women make the rational choice to go into STEM where they have much higher chances of making good money.[496] Other career opportunities which we take for granted in the West are hard to come by in many parts of the Middle East and Asia where options for career choices are scarce.

Thus, rather than disproving the validity of the Things-People dimension across cultures, the low percentage of female science graduates in the West may indicate that many women choose to go into more people-oriented fields rather than things-oriented careers because of greater economic opportunity and gender equality. Put another way, the fact that 67% of science graduates in Iran are female does not show that Iran is leading the way for gender equality, but rather that economic opportunity

---

[496] Stoet, G., Geary, D. (2010). The gender equality paradox in STEM education. *Psychological Science*, 1.

may be low. Women have few options in these traditional, patriarchal, and developing societies, whereas the West has far more opportunities and freedoms which allow people to pursue their own innate interests. **Therefore, it is this critical factor of *interest*, not differences in *ability* or *discrimination*, which can largely explain the gender disparity in Western STEM fields.** It's time for the leaders of our Western nations to realize this, and yet such a realization is unlikely to occur.

Researchers and public policy-makers are hell-bent on getting as many women as possible into Western STEM. Put aside any altruistic motivations for helping women get into STEM, which certainly abound, and instead focus on the fact that social, economic, and political incentives drive the push for women's equal representation across science, technology, engineering, and mathematics. Public policy-makers and researchers can get more grants, more money, and more power and influence by adopting policies which champion women's equality in the STEM fields. This is why such an exploration of sex differences in the Things-People dimension is critical for understanding women's continual underrepresentation in fields like engineering. If we can understand why there is still a lack of women in Western STEM, perhaps we can develop more effective strategies for increasing women's representation.[497]

One such strategy involves socializing our boys and girls in ways that ensures their innate interests can be fully supported and developed. Equally supporting them in their skills and interests can potentially increase women's representation in things-oriented careers and increase men's representation in people-oriented careers. This is not to say that innate interests should be molded for ideologically-desirable outcomes, but rather that innate interests should be bolstered and championed, regardless of whether the outcome between boys and girls is equal. Many studies focusing on such strategies have shown that, while there is a Things-People difference in interest between boys and girls, there is also an average gender difference in mathematics self-concept.

---

[497] Su, R., & Rounds, J., et al. (2009).

Self-concept is the internal understanding of your own ability, an internal structure which often increases in strength as you become more competent in the skill you're developing. Girls tend to have a lower self-concept in math ability than boys, but not for the reason we commonly think. Social constructionists are partially right in the fact that the difference in math self-concept relates to socialization pressures which internalize gender-appropriate abilities in girls and boys. For instance, a boy's skills in math may be especially encouraged and supported, while a girl's skills in math may be discouraged or suppressed. However, socialization is not the sole force at play. Something more psychological is happening underneath the surface which can explain why more males tend to go into STEM.

Social constructionists who decry patriarchy for decreasing girl's confidence in mathematics miss a critical link between *math ability* and *verbal skills*. In turns out that self-concept in math is directly affected by one's self-concept in verbal skills. Because the hemispheres in girl's brains are usually more interconnected, girls tend to excel in *both* math and verbal skills. This is because the abstract logic and concrete linguistic regions are heavily interlinked in females, as opposed to the more lateralized regions in males which show less linkage. Males excel at math just as much as females but, in large measure, lack in verbal skills. Thus, males tend to have a higher math ability self-concept. Let's think about why this might be the case: if one skill is lacking, you are likely to emphasize the one which isn't. If you suck at golf but excel at tennis, wouldn't you put more of your efforts into developing your tennis abilities? In the case of males, verbal skills are lacking, and so mathematics is put on a pedestal. And this process is exemplified in the correlation between self-concept in math and enrollment into STEM: the stronger your math ability self-concept is, the more likely you will enter STEM fields.[498]

For females, however, such a difference in self-concept across math and verbal skills usually doesn't happen, as females excel at both the mathematics side of things and the verbal side of things. Because of this,

---

[498] Lindberg, S., Hyde, J. (2010).

they are more likely to view their abilities in math and linguistics as equally strong. If you view both these skills as equally developed, then what factor will you utilize to decide on career choice? The answer is pure *interest*. If you believe yourself to be equally qualified for math-based and linguistic-based fields, you will decide based on your innate interests and desires, not an internal sense of your ability.

Such claims are backed up by many studies. One study from 2011 focused on differences in interests and how they related to math and verbal skills using 1,490 high school students across the United States.[499] The researchers concluded with four interesting findings which show that the key to understanding the Western STEM disparity is understanding the link between math ability and verbal skills:

> "1) 70 percent more girls than boys had strong math and verbal skills.
> 2) Boys were more than twice as likely as girls to have strong math skills but not strong verbal skills.
> 3) People (regardless of male or female) who had only strong math skills as students were more likely to be working in STEM fields at age 33 than were other students.
> 4) People (regardless of male or female) with strong math and verbal skills as students were less likely to be working in STEM fields at age 33 than were those with only strong math skills."[500]

One can see that, rather than being a female-deficiency in math ability which explains the STEM disparity, it is actually a male-deficiency in verbal skills which somewhat explains men's higher representation in more things-oriented careers. Females are equally capable of mathematics and verbal-based tasks, whereas males tend to be specialized for visuospatial tasks, leaving the verbal and linguistic skills underdeveloped.

---

[499] Wang, et al. (2013). Not lack of ability but more choice: Individual and gender differences in choice of careers in STEM. *Psychological Science*, 1-6.
[500] Jussim, L. (2017). Why Brilliant Girls Tend to Favor Non-STEM Careers. *Psychology Today*.

Knowing this sex difference in self-concept and the large sex difference in the Things-People dimension, we can develop strategies for increasing women's participation in STEM fields and for increasing men's participation in people-oriented careers. For females, one such strategy might involve emphasizing the utilization of verbal and linguistic skills in STEM careers rather than prioritizing things-oriented tasks. If women can realize that STEM careers also integrate heavy amounts of linguistic skills, then their interests in such fields may increase. Therefore, increasing women's interest in STEM is not about increasing their math ability self-concept, which is often already strong, but rather increasing their understanding that verbal and linguistic-based tasks are critical to success in STEM. For males, increasing men's representation in people-oriented careers might involve a discussion on the relationships between people-oriented tasks and things-oriented tasks such as visuospatial cognition and manipulation of objects, ordering systems, and organizing data.

Thus, the large STEM disparity is not from women's lack of ability or their low self-concept in mathematics; rather, it is from men's underdeveloped verbal skills combined with women's greater interest in people-oriented careers. Such differences create the large disparities we see across Western science fields, and before we recognize these factors of differences in interests and unbalanced self-concept in men, we are unlikely to see a change in women's representation in STEM.

SHATTERING THE GLASS CEILING

Beyond the disparity between men and women in STEM, there are additional examples of gender inequality throughout the upper echelons of the West's social, economic, and political hierarchies. As discussed in Chapter 4, women continue to be underrepresented in high-level, highly paid positions and are overrepresented in low-paying jobs; they earn about 82 cents for every dollar a man earns (when only accounting for average earnings across the population by sex), make up only 20% of the top 1% income group, 256 of the world's 2,208 billionaires, and comprise only 5%

of Fortune 500 CEOs, 22% of Fortune 500 board members, and 30% of university presidents.[501]

Such continual lack of equity in the upper strata of the West is known as the *Glass Ceiling,* a term to describe the illusive barriers to women's advancement in a given hierarchy. In 1991, the United States Department of Labor defined it as "artificial barriers based on attitudinal or organizational bias that prevent qualified individuals from advancing upward in their organization into management-level positions."[502]

Social constructionists, radical feminists, researchers, and public-policy makers attribute the Glass Ceiling to bias, discrimination, and injustice; if all societal forces of sexism, bigotry, and misogyny were removed, they say, then equity across all of these hierarchies could be reached, with equal amounts of men and women. If one views the differences between men and women as products of socialization and gender stereotypes, then why wouldn't such a push for equity be, not just correct, but *moral?* Such a belief that men and women should be equally represented across every strata of society is moral insofar as there are no innate differences among men and women, which is to say, that their behaviors, personalities, and interests have no biological and psychological foundations. If, however, there *are* average differences between men and women in their behaviors, personalities, and interests, then such social engineering towards equity would be immoral, as people's individual skills, merit, and choices would have to be squashed through the force of the state in exchange for equal outcomes.

It's true that women are still discriminated against in the workforce, and it's true part of the Glass Ceiling is the result of bias. And yet, it's also true that men and women share average differences in their personalities and interests. Because of this, men's and women's different life choices may not result in the equal outcomes so desired by radical feminists and social constructionists alike.

---

[501] (2019). Gender Economic Inequality. *Inequality.org;* (2018). The data on women leaders. *Pew Research Center, Social & Demographic Trends.*

[502] (2012). The Glass Ceiling Commission. *Australian Centre for Leadership for Women.*

While there are certainly things we can do to help women climb the upper strata, there are other factors which cause the Glass Ceiling to remain in place. One factor comes from women's own life choices. Women tend to be more flexible in their schedules and choose to work less hours, they tend to not negotiate as fiercely as men, and they tend to prioritize work-life balance over the workaholic lifestyles required for success at the upper echelons of a profession.[503] When asked about such preferences, women themselves have responded with agreement.

Pew Research Center asked women (mothers and non-mothers) whether their ideal situation would be to work *full time, part time,* or *not at all* based on four different financial situations: *living comfortably, meeting basic expenses with a little left over, just meeting basis expenses,* or *don't have enough to meet basic expenses.* For those who had comfortable financial situations, women preferred full time work at 31%, part time work at 33%, and no work at all at 34%. In contrast, for those just meeting basic expenses, these women preferred full time work at just 37%, part time at 40%, and no work at all at 22%.

These numbers overwhelmingly show that women, on average, prefer to work part time as much as they prefer to work full time.[504] Thus, this large preference for work-life balance may heavily contribute to disparities between men and women in the upper echelons of the most prestigious jobs.

Knowing that women tend to prioritize work-life balance more than men, perhaps it is average differences in *interest*, rather than discrimination and injustice, which contribute to a significant portion of the gender inequalities across social, economic, and political positions in the West. Of course there are exceptions where discrimination and injustice have contributed to such inequality, but in our contemporary liberal democracies and the most gender-equal societies on earth, it is wrong to claim that all disparities between men and women are the result of insidious patriarchal forces.

---

[503] Allen, T. (2018). Six hard truths for women regarding the glass ceiling. *Forbes.*
[504] Financial well-being linked to views on ideal work situation. *Pew Research Center.*

Thus, perhaps shattering the Glass Ceiling should be less about achieving equity across all hierarchies and more about giving women the freedom and power to live their own lives; to pursue intense and demanding careers; or, if they so choose, to spend more time caring for their children. Whichever path they decide to travel, as long as it's chosen for their own interests, is a path social constructionists and gender theorists should be actively supporting. And yet such a strong support for women's choices is unlikely to arise from the gender ideologues, as rejecting the doctrine of equality of outcome would require the ultimate blasphemy: *that men and women are not the same.*

SEX DIFFERENCES IN SCHOOL AND WORK

Such a claim, that men and women differ, is backed up by not just biological evidence from the natural sciences, but psychological and sociological evidence from the social sciences. When it comes to school and work, these differences become clear. Across 66 research studies, females spend more time studying, attending class, and paying attention in class; they work in more female-typical occupations, are more likely than males to be nurses, and are more involved in parenting and childcare activities.[505] Certain psychological and neurological traits predispose females towards these people-oriented and linguistic-oriented careers. A lack of suboptimal arousal and a lack of heavy brain lateralization means that females get bored less easily and will show a stronger linguistic affinity, perfect traits to have for the modern education system which emphasizes the acquisition of language skills.[506] Because female-typical brains are more interconnected between the hemispheres (linking abstractions and creative-thinking with linguistics), females find school more enjoyable and are therefore more attentive.[507] However, such a sex difference can be equalized as males acquire and develop these linguistic skills as well:

---

[505] Ellis, L. (2011). 556.
[506] Ibid.
[507] Ellis, L. (2011). 556.

"As males slowly come to learn language skills to the same extent as females, or as they take courses that rely less exclusively on left hemispheric function (e.g., advanced mathematics), the sex differences in school enjoyment and performance should diminish. All of this reasoning is consistent with the finding that males have more of an interest in physical science and technology courses and with evidence that males are more often bored. Male boredom should be especially high when required to focus on language acquisition tasks."[508]

While females tend to enjoy school more due to its emphasis on linguistic skills and tend to have a greater interest in people-oriented careers, males often show opposite preferences in regards to school and work. Compared to females, males are more likely to be employed full-time, to work in male-typical occupations, to hold more managerial, administrative, and supervisory jobs; they are more likely to hold corporate executive positions, to hold law enforcement jobs, and to be scientists and engineers.

Like their female counterparts, certain psychological and neurological traits predispose males towards pursuing male-typical careers such as law enforcement or STEM. For law enforcement, work that is often confrontational, men tend to enter in greater numbers partially due to higher levels of *suboptimal arousal*, the neurological tendency to get bored with one's environment.[509] People who have high suboptimal arousal tend to seek out thrilling experiences, sometimes even highly risky ones. Such people are more likely to be males, as prenatal androgens directly influence the degree of suboptimal arousal in humans and mammals.[510] Males are often exposed to significantly higher levels of prenatal androgens than females, creating a sex difference in this neurological trait. For science and engineering, professions which use heavy amounts of visuospatial

---

[508] Ibid.
[509] Ellis, L. (2011). 556.
[510] Ibid., 554.

cognition, men tend to enter in greater numbers partially due to heavier brain lateralization.[511] Lateralization prioritizes the specialization of the visual system, leaving the linguistic connections across hemispheres less developed.[512] Once again, this heavy lateralization is directly related to the amount of prenatal androgens a fetus is exposed to in the womb. The more androgens, the more masculinized the brain structure.[513] Overall, therefore, men and women show average differences in interests: in how they view school, how they prioritize work-life balance, and what career paths they choose to go down.

Knowing these average sex differences in school and work interests should help explain why occupational disparities continue to exist across Western societies despite intense socialization pressures to the contrary. It can partially explain why the Glass Ceiling still remains, and it can explain why women are not fully represented in the upper echelons of society's hierarchies. However, such data into the Things-People dimension and other sex differences in life choices and occupational preferences has not affected the beliefs of western leaders, professionals, and teachers who continually push for more women in STEM.

Ironically, such disparities between men and women have not gone away; instead, the more politicians, researchers, and scientists push for gender equality, the larger the STEM disparity grows. This paradoxical finding, that gender differences in personality and interests grow larger as a country becomes more gender-equal, is the core of the Gender Equality Paradox. For occupational disparities, such a paradox works like this: *the more prosperous and gender-equal a country becomes, the less women pursue STEM careers.*[514]

Such incredible findings are counterintuitive to the social constructionist ideology. Why do gender differences in personality and interests grow larger in the most world-renowned gender egalitarian

---

[511] Ibid.

[512] Ngun, T., et al. (2010). 229.

[513] See chapter, "Conception to Birth" and "The Activational Stage."

[514] Stoet, G., Geary, D. (2010). 10; Khazan, O. (2018). The more gender equality, the fewer women in STEM. *The Atlantic.*

societies? And why does such a disparity grow larger in STEM? Shouldn't more gender-equal countries experience less gender disparities in the workforce? Social constructionists attempt to explain these findings by doubling down on their ideology: *perhaps the system is still sexist and patriarchal, perhaps we still treat men and women differently, and perhaps we need to treat our boys and girls even more the same.* And yet, such attempts to make their societies more accepting, more tolerant, more prosperous, and more gender-equal results in simply more of the status quo: less women in STEM and more women in people-oriented occupations.

Thus, the final defeat of social constructionism can be found not through biological or genetic arguments but simply by presenting average sex differences in *interest*: **when people are given the freedom to live their own lives, to make their own choices without societal pressures, the Things-People dimension becomes the largest cause of occupational disparities.** Men and women, on average, show different interests when it comes to career choices. Such a finding means that the Gender Equality Paradox can indeed be solved, not with low-resolution rhetorical arguments from social constructionists, but with multiple scientific disciplines which integrate biology, psychology, and society, giving us a multidimensional perspective on the paradoxical gender triangle.

And it is this final variable of the Things-People dimension, not socialization, cultural practices, discrimination, or injustice, which solves the Gender Equality Paradox by dismantling the ideology of the blank slate at its foundational doctrines.

13 INTERESTS AND INEQUALITY

Graph: The Gender Equality Paradox in STEM.

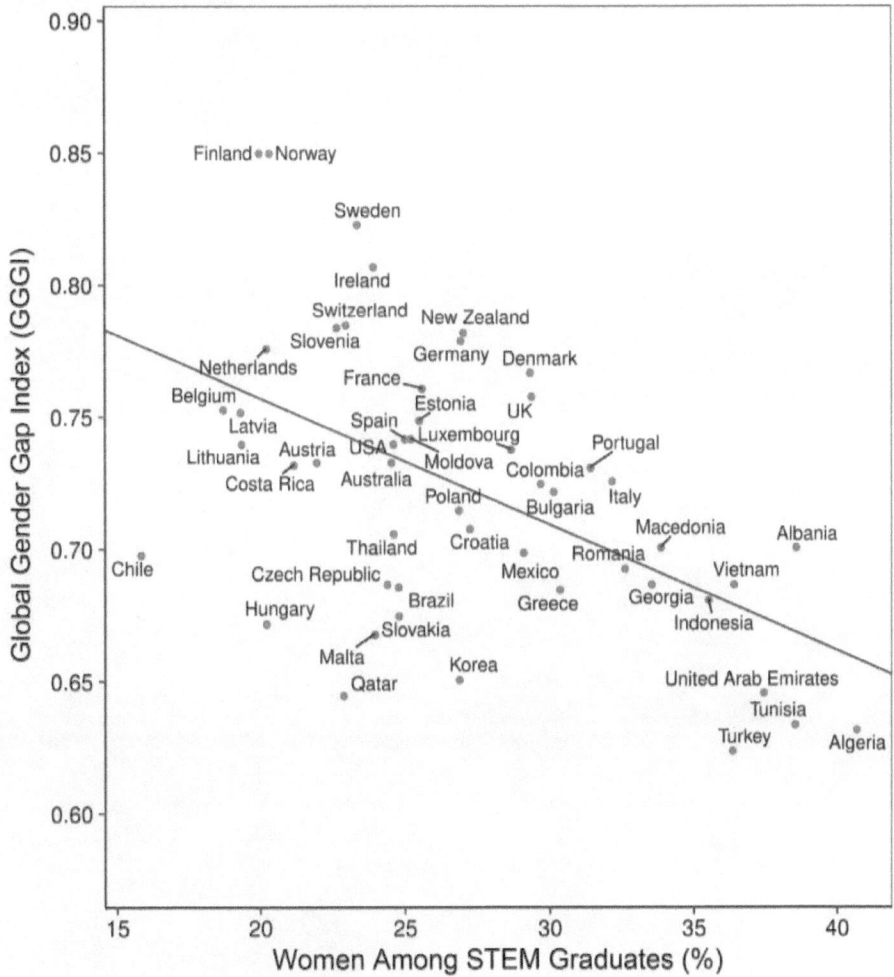

The Gender Equality Paradox in STEM: the more gender-equal the society, the less women in STEM. Source: Stoet, G., Geary, D. (2010).

14

SOLVING THE GENDER PARADOX

Like Sweden and Finland, Norway is among the world's most gender-equal countries, a society which has done the most, in social and economic terms, for women's equality.[515] These factors include reproductive health, proportion of parliamentary seats occupied by females, the proportion of adult females and males aged 25 years and older with a secondary education, and labor force participation rate among females and males aged 15 and older.[516] In fact, Norway is so gender-equal that men and women have the same social and economic foundations for success: "*In Norway, men and women are formally on equal footing: the same laws protect them, they have equal access to education, health, social services, and the same opportunities in the workforce.*"[517]

And yet, despite these landmark steps to becoming the most egalitarian society the world has ever known, giving men and women true equality of opportunity, equity has still not reached the gender-equal utopia of Norway: women comprise only 36.3% of political leaders and 39% of municipal counter leaders, women are still largely working part-time at 41.3%, compared to only 18.8% of men who work part-time.[518] Why does such a disparity still exist in the world's most gender-equal society?

---

[515] Gender Inequality Index (GII). *United Nations Development Programme, Human Development Reports.*

[516] Ibid.

[517] Brother, M. (2015). Gender equality in Norway: Progressive policies and major changes. *HuffPo.*

[518] Indicators for gender equality in municipalities. *Statistics Norway.*

It's even worse than that for the social constructionists who desire to see equal representation of men and women in every field, for the utopia of equity is still light-years away. Despite treating their boys and girls equally, Norway has some of the world's largest gender differences in personality and occupational preferences. In 2016, women made up only 5% of building and construction graduates; 6% of electricity and electronics graduates, 10% of technical and industrial production, 34% of service and transport, 33% of natural science graduates. On the other hand, men made up only 14% of design, arts, and crafts graduates; 17% of healthcare, childhood, and youth development; 32% of music, dance, and drama; 21% of health, welfare, and sport; 27% of education graduates; 37% of social sciences and law, and 40% of humanities and arts graduates.[519]

As these graduates enter the workforce, such disparities between men and women *grow* in magnitude. There are still plenty of male-dominated occupations just as there are many female-dominated occupations. In fact, half of all female employees work in the ten most female-dominated occupations, and one-quarter of male employees work in the ten most male-dominated occupations.[520] A statistical analysis shows us where these disparities exist: women make up only 1% of building frame workers, 1% of building finishers, 3% of moulders and welders, 3% of machinery mechanics and repairers, and 3% of electrical equipment installers. For the male-dominated occupation of engineering, women only make up 20% of engineers.[521] Men, on the other hand, make up only 7% of hairdressers and beauticians, 17% of personal care workers, 17% of child care workers and teaching assistants, and 19% of numerical clerks. For the female-dominated occupation of nursing, men only make up 11%.[522]

Equal representation of men and women was only achieved across five major occupations: medical doctors, legal professionals, university and higher education teachers, architects and planners, and sports and fitness workers.

---

[519] Kristiansen, J., Sandnes, T. (2018). Women and men in Norway. *Statistics Norway*. 11.
[520] Ibid., 16.
[521] Hatfield, S. (2014). Where are all the female engineers? *Independent*.
[522] Kristiansen, J., Sandnes, T. (2018). 16.

What are the typical gender-dominated occupations in Norway?[523]

Occupation Type	Percent Women	Percent Men
Hairdressers, beauticians, etc.	93	7
Nursing and midfwifery professionals	89	11
Personal care workers	83	17
Child care workers and TAs	83	17
Numerical clerks	81	19
Medical doctors	51	49
Legal professionals	51	49
Higher education teachers	49	51
Architects, planners, etc.	49	51
Sport and fitness workers	48	52
Electrical equipment installers, etc.	3	97
Machinery mechanics and repairers	3	97
Moulders and welders, etc.	3	97
Building finishers, etc.	1	99
Building frame workers, etc.	1	99

[523] *Source:* Kristiansen, J., Sandnes, T. (2018). 16.

Such huge occupational disparities between men and women are ubiquitous across Scandinavia and the West, even though our contemporary liberal democracies are some of the most prosperous places for women on the planet. We look out across the Middle East and Southeast Asia to see that women make up the majority of STEM graduates, earning science and technology degrees in record numbers compared to their Western sisters. Even in Iran, an Islamic theocracy which stones women for being raped, women make up 67% of science graduates.[524] Dismayed at the fact that the countries with the most female STEM graduates are the most gender-unequal societies on the planet, social constructionists have become convinced that our liberal democracies are doing something wrong.

What are we doing to cause men and women to segregate themselves into gender-typical professions? And what are the patriarchies of Iran and Malaysia doing right? For social constructionists in Norway, such a question can only be answered through one variable: *socialization*. And it's the one answer we can always count on from the ideologues; it's the univariate answer which can solve all of society's inequalities through the dismantling of men and women as constructed categories. If patriarchal forces are so great and implicit in our unconscious that they have maintained these occupational disparities across the most gender-equal societies, then it will take just as much power to destroy these gender differences.

For the social constructionist, solving the Gender Equality Paradox does not involve allowing men and women to express their innate interests and choices but instead requires a tectonic feat of social engineering: it means more equal socialization, more cultural transformation, and a complete elimination of any and all gender differences. It means treating boys and girls even more equally, even more the same, and socializing them to believe that there are no differences between them, no differences in their personalities, their interests, their desires, and ultimately, their

---

[524] A Complex Formula: Girls and Women in Science, Technology, Engineering, and Mathematics in Asia. *United Nations Educational, Scientific, and Cultural Organization (UNESCO)*. 38.

biology. In the mind of the ideologues, to believe that boys and girls are different reeks of traditionalism, conservatism, and the evils of biological essentialism, and yet such conservative forces are not the leading powers of our time. Social constructionist ideology, from which power and influence accrue, is the dominant force across the upper hierarchies of our liberal democracies. Social constructionist policies and practices are continually being adopted by Western nations to treat boys and girls equally. And ironically, the more the social constructionists try and socialize boys and girls the same way, the more gender differences arise. For example, the sex difference in trait Agreeableness grows larger as a country becomes more prosperous, freer, and gender-equal; in other words, the more free the society, the more men and women express their innate selves.[525]

But instead of adapting to this data, the ideologues dig in their heels; rather than considering that men and women might have average differences in hormone levels, brain structure, interests, and occupational preferences, social constructionists respond to the Gender Equality Paradox by doubling down on their ideology as they sit in the most gender-equal societies on earth: "It shows that the gender roles are still deeply rooted among the young," a Norwegian sociologist said, "We treat boys and girls differently from the very start."[526]

While there is plenty of evidence that we as a society often exaggerate or suppress the innate tendencies of boys and girls, or those who express themselves in non-binary ways, it does not mean that these innate tendencies do not have biological and psychological components to them. To say otherwise is to play into the hands of those who wish to mold individuals in preferential ways, who wish to apply stereotypical and narrow definitions of what it means to be a man or a woman, and who wish to constrain people to the subjective and arbitrary definitions of society. For the health of all individuals, what we need is not a rejection of biology but rather an acceptance of it. As we have explored throughout the book, males and females differ in a variety of ways. From genetics, hormones,

---

[525] Lippa, R. (2010). 628.
[526] Hjernevask (Brainwash), "The Gender Equality Paradox." *YouTube*. 9:00.

neuroanatomy, neurocircuitry, toy preferences, behavior, interests, and occupational preferences, there are real differences between males and females that do not result from socialization pressures alone--differences which can be seen in nonhuman mammals such as mice and monkeys. And while there is considerable overlap between men and women and boys and girls in many traits, there are also areas of critical differences in occupational preferences such as men's interest in things-oriented activities and women's interest in people-oriented activities, in predispositions towards certain neurological diseases, and in tendencies to perceive the world through different perceptual structures.

Thus, the solution to the Gender Equality Paradox requires that we gain more than just one perspective. Social constructionists use only one perspective to solve the paradox. By saying that all gender differences are the result of sociocultural forces, even the most basic phenomena of sex and gender differences fail to be explained. And so if we reject the univariate position of the social constructionists and instead adopt the sophisticated biopsychosocial model, which integrates the forces of biology, psychology, and society, then we can gain a more accurate picture of the paradoxical gender triangle. Adopting such a high resolution model for sex and gender differences allows us to solve the gender paradox quite easily, for we can see the illusion from multiple perspectives.

Solving such an illusion requires no grand theories of insidious patriarchal forces, no social engineering, and no societal-wide transformations. Instead, it requires an investigation into how the intricacies of biology, psychology, and society shape the differences between men and women. Therefore, to solve the Gender Equality Paradox, I will present a review of all the evidence thus far as we condense the biopsychosocial model into its most basic attributes.

THE SOLUTION: A BIOPSYCHOSOCIAL MODEL

Integrating the three major factors of the biopsychosocial model allows us to create an accurate framework for the existence of sex and gender differences, and therefore, solve the paradox. Using this model, we can define biological sex as two functionally unique reproductive capacities (male and female) which create the formwork from which gender differences evolve. Gender can be thought of as the behavior-based mixture which is poured into the mold of biological sex. The biological sex formwork controls the boundaries of gender differences, and the composition of gender differences is created through a mix of the three factors:

1) **Biology** (chromosomes, genes, hormones)
2) **Psychology** (neuroanatomy/brain circuitry, perceptual structure)
3) **Society/Culture** (socialization, cultural practices, social learning)

Throughout the book, following the chain of events which lead to gender differences in behavior allowed us to explore how these three factors interact.

The first factor was **biology**: a person's sex is determined at conception through chromosomes and genetics. These genetics sexually differentiate males and females through the production of sex hormones. Exposure to these sex hormones, combined with sex-specific genes, causes physical and neurological changes which further differentiate males and females. The combination of hormones and genes creates a unique neural structure in the fetus, from which a proto-personality will develop.

The second factor was **psychology**: a person's neuroanatomy and neurocircuitry, wired *in utero*, forms the initial structure from which one's personality can develop and mature. From this personality, certain psychological traits will arise, influencing one's behaviors and interests through genetic and hormonally-defined perceptual structures.

The third factor was **society**: socialization, social learning, and cultural practices influence these personality structures through external behavioral reinforcements and punishments combined with internalization of learned behavior. Here, the innate biological and psychological frameworks formed *in utero* can be exaggerated or suppressed, but not eliminated. Gender stereotypes, generalizations of male-typical and female-typical behavior, can further influence a person's behaviors, either exaggerating innate psychological structures or suppressing them. Therefore, because each individual is born with a pre-existing personality structure modulated by a unique neuroanatomy and physiology, socialization is the tertiary force of many individual differences in behavior and interests, not the primary or secondary.

Because individual differences are the result of biological, psychological, and social forces, the Gender Equality Paradox can be explained in simple terms: *if we eliminate social forces as a cause for gender differences, we allow biological and psychological traits to fully express themselves.* Or, put another way, when you flatten out the sociocultural effects, you maximize the biological and psychological effects. Such a relationship between society and biology can only exist if biology exists.

Thus, we began the book by looking at how social constructionists defined biological sex as an arbitrary and subjective category. The social constructionists used arguments from indefinite variation which show chromosomal, hormonal, and gonadal abnormalities; they used arguments from pseudo-hermaphroditism, showing that non-functioning vestigial reproductive structures can exist in the opposite sex; and they used hormonal variability to claim that male and female cannot be easily defined. However, despite the claims of social constructionists, we showed that male and female *can* be reliably sorted into two distinct categories. While there is plenty of variation in sexual characteristics, there are only two reproductive capacities (large gametes and small gametes). No matter how you look at it, there are only two capacities for sexual reproduction, defined as the large gametes known as eggs and the small gametes known as sperm. In genetics, we found that the SRY gene is the primary activator which initiates male sexual development from the default female pathway, forming testes from the ovaries. Sex hormones produced in the womb further differentiate the fetus by promoting the development of male or female reproductive structures. Sometimes, as in the case of intersex, these pathways are slightly altered, creating abnormalities in external genitalia and hormone levels. We explored such cases throughout the book, looking at how girls with Congenital Adrenal Hyperplasia exhibit ambiguous genitalia because of high prenatal androgen exposure. Furthermore, we showed that removing the testes from a developing male in utero causes him to develop as a female, providing evidence for the importance of the gonads for sexual development. We showed how sex hormones such as testosterone create a uniquely masculine neural structure, increasing suboptimal arousal and lateralization, which then increases risk-taking behavior and affinity for visuospatial tasks.

Next, we looked at how these sex hormones affect the gender development of the individual and their preference for male-typical versus female-typical toys. We rejected the ideas of Dr. John Money, who believed infants were born psychosexually neutral and could be molded at whim by societal forces, and instead, we championed the idea that one's gender identity is directly linked to genes and hormones. Separating genes and hormones from gender identity and instead linking gender identity to the appearance of the genitals was shown to be harmful and dangerous to a child's healthy development. It is not the appearance of the genitals which makes a boy and girl who they are; rather, it is their genes and hormones, as Dr. Money's failed experiments showed.

We proved that sex hormones have a direct effect on a child's toy preferences regardless of what sex they were. Males exposed to normal levels of androgens exhibited male-typical play preferences and behaviors; perhaps more importantly, we showed that females exposed to abnormally high levels of androgens also exhibited the same male-typical preferences and behaviors despite socialization pressures to act like a normal girl. We also showed that these girls with Congenital Adrenal Hyperplasia rarely experience gender identity disorder and instead identify as normal girls. Additionally, evidence from rhesus macaques and vervet monkeys showed parallel toy preferences with those of children, providing another piece of evidence that sex hormones and one's biological sex influence male-typical versus female-typical interests and behaviors.

On the psychological front, we showed how gender stereotypes can turn from mere generalizations of behavioral patterns and into pathological ways of thinking. We provided evidence that our Western societies still proliferate many caricatured stereotypes of the masculine and feminine archetypes. These stereotypes influence an individual's behavior to some degree, potentially exaggerating or suppressing innate tendencies. We explored how integrating the masculine and feminine archetypes into an individual's psyche can dismantle the narrow stereotypes of what it means to be a man or a woman, and therefore, relink the healthy syzygy of the masculine and the feminine which our culture has so heavily tried to separate. Lastly, we argued that no matter what sex you are, you should be able to express your innate interests and desires. Girls with CAH, for example, mostly pursue male-typical careers, even though they have two X chromosomes. The higher exposure to prenatal androgens means they will forever have male-typical psychologies. Attempts at using the tertiary factor of socialization to change their behavior towards female-typical interests almost always fails. Because of this, such an abnormality should be accepted rather than suppressed. Rather than trying to socialize them into female-typical interests, we argued that they should be allowed explore what they're interested in with the amount of freedom necessary to produce healthy psychological development.

Next, we looked at how the stage of puberty activates dormant neural circuits formed *in utero* to create additional sex differentiation between males and females. Specifically, we looked at how sex hormones during puberty initiate physical changes in the body and the brain, and how these changes influence behavior and interests. We showed that there is indeed a male-typical and female-typical brain using evidence from neuroendocrinology and brain mapping technologies. These fields of research showed us that males and females have different levels of gray and white matter, different amounts of sex hormone receptors in specific areas of the brain such as the amygdala and the hippocampus, and we showed that exposure to sex hormones can rewire entire brain regions.

Furthermore, we showed that the male and female brain is lateralized differently. Males showed heavy lateralization, or connection within each hemisphere, and females showed heavy interconnectivity, or connection between the hemispheres. These sex differences in neurological connectivity altered how the brain emphasized tasks. In males, spatial ability was heavily specialized, whereas in females, linguistics and abstractions were heavily linked across the hemispheres. Lastly, we looked at how sex can now be predicted through looking at the brain with up to 97% accuracy.

As we moved further into the psychological factors of gender differences, we found that cross-cultural meta-analyses showed large gender differences favoring women in trait Extraversion, Agreeableness, and Neuroticism, and that these differences tend to grow larger as a country becomes more gender-equal. We looked at how these gender differences in personality traits might be formed through both biological and social factors. Evolutionary biology explains these differences through males and females unique reproductive capacities and sex roles which, over time, formed different perceptual structures. And yet, we showed that sociological factors also played a role in the development of gender differences in personality, as society's division of labor between men and women may have also exaggerated trait-based differences.

Finally, after integrating the personality psychology research, we studied gender differences in occupational preferences across the world and how these differences might influence gender disparities in occupations. We found that the largest psychological sex difference between men and women was the Things-People dimension: *men tend to be more interested in things-oriented activities and women tend to be more interested in people-oriented activities.* The effect size was $d = 0.93$, one of the largest effect sizes in the social sciences, showing that only 18% of women were as interested in engineering as the average man. Such a large difference was explained psychology and neurologically through the wiring of the brain: men tend to be more lateralized and developed for visuospatial tasks, systematizing, and organizing data, while women tend to be more interconnected and developed for both verbal tasks, linguistics, interaction with people, and logical analysis.

Lastly, we showed that the disparity in Western STEM fields is the result of two major factors. The first factor is women's much lower interest in things-oriented activities such as those in STEM. The second factor is women's equally strong self-concept in math ability and verbal skills compared to men's low self-concept in verbal skills and high self-concept in math ability. We showed that the more you are developed across two skills, the more likely you are to pick the skill you are most interested in, regardless of ability. Therefore, we found that women tended to pick people-oriented careers out of pure interest, even though they believed themselves to be, and were, equally good at mathematics. We found that men's perceived verbal deficiency combined with their interest in things-oriented activities led men to pursue STEM careers in much higher amounts than their women counterparts. We showed that socialization can certainly play a role in whether women and men pursue gender-atypical occupations. For increasing women in STEM, we proposed that people-oriented tasks be emphasized to show that verbal skills and interacting with people is a critical aspect of STEM. Likewise, for increasing men in people-oriented careers, we proposed that things-oriented tasks be emphasized to show that visuospatial cognition and data systematizing may be incredibly useful for certain people-oriented occupations. And finally, we argued that disparities in Western STEM can be largely attributed to average gender differences in occupational preferences, not merely discrimination, injustice, or stereotyped gender roles.

To condense our biopsychosocial model into a concise paragraph, we can say that *a person's sex is determined by their genetics, which then activates hormones. The hormones then form physical and neurological differences between males and females. These neurological differences produce different perceptual structures, or personalities, through which the facts of the world are perceived. These personalities then exhibit unique interests, desires, and values which can be exaggerated or suppressed through social forces. These unique interests then produce disparities in society as individuals choose their own unique life-paths. Average gender differences in interests and life choices then produce many of the disparities we see across the most gender-equal countries on the planet.*

It is this multivariate solution, not the univariate explanations of social constructionists, which accurately attributes the causes of the Gender Equality Paradox to each individual's own innate interests and choices-- choices which are influenced through biological, psychological, and sociocultural forces.

And yet, despite the mountain of evidence we presented for the influences of biology and psychology in the production of gender differences, it is only a small fraction of the entire field of research. From evolutionary biology and neuroendocrinology to psychology and sociology, there are thousands of additional studies, experiments, and meta-analyses which elucidate the complex interaction of the biopsychosocial factors in sex and gender differences and similarities.

To put an end to the Gender Equality Paradox for good, I will dismantle the last remnants of strict social construction ideology by showing 64 universal differences between males and females across the world--differences which originate from the complex interaction of biology, psychology, and society. These sex differences are observed across almost all cultures and societies, showing that many differences between males and females in interests and behavior are not mere social constructs but are products of evolutionary and psychological forces extending across the eons. Many of these sex differences grow larger as a society becomes more prosperous, freer, and more gender-equal because people are given more freedom and opportunity to pursue their own interests and express their innate selves; such is the solution to the Gender Equality Paradox.

The following tables were produced by anthropologist and sociologist Lee Ellis, who developed an integrative model known as the Evolutionary Neuroandrogenic Theory (ENA) to explain over 60 universal sex differences in perception, cognition, behavior, and interests. ENA states that sex and gender differences arise through genetic, hormonal, and social learning factors which shape male and female behavior. Similar to the biopsychosocial model, Ellis's theory integrates many factors into a complex and multidimensional analysis. With the continual evidence from biology and psychology, social constructionists cannot continue to argue that sex and gender differences are the result of social forces alone.

UNIVERSAL SEX DIFFERENCES

Table 1. Universal sex differences involving stratification and work and occupations.[527]

Universal sex difference	# of studies	# of nations
1) Females spend more time studying, attending class, and paying attention in class.	15	7
2) Males are more likely to be employed full-time.	31	11 (6)
3) Among employed persons, males work longer hours.	19	5 (1)
4) Males work in male-typical occupations more.	42	12 (2)
5) Females work in female-typical occupations more.	22	8 (2)
6) Males hold more managerial, administrative, and supervisory jobs.	58	8 (4)
7) More corporate executive officers are male.	18	7 (1)
8) Males hold more law enforcement jobs.	14	4
9) Females are more likely to be nurses.	12	7
10) Males are more likely to be scientists and engineers.	21	4
11) More college professors are males.	18	7 (2)
12) Females are more involved in parenting and childcare activities.	17	7 (1)

*The number in parentheses refers to the number of individual studies which sampled from more than one country.

---

[527] Ellis, L. (2011). 557.

Table 2. Universal sex differences in consuming behavior and interactions between individuals and social institutions.[528]

Universal sex difference	# of studies	# of nations
13) Males begin alcohol consumption at an earlier age.	12	2
14) Males are more likely to binge drink.	15	6
15) Self-reported property crimes are more common among males.	20	6 (1)
16) Males carry weapons more.	11	4
17) More males engage in sexual assault.	10	1

---

[528] Ellis, L. (2011). 557.

Table 3. Universal sex differences in social and play behavior.[529]

Universal sex difference	# of studies	# of nations
18) Females are more likely to establish and maintain eye contact in social interactions.	10	2
19) Females provide more care to their offspring.	38	9 (6)
20) In the case of divorce, females retain primary custody of their children more.	14	4
21) Females interact socially in small groups more.	23	6
22) Females are more likely to have intimate friendships.	35	5 (1)
23) Females confide/share secrets with friends more.	31	6
24) Males engage in more conversations focused on leisure and work.	14	2
25) Males play more with masculine toys.	13	2
26) Females paly with dolls and stuffed animals more.	16	3
27) More females play house and parenting.	12	3
28) Males are involved in directing competitive types of sports more.	30	5
29) Males initiate sexual intimacy more.	13	2

[529] Ellis, L. (2011). 557.

Table 4. Universal sex differences in personality and general behavior.[530]

Universal sex difference	# of studies	# of nations
30) Males explore their environments more.	24	8 (1)
31) Females are friendlier.	11	4
32) Males are more hostile.	14	2
33) Males take more risks in career and business decision making.	17	3
34) Females are more feminine.	30	9 (1)
35) Females are more likely to use a "cradling" book-carrying style.	11	1
36) Females diet more.	40	9 (1)

---

[530] Ellis, L. (2011). 557.

Table 5. Universal sex differences in preferences, attitudes, and interests.

Universal sex difference	# of studies	# of nations
37) Liking school in general is more characteristic of females.	21	8
38) Males have greater interest in physical science and technology.	16	4
39) Males are more accepting of aggression as a response to social problem solving.	23	3
40) Males have more interest in watching sporting events.	15	6
41) Males have more interest in participating in sporting activities.	18	6
42) Males have more interest in athletic activities in general.	11	4
43) The desire for promiscuous sex/numerous sex partners is greater for males.	15	4 (1)
44) Males have a stronger sex drive/are more interested in engaging in sex.	30	7
45) Females are more inclined to prefer mates who are taller than themselves.	11	2 (1)
46) Females are more likely to prefer mates who are high in social status.	20	5 (3)
47) Males prefer mates who are younger than themselves more.	38	15 (1)
48) Females prefer mates who are older than themselves more.	11	8 (3)

Table 6. Universal sex differences in cognition and mental illness.[531]

Universal sex difference	# of studies	# of nations
49) Males have more learning disabilities/difficulties.	12	5
50) More females assess their weight to be excessive (or want to lose weight).	39	7
51) Females more frequently attribute failure to "internal" factors (i.e., their own shortcomings).	10	2
52) Females ruminate over negative emotional experiences more.	21	5 (1)
53) Females experience more psychological distress/mental problems in general.	14	4
54) Problem drinking/alcohol dependence is more common among males.	14	3
55) Anorexia nervosa is more common among females.	25	8
56) Females have higher rates of bulimia.	30	9
57) Panic disorders are more common in females.	29	5
58) Males exhibit higher rates of ADHD/hyperactivity.	51	8
59) Males have higher rates of autism and Asperger syndrome.	28	8
60) Psychoticism is higher among males.	12	8

[531] Ellis, L. (2011). 558.

Table 7. Universal sex differences involving perceptions and emotions.[532]

Universal sex difference	# of studies	# of nations
61) Females estimate/perceive greater hazards and injury risks in their environments.	14	6
62) Females have more stress/anxiety associated with providing care to others.	10	2
63) Males are more prone to boredom.	11	4
64) Females cry more than males as adults.	12	4 (1)

---

[532] Ellis, L. (2011). 559.

IDEOLOGY AND THE DENIAL OF SCIENCE

When confronted with these 64 universal sex differences in behavior, interests, cognition, and perception, strong social constructionists will not accept the fact that a considerable number of these sex differences may have originated from more primordial sources of biology and psychology. Instead, the social constructionist doubles down by claiming that these differences are simply the result of socialization forces. Instead of honestly considering the evidence for sex and gender differences and its profound implications for understanding individual behavior, the strong social constructionist and gender theorist sees these differences between men and women as resulting not from biological, psychological, and sociological forces but instead from linguistic power structures which maintain and perpetuate men's domination and control over women.

And yet, as medical technology continues to advance and researchers continue to adopt interactionist approaches for understanding the complexity of sex and gender, approaches which integrate a variety of factors, the theory of social constructionism and the ideology of the blank slate continue to fade into irrelevance in the natural sciences. Unfortunately, such ideological theories are not fading away in the social sciences, media, and the public. In fact, the doctrines of social constructionism maintain their grip over our society's social, economic, and political institutions, and this grip has only gotten stronger.

As our liberal democracies rightfully attempt to protect the rights of all individuals from discrimination, social constructionist ideologues have gotten ahold of our culture's zeitgeist. As these ideologues grab onto the fabric of our institutions and pull them into the darkness of ideological rigidity, the extreme social constructionists and gender theorists continually gain more and more power in both scientific and political positions, instituting laws and policies which enact the very beliefs of Dr. John Money: that boys and girls are not born different, but *made* different.

Rather than simply giving men and women the equal opportunities to make their own choices and express their innate selves, social constructionist ideologues choose to not let this happen: *there must be equal representation of men and women in every hierarchy of society; the doctrine of equity must be achieved.* No matter how damaging or regressive to society, the doctrines of social constructionism must be instituted. If quotas must be implemented to get equal representations of women in STEM, so be it. If we must discriminate against men in favor of women so that we can eliminate any and all gender gaps, so be it; if we must redefine what a man and a woman is, dismantle the category of biological sex, and force our manipulative ideology down the throat of the public, then so be it. If we have to gaslight society, make people think that sex and gender are constructed categories which can be changed at whim, then perhaps we can finally achieve our great utopia free of discrimination, injustice, and oppression.

Like any ideology, extreme social constructionists, postmodernists, and radical gender theorists use name-calling, verbal attacks, manipulation, and gaslighting to get what they want, and what they want is *you* to shut up--to stop saying that male and female exist, to stop believing that gender differences are partially biological, and to stop thinking that the Gender Equality Paradox is the result of innate interests unimpeded by social forces.

Whatever you do, do not shut up, and do not stop speaking the truth about sex and gender differences and similarities. This should be done not just for the health of society, but for the psychological and sexual health of children who have been damaged and abused by the ideology of social constructionism. Without an understanding of the science of sex and gender, our society will continue to descend into the depths of this confused sexual mess, and the ideologues will continue to gain more power, more control, and more influence over our institutions, our laws, and our societies.

15

CRITIQUING SOCIAL CONSTRUCTIONISM

Before we delve into the case studies of how postmodern social constructionists utilize gaslighting tactics to argue that both sex and gender are social constructs, we must review the major claims social constructionists make about sex and gender. We should not ignore the fact that social constructionists communicate important truths in every topic about sex and gender, and while some of the things they say are true, these truths can only go so far. For this section, I will review seven social constructionist claims we explored throughout the book and briefly analyze where they're right and where they go wrong.

BIOLOGICAL SEX IS A SOCIAL CONSTRUCT

For social constructionists, the first step to eliminating all gender gaps seems to be the deconstruction of biological sex as a category. In some sense the binary definitions of what it means to be male and what it means to be female are, as the social constructionists say, *bullshit*. In the biological sex realm, we began the book by discussing intersex infants and the argument from indefinite variation.

Social constructionists utilize the cases of intersex individuals to argue that biological sex is a subjective category defined by society; they argue that sex exists on a spectrum, not a binary; and finally, they claim that the existence of intersex proves their arguments. Social constructionists are right that there is a variation of chromosomal, hormonal, and genital abnormalities and that these abnormalities can be surprisingly common compared to other conditions.

"Science is constantly evolving and learning," said one enthusiastic Twitter user, "There are a number of chromosomal disorders and we are discovering more every year. We haven't even scratched the surface of understanding the effects. People can be XY, X, XX, XXY, XYY, XXX, XXXX, XXXXY, XXXXX, and many more."[533]

Social constructionists are also right that societally-defined criteria for what constitutes a male and what constitutes a female has been unwillingly pushed onto intersex individuals. For example, medical practitioners of the past often castrated young boys who had ambiguous genitalia and raised them as girls. They thought that healthy sexual development was independent of genes and hormones and was instead directly tied to whether someone had "normal" looking male or female genitalia. Because of this, intersex infants faced continual surgery to correct their genitals, even though in some instances, these genitals did not need to be corrected. It was therefore the definitions of society, not of biology, which relegated a considerable amount of intersex people to narrow definitions of what it means to be a man and what it means to be a woman. Just because your genitals are different than a normal man or woman does not make you less of a male or female. And here the social constructionists are partially right: there is a level of variation when it comes to sex, and societal definitions for male and female have often been too narrow.

Where social constructionists go wrong is not in the recognition of sex variation or the fact that disorders of sexual development exist; **they go wrong when they claim that the existence of these abnormalities proves that biological sex exists on a spectrum.** Here there is a failure of definitions. Biological sex, by definition, can only function as a binary system: you either produce large gametes (eggs) or small gametes (sperm). The sperm is a small mobile cell which fertilizes the egg. Therefore, if you have the phenotypic structures to produce large gametes, you're a female, and if you have the phenotypic structures to produce small gametes, you're

---

[533] The Mad Scientist. (2019). In response to intersex infants. *Twitter*.

a male. Within these two categories of male and female, you have a wide range of sexual characteristics that can develop from genetic and hormonal differences (such as different variations of sexual characteristics, i.e., breasts, facial hair, and genitals).

It is therefore sexual *characteristics* which exist on more of a spectrum, not biological sex. The category of biological sex is by definition a binary system. We can only say it is not binary if intersex people constitute new sexes, and to constitute new sexes, you'd have to produce a new type of gamete outside of eggs and sperm. And unfortunately, for social constructionist arguments, no such third gamete has ever been formed.

Like everyone else, the sex of intersex individuals is determined at conception. Thus, they are not new sexes but are rather males or females whose development pathways are slightly altered by either chromosomal or hormonal abnormalities--conditions which produce somewhat ambiguous structures such as a fused labia and enlarged clitoris in girls with Congenital Adrenal Hyperplasia (XX with abnormal levels of prenatal androgen). Just because they have genitals which have been masculinized does not mean they are not girls; it just means they have unusual genitals. These girls still have female gonad structures which can produce eggs from the ovaries.

Therefore, intersex infants are still male or female; they just have abnormal genotypic and phenotypic structures, and that's okay! We as a society should not judge individuals who have experienced unusual development paths, nor should we preemptively relegate them to a strict male or female identity through genital surgery. Instead, we should recognize that they are abnormally developed females or abnormally developed males who may exhibit contrasexual psychological qualities and interests. For example, girls with CAH usually exhibit male-typical interests, and people still correctly recognize them as girls. Just because they have abnormal genitalia and prefer male-typical things does not make them *less female*.

In the past, abnormal genitalia was viewed as hermaphroditic (outside of male or female), and thus something which must be medically corrected. Because of this, abnormal genitalia was dealt with through surgical intervention, sometimes 'assigning' an individual to the wrong sex. Or, in the words of an intersex person:

> "In the past, the idea that people with DSDs were 'not male or female' was the justification used for unnecesary surgeries to assign a sex. Back then, the word used was 'hermaphrodite,' these days 'the sex spectrum' is just the same old nonsense repackaged."[534]

Finally, when it comes to how common these disorders of sexual development are, it really does not matter if intersex conditions constitute a small or large proportion of the population. Regardless of what genotypes or phenotypes they have, they do not form new sexes. Despite this, social constructionists continue to argue that there are more sexes than male and female as if they have made some profound scientific discovery; such a discovery, of course, does not exist. Category errors like this are a common thing to see from those who do not fully understand what biological sex is. Saying biological sex is binary does not deny the variation within males and females when it comes to chromosomes, hormones, and genitals; or, as developmental biologist Dr. Emma Hilton describes it:

> "Nobody is saying that sex development is not complex. Nobody denies molecular and anatomical variation within sex. To interpret physical variation (normal/pathological) within sex as additional or multiple sexes is a category error. Is there a third gamete, third reproductive role, and third sex?"[535]

The second problem of saying that biological sex exists on a spectrum results from the conflation of gender identity variation with disorders of sexual development. Though intersex infants have chromosomal, hormonal, and genital abnormalities, most intersex individuals identify as

---

[534] Intersex Facts. (2020). In response to 'sex is a spectrum.' *Twitter*.
[535] Hilton, E. (2019). In response to intersex infants. *Twitter*.

male or female and exhibit either male-typical or female-typical psychological traits; most do not identify as transgender. Most transgender people, by comparison, do not have disorders of sexual development (such as intersex) but rather have disorders of gender identity. Some may not have such disorders and may simply want to present themselves in gender-atypical ways. Other trans people have actual psychological disorders, similar to anorexia, that interfere with healthy development. Therefore, because you cannot conflate intersex individuals who have disorders of sexual development with transgenderism, it is time for social constructionists to stop using intersex infants for their own ideological agenda.

To be accurate when it comes to the topic of biological sex, social constructionist arguments for the spectrum of sex should be focused on the variation of sexual characteristics such as chromosomes, hormones, and genitals, not biological sex itself, which is always either male or female through the two unique gametes. Thus the social constructionist argument can be refined to this: *biological sex is binary, whereas sexual characteristics within the binary exist as bimodal distributions.* **You can therefore simultaneously say that biological sex is binary while recognizing that plenty of variation exists within the two categories.**

> "We can create as many sex spectrums as we like by taking any arbitrary characteristic that varies within and between sexes. Hormones, height, beard length -- but all we are doing is showing a distribution of these characteristics, not a distribution of sex."[536]

And here we should recognize the major problem with conflating variations of sexual characteristics with sex itself. **If we define biological sex through variations in sexual characteristics, we can easily relegate people to sexist definitions of what it means to be a man and what it means to be a woman.** If you have unusual male genitals, does that make you a different sex altogether? Does that make you less of a male? If you have

---

[536] Lewis, A. (2019). In response to intersex infants. *Twitter*.

smaller breasts and a larger bone structure as a female, does that make you less of a woman? Or even more offensive: if you lost your breasts to cancer, does that put you closer to the male end of the spectrum?

> "Each of those characteristic spectrums would be extremely offensive if taken as a sex spectrum. Is a man a female because he is short in stature? Are you less of a woman if you have small breasts?"[537]

By conflating sexual variation with biological sex itself, social constructionists ironically relegate people, and intersex individuals, to rigid definitions of what it means to be a male or female, thereby reinforcing the stereotypes they supposedly wish to dismantle. Because of this, social constructionists are stuck between a rock and a hard place: if they wish to dismantle the category of biological sex and replace it with spectrum-based definitions, they put themselves in danger of 1) reinforcing sexist stereotypes and 2) essentializing people's biological sex to mere sexual characteristics.

Thus, attempts to deconstruct the category of biological sex by arguing from indefinite variation will always end up failing. *Sex characteristics vary within males and females, but the category of male and female does not.* In light of these arguments, consider the three questions below. If biological sex is a social construct, then…

> 1) What is the 'sex' with which a trans person's 'gender identity' does not align?
> 2) Why take hormones, have surgery, and perform other measures to mimic the opposite sex?
> 3) How is it that only females produce ova and only males produce sperm?[538]

---

[537] Kiernan, K. (2019). In response to intersex infants. *Twitter*.
[538] KProtein19. (2019). In response to intersex infants. *Twitter*.

GENDER IS A SOCIAL CONSTRUCT

This is perhaps the most overused and clichéd trope of our contemporary culture. Of course gender is socially constructed! But such a statement is too broad and ill-defined. It's spoken as if it has some profound meaning, yet the exact meaning of the phrase disintegrates immediately after being uttered. We need to be more specific what we mean when we say *gender is a social construct*. Gender can be divided into many things, from sex roles and self-expression to beliefs and attitudes around masculinity and femininity.

Social constructionists are correct when they show specific elements of gender which are socially constructed. Gendered fashion is one example, where differences in clothes between men and women change over time and across cultures. In fact, at one point during the Enlightenment Age, high heels were predominately worn by men to signify nobility and status, and these high heels were not symbols of femininity but of masculinity and power. Beliefs and attitudes about gender roles and gender expression also change over time and across cultures. Cultural practices continually regulate what it means for males and females in a specific culture to express their gender. No person who understands the diversity of cultural practices would deny the fact that the features and aesthetics of gender expression can change across cultures.

Social constructionists are correct when they show the ubiquity of gendered products--products which are often unnecessarily gendered. For example, selling large masculine earplugs versus small feminine ones would be ridiculous, and yet such practices can be easily found across our societies where gendered toys are commonplace. These products are often not based in actual sex differences, but are rather based in stereotypical and overly gendered notions of male and female behavior.[539]

Social constructionists are correct when they show how behavioral reinforcement and punishment can affect certain behaviors. It would be wrong to say that boys and girls are treated the same in every aspect of their lives. Some supposedly 'innate' differences between boys and girls are

---

[539] See chapter, "Philosophy of Gender."

indeed socially constructed rather than the result of genetics and hormones. Famous researchers into the sociology of gender, West and Zimmerman, called this process the 'construction of gender essentialness.'

> "Doing gender means creating differences between girls and boys and women and men, differences that are not natural, essential, or biological. Once the differences have been constructed, they are used to reinforce the 'essentialness' of gender."[540]

This construction of gender essentialness socializes boys and girls to think that the differences between them are innate. Certain aspects of behavior are reinforced and solidified in this way, and these reinforcements often show up through gender stereotyping, which judges the behavior of individuals through generalizations of behavioral patterns. We can see this through how boys and girls often socialize one another to stay within gender-typical boundaries. If one dares to stray from typical gender norms, peers will often marginalize those who differ. Thus, believing that gender differences in behavior are all fixed and binary ignores the wide diversity of behavior *within* and *between* males and females, and it ignores the fact that men and women overlap in many traits.

When it comes to whether men and women are more alike than different, the social constructionists are partially right. On an aggregated level, 78% of gender differences between men and women are zero to small, and only 22% of gender differences are moderate to large.[541] Gender and Women's Studies professor Janet Hyde produced the meta-analysis which provided these statistics, and she calls it the *gender similarities hypothesis*.[542] I agree with this hypothesis. Across all the categories, men and women are more like than different.

However, in saying that gender is socially constructed and completely separate from sex, the social constructionists miss the mark. While certain

---

[540] West and Zimmerman, (1987). 138.
[541] These are *effect size* metrics which are statistically significant.
[542] See chapter, "Exploring Gender Differences."

aspects of *gender expression* are socially constructed, there are important gender differences between males and females which exhibit themselves through behavior, personality, interests, and psychology. These differences are often more apparent in the extreme ends of the distribution than the average and grow larger in more gender-equal nations.

The first major category, gender identity (sense of one's own gender), is not socially constructed nor is it independent of genetics and hormones. Social constructionists believe that boys and girls are not born different, but are rather *made* different. Such statements were backed up by Dr. John Money, a sexologist who believed infants were born psychosexually neutral and that the appearance of their genitals determined their gender identity. His experiments on intersex infants and on a normal and healthy male infant proved his theory wrong. Though the boy was raised as a girl from infancy with a fake vagina, the boy never accepted his socialized gender identity regardless of all the behavioral reinforcement from physicians, parents, and psychiatrists who said he was a girl, who gave him girl-typical toys, and who forced him to dress in girl-typical clothes. He never felt like a girl and always thought he was a boy, preferring male-typical toys and clothes. Thus, the boy's gender identity, rather than being socially constructed, directly related to his genetics and hormones, which were all male-typical. It turned out that genetics and hormones were not independent from gender.[543] Believing that gender identity was socially constructed led to the mutilation, castration, and psychological abuse of thousands of children throughout the 1970s and 1980s.

Social constructionists get it wrong when they say that gender identity is a social construct. Nonhuman mammals show toy preferences which parallel those of children. Rhesus macaques, for example, show large sex differences in their preferences for toy trucks versus dolls, providing evidence that gendered toy preferences may be the result of more biological factors. Further evidence for the effects of genes and hormones on gender identity comes from girls with Congenital Adrenal Hyperplasia. These

---

[543] See chapter, "Development of Gender Identity."

girls, as I discussed throughout the book, were exposed to high levels of prenatal testosterone. Throughout infancy, childhood, and adulthood, these girls consistently exhibited male-typical psychological traits such as an interest in things-oriented activities and a higher spatial ability than their unaffected sisters. In fact, in almost every aspect these girls with CAH were much more like males than females, and yet, most still identified as females. This shows that one can have male-typical traits and still be a female, and vice-versa; prenatal hormones have a large organizational effect on the brains of boys and girls, which then produce gender differences in toy preferences, behavior, interests, and occupational preferences.

Thus, rather than saying *gender is a social construct*, social constructionists would do better by being more specific. What aspects of gender are socially constructed (such as fashion, expression, and attitudes) and what aspects of gender are more based in genetic and hormonal factors (such as specific toy preferences, interests, and occupational preferences)? And so, knowing this data, social constructionists could instead say "certain aspects of gender are socially constructed; other aspects are not." But, unfortunately, such a statement doesn't have quite the ring to it.

While it is true social systems have been constructed around gender expression and behavior, this does not mean that all aspects of gender are socially constructed. Rather, as I have argued throughout the book, gender is the mixture which is poured into the mold of biological sex. It's how biological sex expresses itself through behavior. Or, in the words of evolutionary biologist Heather Heying:

> "No, gender is not a human construct; it's not something we created just for us. It's the behavior that goes along with all of the foundational, anatomical, physiological, chromosomal, and genetic factors that came before. Is it much more fluid? Is it much more variable, especially in modern times with modern economies? Are some of the traditional expectations of gender roles outdated, unnecessary, and frankly, alarming? Yes, for sure.

We can throw a bunch of those out, but pretending that they are not based on millions of years of evolution is *batshit* crazy.

Now are there as many ways to express gender as there are humans on the planet? Yes. But then that's not useful as a category; categories are about generalization. Categories in biology always have fuzzy boundaries, but it doesn't render the categories untrue or meaningless."[544]

Psychology professor Christopher Ferguson also reiterates the fact that gender differences are shaped by genetic, hormonal, and social forces, and not socialization alone:

> "Ultimately, the mantra that 'gender is a social construct' is misleading and may cause significant confusion and unnecessary acrimony. It is more reasonable to suggest that gender is an internalized sense of masculinity/femininity that is shaped by a complex interaction of genetic, hormonal and social forces. Granted, that's probably harder to fit on a coffee mug. But I remain optimistic that if we are realistic about the complex interplay of biology and environment, we can work toward an egalitarian and open society that allows individuals to express their individuality whether or not they conform to traditional (or progressive) gender role norms."[545]

GENDER STEREOTYPES ARE PATHOLOGICAL

It is in gender stereotype pathology where the strongest social constructionist arguments reside. Gender stereotypes are often harmful phenomenon where gendered behavioral patterns are generalized and viewed as fixed and binary. Male and female behaviors are then judged according to these patterns. Individual behavior, being more variable

---

[544] Heather, H. (2019). Bret Weinstein and Heather Heying: Gender Ideology vs Biology. *Rebel Wisdom, YouTube*, 10m00s.
[545] Ferguson, C. (2019). Is gender a social construct? *Quillette*.

within populations than *between* them, makes stereotyping a slippery slope.

Social constructionists are correct when they say that gender stereotypes can be harmful. Male and female behavior has much more variation within males and within females than between males and females. Because of this, gender stereotyping occurs when we apply generalizations to individuals. For example, telling a girl she shouldn't go into engineering because she's female is *stereotyping*; it's relegating an individual to narrow views of male or female behavior. Gender stereotypes have historically held people back from expressing their true selves, and such stereotypical ways of thinking about male and female behavior can quickly turn pathological.

Social constructionists are correct when they say that gender stereotypes are common throughout our culture. The media and entertainment industries continue to push the ideology that masculinity and femininity are completely separate and distinct categories. Masculine behavior in females is punished, whereas feminine behavior in males is often punished even more severely. These societal stereotypes of what it means to be a man and what it means to be a woman are naturally constricting, prejudicial, and sexist. And it's damaging to an individual's psychological development.

Social constructionists are correct when they say that certain gender stereotypes are culturally constructed. Certain aspects of gendered behavior are defined through culture, and the more superficial gender roles and stereotypes change across time and culture. Stereotypes about gendered fashion, gendered roles, and gendered expression are all defined through cultural norms and must be ignored by those who wish to live their own lives.

However, social constructionists get it wrong when they say masculinity and femininity are defined through culture alone. Archetypes of the masculine and the feminine have existed across cultures and across time in the collective unconscious of human beings. These archetypes, or ideals of the masculine and the feminine, can be observed through studying images, symbols, and mythologies of past cultures. The masculine and feminine archetypes arise throughout cultures and throughout time as interconnected syzygies. A syzygy is an interconnected and corresponding pair of something, often occurring in nature. Examples of syzygies are the moon and sun, yin and yang, chaos and order, and even the feminine and the masculine.

Analyzing the stories, images, and symbols across human history gives us actual definitions for femininity and masculinity which are rooted much deeper than culture. The feminine is what renews, transforms, and brings new life to old structures, yet it can also be what devours and what kills.[546] On the other hand, the masculine is that which orders, creates through spirit, and forms unrealized potential into actualized reality through *logos,* or logic, and yet, it too has its pathological sides. The masculine can turn constrictive, hierarchical, and tyrannical.[547] Both males and females exhibit qualities of the masculine and the feminine. And our culture often relegates masculine qualities to males and feminine qualities to females. In reality, the feminine must be integrated into a man for him to become a complete individual, and reversely, the masculine must be integrated into a woman for her to become complete as well. Rather than separating the masculine/feminine syzygy, our culture needs to embrace the fact that they must be integrated together. Otherwise, breaking the syzygy will result in a pathological *split-brain,* where you get too much of either masculine or feminine qualities.

Social constructionists get it wrong when they themselves apply stereotypes to masculinity and femininity. Social constructionists embrace the very stereotypes they claim to want to dismantle; namely, that masculinity is about power and dominance and femininity is about submissiveness and timidity.[548] By conceptualizing the feminine as a submissive behavioral pattern in response to masculinity's domination and power, social constructionists end up reinforcing stereotypical beliefs about men and women, thereby hurting the very cause they are supposedly championing.

---

[546] See chapter, "Gender Stereotype Pathology."

[547] Ibid.

[548] Connell, R. W. (1987). *Gender and power: Society, the person and sexual politics.* Cambridge: Polity, 188.

Social constructionists get it wrong when they apply gender stereotypes to individuals. A girl who likes male-typical toys and dresses like a boy is a *tomboy,* not a *boy.* Just because a female likes male-typical things does not make her a male, and it does not make her less of a female. The same is true for boys: just because a male likes female-typical things does not make him a female, and it does not make him less of a male.

"In this context..." said evolutionary biologist Heather Heying, "if I were growing up now the way I grew up, someone would be labeling me *trans,* because I was a tomboy who far preferred dirt to dresses and math to spending time in museums. '*Oh, well that's gendered...*' Since when is math a 'male' thing? That's not okay. Yes, more *men* end up choosing mathematically-intensive careers, but are you now going to say that if I, as a woman, am interested in math, I am therefore acting *male*? That's actually regressive, sexist, and backwards. It's actually regressive, sexist, and backwards to make a claim that by not wanting to wear a dress or girly clothes you are secretly a boy. No you are not."[549]

SOCIALIZATION AT PUBERTY MAKES BOYS AND GIRLS DIFFERENT

Puberty is a time of increasing changes in boys and girls as they mature to become men and women, and it's also a time where socialization pressures and hormones make boys and girls more different. Social constructionists use the activational stage of puberty to argue that increasing gender differences are due to socialization.

Social constructionists are correct when they say that socialization at puberty influences the behavior of boys and girls. Certain aspects of behavior at puberty can be heightened through socialization and peer

---

[549] Rebel Wisdom. (2019). Bret Weinstein and Heather Heying: Gender Ideology vs Biology. *YouTube.* 13m20s.

pressure. Guys may pressure other boys into doing aggressive things, while girls may pressure other girls in similar but distinct ways. As boys' and girls' bodies become sexually mature, this further differentiates them, and socialization can easily affect certain gender differences in behavior at puberty.

Social constructionists, however, are wrong when they claim that there are no differences between boys and girls which arise through genetics and hormones at puberty. Besides the obvious morphological differences in physical characteristics, there is increasing evidence from the frontiers in neuroendocrinology that sex hormones during puberty activate dormant neural circuits for sex-differentiated behavior. These neural circuits were organized *in utero* through exposure to sex hormones and activation of sex-specific genes.

Activational effects at puberty cause structural differences in the brains of boys and girls. These structural differences involve heightened levels of suboptimal arousal and lateralization in men, which represents itself through men's higher interest in risky occupations such as law enforcement or visuospatial/things-oriented careers such as engineering. The structural sex differences in the brain are documented through differences in gray and white matter volume, amygdala and hippocampus size, cortical morphology, and differential activity of brain regions. After puberty, someone's sex can be predicted through the brain with 97% accuracy.

GENDER DIFFERENCES IN PERSONALITY DO NOT EXIST

When it comes to personality, social constructionists are wary to say that differences between men and women in personality traits exist. If they agree that some differences exist, they will claim that most or all these traits are socially constructed.

Social constructionists are correct when they say that men and women are more alike than different across personality traits. Each of the Big Five traits (Extraversion, Agreeableness, Conscientiousness,

Neuroticism, and Openness to Experience) show significant amounts of overlap between men and women. Because of this, predicting the levels of any given trait in an individual simply through gender is not entirely reliable. (60% of women have higher trait Agreeableness than the average man, and 40% of men have higher trait Agreeableness than the average woman.) Thus, there are plenty of exceptions to the general patterns of personality traits that it is often hard to nail down the traits of an individual based on gender. On almost all the distributions of traits, men and women show similar average levels and plenty of overlap.

Social constructionists are wrong when they say that gender differences in personality are not important. While men and women share significant overlap in personality traits, it is often at the extreme ends of the bimodal distributions where the differences are most apparent. For trait Agreeableness, the distributions of men and women are shifted so that, on the tail ends, the most agreeable people are almost all women and the most disagreeable people are almost all men. These differences in personality cannot be ignored because, when examined on a society-wide level, they can help to explain generalized patterns of behavior between men and women on the extremes. For example, 93% of prison inmates are men. This disparity is, in large part, due to the fact that the most disagreeable people (defined through trait Agreeableness) are almost always men.

Social constructionists are wrong when they say that gender differences in personality vary significantly across cultures. Gender differences in each Big Five trait stays consistent in favoring men or women across cultures. For example, Extraversion, Agreeableness, and Neuroticism are all higher in women across all cultures. However, these gender differences in personality do not minimize as countries become more gender-equal; *rather, these personality differences grow larger in magnitude the more gender-equal the country becomes.*[550] This fact alone disproves social constructionist arguments which say that all gender differences in personality are culturally constructed.

---

[550] See chapter, "Personality and the Big Five."

GENDER DIFFERENCES IN OCCUPATIONAL PREFERENCES DO NOT EXIST

If gender differences in occupational preferences can be shown to not exist or to result from sociocultural forces, then perhaps the gender gaps in Western STEM fields can be closed. Because of this, gender differences in occupational preferences are critical to understanding gender gaps across career fields.

Social constructionists are correct when they say males and females show considerable overlap in occupational preferences. Men and women do not all go into different fields. There are women who are interested in more things-oriented careers such as engineering just as there are men who are interested in more people-oriented careers such as nursing. Thus, there are plenty of exceptions to the generalized patterns of occupational preferences. However, the fact that the exceptions exist do not disprove the fact that gender differences in occupational preferences are significant.

Social constructionists are wrong when they say men and women do not show significant differences in occupational preferences. Analyzing the occupational interests of over 500,000 individuals, researchers concluded that one of the largest differences between men and women was the Things-People dimension. Men tend to be more interested in things-oriented activities and women tend to be more interested in people-oriented activities. While there is overlap between men and women in this dimension, the sex difference remains large:

> "Men have stronger Realistic, Investigative, and STEM interests, and women have stronger Artistic and Social interests that parallel the Things-People sex difference. These differences were large, with the mean effect size of 0.84 for Realistic interests and 1.11 for engineering interests, equal to a 50.9% and 40.7% overlap of male and female distributions, respectively. The mean effect size for Social interest ($d = -0.68$) was moderate, equal to a 58.4% overlap of distributions. In other words, only 13.3% of female respondents were more interested in engineering than an

average man, whereas 74.9% of female respondents showed stronger Social interests than an average man.[551]

Social constructionists are wrong when they say that these differences in occupational preferences cannot explain the gender disparities in occupations. In the West, people have the highest economic opportunities in the world. Western societies are ranked among the world's top gender-equal societies, providing equal support and encouragement to men and women to pursue what they're interested in. The reason why gender disparities in occupations are larger in more developed nations is because the more freedom you give to individuals, the more they will go into careers that interest them rather than careers that will simply earn them money. Thus, occupational preferences between men and women largely explain the gender disparities in occupations across Western societies.

GENDER GAPS ARE INDICATIVE OF INJUSTICE

Gender gaps in occupations and earnings are often used by social constructionists to argue for the existence of systemic discrimination against women. Gender gaps are always seen as indicators of injustice, no matter the actual origins of the gaps.

Social constructionists are correct when they say that some gender gaps are indicators of injustice. Discrimination against women certainly existed and continues to exist even in Western societies. For instance, some occupational disparities may be the result of men not hiring women in certain fields.

Social constructionists are wrong when they say that *all* gender gaps are indicative of injustice. Gender gaps should be evaluated with context. Many gender gaps across the West (such as the wage gap and STEM gap) are mostly examples of women's free choices when it comes to life and work, not injustice committed against them by employers or society.

---

[551] Su, R., & Rounds, J., et al. (2009). 873.

Rather, combining women's different preferences for work-life balance and their average interest in more people-oriented careers explains much of the disparity in occupations across Western nations. For example, Norway is ranked among the world's top gender-equal nations, and yet, women make up minorities of engineers. Such a gap can be mostly explained through the sex difference in the Things-People dimension of occupational interests, not socialization or discrimination more broadly.

In review, **biological sex is not a social construct**. Variation in sexual characteristics does not constitute new sexes. **Gender is not entirely a social construct.** Certain aspects of gender expression are constructed, but other aspects, such as an interest in male-typical versus female-typical toys, are often formed from genetics, hormones, and brain structures. **Gender stereotypes can be pathological and should be used with extreme caution.** The masculine and feminine should be integrated, not separated. Furthermore, **personality differences between men and women *do* exist and are cross-culturally universal (they grow larger the more gender-equal the country becomes). These personality differences inform the gender differences in occupational preferences and the Things-People dimension, from which many gender disparities in occupations originate**. Despite all the evidence I have provided throughout the book, strong social constructionists tend to ignore scientific evidence completely. Instead of adjusting their beliefs to the evidence, the social constructionist doubles down on their ideology: w*e must try harder to eliminate any and all gender gaps.*

When each of these seven arguments for the deconstruction of sex and gender fails, radical social constructionists switch to more *postmodern* tactics of manipulation and control to make their opponents shut up: rather than using rational discourse backed up by evidence, the gender ideologue employs gaslighting strategies to make you think that the realities of sex and gender do not have any biological origins.

16

THE GASLIGHTING IDEOLOGY

When biologists, psychologists, and even sociologists show you that your ideology is devoid of logic, reason, and evidence, what do you do? If you cannot use the tools of science to argue for the validity of your beliefs, then you must employ other strategies to convince people your beliefs are true descriptions of reality. Distorting science to fit your agenda, claiming the West is an oppressive patriarchy, and shutting down voices of dissent by slinging meaningless buzzwords of sexism, homophobia, and transphobia are all strategies used to confuse, manipulate, and silence the public into accepting your ideological beliefs. The postmodernists believed that such *fatal strategies* must utilize rhetoric, aesthetics, paradox, and contradiction to overcome the reason and science of the logocentric Western ethos. If our society's faith in the power of reason, logic, and science can be destabilized, then perhaps the patriarchal power structure can be overthrown. It's not hyperbole to say that this is the mainstream view of the postmodern social constructionists and the zeitgeist of our time.

Postmodernists believe that such fatal strategies must be used to restructure power hierarchies along more equitable lines, thereby eliminating all gender gaps between men and women:

> "Men's lives are structured around relationships of power and men's differential access to power, as well as the differential access to that power of men as a group," wrote sociologist Michael Kimmel, "Our imperfect analysis of our own situation leads us to believe that we men need more power, rather than

leading us to support feminists' efforts to rearrange power relationships along more equitable lines."[552]

It does not matter whether these gaps originate from the free choices of individuals; what matters is that these gender gaps continue to exist across Western societies, and for the postmodern social constructionist, such gaps must be destroyed. Equality of outcome must be achieved. And yet, as I have shown throughout the book, achieving equality of outcome is impossible because individuals are not blank slates. We cannot mold people in ways we see fit; instead, we have to give people the equal opportunity to live their best lives and express themselves for who they are. Thus, some gender gaps in the most gender-equal societies, like Scandinavia, are not due to patriarchal oppression but rather due to men's and women's own choices and inclinations, the freedom with which they are able to live their lives and pursue their interests.

And yet, the idea that men and women might have average differences when it comes to life choices and career preferences is akin to blasphemy for the social constructionist who, while using rhetorical strategies and aesthetic arguments, believes that all gender gaps are indicative of injustices which must be eliminated. Such a view is anathema to the idea of individual liberty--that through equal opportunity, people will often make different life choices. However, for the social constructionist ideologue, there is no room for the effects of more primordial sources of biology and psychology when it comes to explaining differences in individual choice. Instead of adapting their theory to the mountain of scientific evidence I have presented throughout the book, the ideologue will switch to different tactics altogether.

When you have abandoned reason and discourse in favor of your ideological beliefs, you are not left with much to base a philosophy off of, and it is here, in the depths of ideological deception, where postmodernism

---

[552] Kimmel, M. (2000). Masculinity as homophobia: Fear, shame, and silence in the construction of gender identity. 218.

comes to save you. In Chapter 3 (The Postmodern Tyranny), we explored the philosophical framework of postmodernism, the ultimate theory of rejecting theories, the judgment rejecting judgments, and the truth rejecting truths. We discovered that postmodernism was the underlying philosophy of strong social constructionist arguments--especially the arguments which claim that the differences between males and females are the result of an unequal power structure designed for domination and control. For the postmodernist and their social constructionist ideologues, everything is about *power*. And when rational discourse is abandoned for ideology and ignorance, then *power* is all that's left.

When social constructionists abandon reason, logic, and science, they turn to the postmodernist values of power, social control, and gaslighting to shut down disagreement, to stifle any voices of dissension, and to twist scientific evidence in their favor so that they can convince you your beliefs are not just wrong, but *bigoted, abusive,* and *immoral*. These attacks are not simply directed at people who share actual hateful views; rather, these attacks are often directed at those who champion the very causes the strong social constructionists say they believe in. Why? Because ideologues often turn on their own. *Everyone must believe the Emperor is fully clothed.*

If you cannot use rational discourse and scientific evidence to bolster your beliefs, then you must turn to the gaslighting tactics of the postmodernists. Gaslighting is a form of psychological manipulation used to gain power in which a person, or entity, makes someone question their own reality. Gaslighting tactics are common among abusers, narcissists, cult leaders, and even dictators. People who are possessed by an ideology will often use gaslighting to get what they want, and it is in such ideological possession where the true dangers lie. Whether these tactics are utilized with conscious awareness or remain in the unconscious depends on the person and the context, but many ideologues nevertheless gaslight their victims into believing that their own perceptions are wrong.

Gaslighters use many tactics to manipulate. They might tell outright lies; deny they ever said something; use your identity against you; wear you down over time; do things which don't match their words; utilize positive reinforcement to confuse you; use their knowledgeable disposition to make you rely on them; accuse you of the very things they are doing; align others against you; cultivate distrust; make you question your own sanity; and get you to think that everyone else is a liar.[553] In the 1944 movie, *Gaslight*, a man manipulates his wife to the point she thinks she's losing her mind.[554] Those possessed by an ideology also utilize such tactics, and no one is immune to it. Thus, it is critical to understand how radical social constructionists gripped by ideology may consciously or unconsciously employ *gaslighting* when discussing the realities of sex and gender. One would think such tactics are uncommon, and yet, as I will show, social constructionist ideology is so pervasive in our institutions that gaslighting, rather than rational discourse, has become the cultural norm. We will now explore this phenomenon through three unique case studies.

J.K. ROWLING AND THE DENIAL OF SEX

Denying that male and female exist as biological realities has quickly become a common gaslighting tactic among social constructionist ideologues. If they can make you think that biological sex is as socially constructed as gender, then their ideas have rhetorical power. They argue that biological sex exists on a spectrum by pointing to examples of intersex infants who have chromosomal, hormonal, and genital abnormalities. This gambit is designed to direct your attention away from the central issue (that you cannot change biological sex) and make you think that biological sex is much more complicated than simply male or female. To bolster the effectiveness of such an argument, social constructionist ideologues will utilize the complexity of biological sex determination to make you question whether male and female are real categories. After hearing such an argument from social constructionists, a person unfamiliar with the science of sex determination may conclude, out of fear of being seen as a science-denier, that biological sex exists on a spectrum and not a binary (male-female). However, such a conclusion is not backed up by any science and

---

[553] Sarkis, S. (2017). 11 warning signs of gaslighting. *Psychology Today*.
[554] Ibid.

is instead supported through conflating the variation of male and female sexual characteristics with biological sex itself.

This tactic of utilizing intersex individuals to dismantle biological sex originated in queer theory of the 1990s, especially in the writings of postmodernists Judith Butler, Fausto-Sterling, and Michel Foucault, where biological sex was defined as a constructed linguistic category used to oppress marginalized groups.[555] Such a theory is now being used to gaslight the public into believing that biological sex, not gender, is socially constructed.

Take, for example, the case of Maya Forstater and J.K. Rowling. In late 2019, a feminist named Maya Forstater lost an employment tribunal case which would have allowed her to take legal action against the global think tank who fired her in early March. Forstater was terminated because she had written multiple articles critical of the proposed reformations to the UK's Gender Recognition Certificate (GRC).[556] In the past, the GRC was used for transgender individuals who were diagnosed with gender dysphoria and needed to change their sex for legal recognition.[557] It allowed them to change their sex after going through the necessary medical and legal processes.

However, the GRC never reached the level of signatories the UK government had predicted, and so, to increase the amount of GRC applicants, the UK proposed a reformation to the GRC which would allow anyone who "self-identifies" as a woman or a man to be legally recognized as a female or male.[558] This meant that anyone, regardless of biological sex, could legally change their sex without medical documentation.

---

[555] Fausto-Sterling, A. (2000). *Sexing the body: Gender politics and the construction of sexuality*. Basic Books.

[556] Forstater, M. (2019). International development: Let's talk about sex. *Medium*.

[557] Hinsliff, G. (2019). Maya Forstater's case was about protected beliefs, not trans rights. *The Guardian*.

[558] Government Equalities Office. (2018). Reform of the Gender Recognition Act 2004. *Gov.uk*.

Forstater argued this change would radically expand the legal definition of 'women' to include anyone who wishes to identify as one, regardless of any medically documented gender dysphoria:

> "I share the concerns of @fairplaywomen that radically expanding the legal definition of 'women' so that it can include both males and females makes it a meaningless concept, and will undermine women's rights and protections for vulnerable women and girls."[559]

While arguing that this expansion of the GRC was too broad and could affect the rights of females, Forstater still affirmed that no one should be discriminated against for their identity:

> "One argument for changing the definition of women from being about sex to being about gender identity (or gender expression) is one of compassion and inclusion. It is polite to treat people in the way they feel comfortable. People should be free to express their identity. No one should be forced to conform to gender stereotypes or victimized for the way they dress."[560]

While recognizing the importance of treating everyone with respect and dignity, Forstater argued that biological sex should not be ignored. She believed that biological sex is a material reality which should not be conflated with gender; that being male or female is an immutable biological fact, not a feeling or an identity; that sex matters for the legal protections of females; and that it is important to be able to talk about sex so that discrimination, violence, and oppression of women and girls can be fought.[561] Because recognizing biological sex matters for the legal protection of females, Forstater further argued that the GRC's loose definition of gender identity was problematic:

---

[559] Forstater, M. (2019). In response to the GRC reform. *Twitter*.
[560] Forstater, M. (2019). International development: Let's talk about sex. *Medium*.
[561] Forstater, M. (2019). Claimant's witness statement. *Medium*.

> "Someone who has simply declared a self-identified gender identity remains physically no different from any other person of their sex--for example in relation to pregnancy risk, sex-based patterns of offending, and the privacy and dignity of others. It is important to recognize that people cannot literally change sex, in the safeguarding of children and vulnerable people."[562]

For saying such things, she was fired. The reason: *for denying the legal recognition of trans-people's identity*. Once fellow staff found the content of her criticisms offensive, Forstater's employer decided not to renew her contract.[563] To them, she was erasing the existence of trans identities by saying that they could not change their biological sex; while Forstater agreed that trans individuals can still *identify* as the opposite sex through proper medical and legal means, she did not agree that biological sex could be changed through mere subjective feeling.

After presenting her case at the employment tribunal, the Tribunal deemed Forstater's opinion that biological sex cannot be changed as a belief which is "not worthy of respect in a democratic society."[564] The Tribunal said that such a belief that there are "only two sexes, male and female...is incompatible with the human rights of others that have been identified and defined by the ECHR (European Court of Human Rights)."

In other words, to say that *male and female are biological realities which cannot be changed* is not just incorrect, but an idea which is *not worthy of respect in a democratic society*. Because of this, Forstater's claim of wrongful termination was denied. And this is where the ideology of social constructionism begins to gaslight the public: in the realm of biological sex. After the outcome of the case broke the news, the author of the Harry Potter series, J.K. Rowling, tweeted her support:

---

[562] Forstater, M. (2019). Sex and Gender, What I Believe. *Hiya Maya*.
[563] Hinsliff, G. (2019).
[564] Newman, D. (2019). Forstater vs CGD Europe, What the tribunal actually found. *A Range of Reasonable Responses*.

"Dress however you please. Call yourself whatever you like. Sleep with any consenting adult who'll have you. Live your best life in peace and security. But force women out of their jobs for stating that sex is real? #IStandWithMaya #ThisIsNotADrill."[565]

Rowling didn't know it yet, but she had just kicked a hornets' nest. In just a matter of hours, her tweet received hundreds of thousands of likes, tens of thousands of retweets, and tens of thousands of responses. And yet, while such a tweet seemed innocuous enough, this was the tweet which launched Forstater's entire case into the mainstream. It was here where the social constructionist ideologues launched their first strike.

The Human Rights Campaign quickly responded to J.K. Rowling with the first gaslighting tactic: *make you question your own belief.* "Trans women are women," the HRC wrote, "Trans men are men. Non-binary people are non-binary. CC: JK Rowling."[566] By saying transwomen are women and transmen are men, the Human Rights Campaign utilized the manipulation of language to make people question their own belief that biological sex cannot be changed. Another human rights group, GLAAD, responded by saying, "JK Rowling has aligned herself with an anti-science ideology that denies the basic humanity of people who are transgender. Trans and non-binary people are not a threat to women, and to imply otherwise puts trans people at risk."[567] Here, GLAAD utilized an *appeal to authority* gaslighting tactic to convince others that Rowling's innocuous statement was actually scientifically illiterate and bigoted.

Forstater and JK Rowling were not denying that individuals can identify as the opposite sex. And in this sense, transwomen *are* women insofar as they *identify* as one. What Forstater and Rowling did deny, however, was that biological sex could be changed--and it was this belief that directly challenged the social constructionist ideology of the blank slate. As Rowling's tweet in support of the reality of biological sex got more and more attention, the ideological gaslighting continued.

---

[565] Rowling, J.K. (2019). In response to the Maya Forstater case. *Twitter.*
[566] Human Rights Campaign. (2019). In response to J.K. Rowling. *Twitter.*
[567] GLAAD. (2019). In response to J.K. Rowling. *Twitter.*

In response to Rowling's tweet saying 'sex is real,' a handful of doctors, scientists, and biologists said that those who believe biological sex is binary (male and female) are either ignorant or lying. Here the social constructionist ideologues utilize appeals to authority to convince you that your commonly held belief is actually scientifically illiterate. Here are two examples:

1. "Those who believe that biological sex is binary are ill-informed and ignorant of an overwhelming body of clinical research that proves it is an outdated and false belief," wrote one doctor.[568]

2. "Biological sex is very complicated, not binary," a biologist wrote, "and there is extensive variation within the sexes (this was literally the topic of my dissertation). Trans women are women."[569]

While it is reasonable to use appeals to authority if you can integrate scientific evidence, such a tactic can become a form of gaslighting in the absence of evidence. Thus, *there is no peer-reviewed scientific paper which will describe biological sex as a 'spectrum.'* While there is certainly a variation of sexual characteristics within the category of male and female, there are still only two sexes (one produces ova, the other produces sperm). When it comes to disorders of sexual development (DSDs), intersex infants are simply males or females who have had developmental abnormalities. Each intersex infant still has either a male internal reproductive structure or a female internal reproductive structure, and most of them identify as male or female as they grow up, since abnormalities such as masculinized genitals or androgen excess have no effect on their gender identity.[570]

---

[568] Cilona, J. (2019). In response to J.K. Rowling. *Twitter.*
[569] Danish, L. (2019). In response to J.K. Rowling. *Twitter.*
[570] Berenbaum, S., Bailey, J. (2003). 1102-1106; Dessens, A.B., Slijper, F.M.E., Drop, S.L.S. (2005). Gender dysphoria and gender change in chromosomal females with congenital adrenal hyperplasia. *Archives of Sexual Behavior, 34,* 389-397.

Therefore, conflating intersex infants with transgenderism actually does nothing for the social constructionist argument, but the conflation's power relies on you thinking that your scientific knowledge is 'behind the times.' *Perhaps you are not up-to-date on the recent science.* But this tactic only works if you accept that a third biological sex (outside of male and female reproduction) exists.

Ironically, such a tactic relies on your own scientific ignorance to be effective. If you do not know that biological sex is defined through the two gametes required for sexual reproduction, you are likely to buy into the social constructionist's argument that biological sex exists on a spectrum. Yet this tactic does not work on everyone, especially those familiar with how sex is determined in the womb.[571]

Using intersex infants to argue that sex exists on a spectrum was described by UK researcher Andy Lewis as a "Flapdoodle Gambit," which is to say, a technique where you introduce a "complex tangential subject to your argument, such as quantum physics, neuroscience, or genetics, that you are sure your opponent cannot spend the time unpicking, so that you can make wild, arbitrary claims about your own beliefs."[572]

If you remember from Chapter 1 (Sex as a Social Construct), I labeled the intersex argument as the *argument from indefinite variation*, which is used to claim that the male-female category is unreliable and defunct; Lewis conceptualizes it this way:

> "The intersex gambit in gender theology is a flapdoodle gambit that tries to create an equivalence between the complexities of sex determination mechanisms and processes and there being a continuum of complex sexes. In all, the flapdoodle gambit is a bait and switch technique that relies on equivocation and conflation in the hope that you do not spot the switcheroo."[573]

---

[571] See chapter, "Sex as a Biological Mechanism."
[572] Lewis, A. (2019). The Intersex Flapdoodle Gambit. *Twitter*.
[573] Ibid.

16 THE GASLIGHTING IDEOLOGY

Responding to such flapdoodle gambits of conflating intersex with transgenderism, one Twitter user correctly defines male and female through the gametes:

> "Sex is binary. Females produce eggs and bear offspring, while males produce sperm and impregnate females. The existence of disorders of sexual development (or, more imprecisely, of 'intersex' persons) in no way collapses this distinction."[574]

If, however, the appeal to scientific authority does not work, the ideologues will combine these tactics with bullying and name-calling to make you think your beliefs are not just scientifically illiterate but *bigoted* and *stupid*:

> "Every single scientist I know: *sex can't be called a binary because we know it exists on a broad spectrum controlled by gene expression.* Transphobes: *excuse me, but when I was 8 someone told me about X and Y chromosomes.*"[575] (This tweet received over 100,000 likes and 20,000 retweets.)

This tactic of bullying, name-calling, and labeling your opponents as 'transphobes' is the next popular step for ideologues who cannot use scientific evidence and reason to argue their points. Such manipulation is designed to get *you* to shut up and to stop you from believing that biological sex is real. As such gaslighting tactics were being used to confuse and manipulate, the media quickly picked up the story:

> "The Harry Potter author is being called a transexclusionary radical feminist (TERF), after she defended a

---

[574] Kearns, M. (2019). In response to J.K. Rowling. *Twitter*.
[575] Zeru. (2019). In response to J.K. Rowling. *Twitter*.

researcher who was fired for saying that people can't change their biological sex," Variety Magazine reported.[576]

Other than the often-used label 'transphobe,' *trans-exclusionary radical feminist* is yet another popular term used to bully and silence voices who dare question social constructionist dogma that one can change biological sex, and yet, such a term is used almost across the board on social media to describe feminists who are critical towards transgenderism and social construction ideology.

Saying *transwomen are women* and labeling your opponents as *TERFs* are not just gaslighting tactics used to silence dissent but are tactics designed to "make it impossible to state any distinction between transwomen and cis ones."[577] For the radical trans activists and gender ideologue, there can be no distinction between a biological female and a male who identifies as one. Labeling anyone who dissents as a *TERF* or *transphobe* is ultimately not about protecting trans people; rather, it's a way to manipulate, control, and distort the truth. Robert Jay Lifton calls phrases such as *transwomen are women* and labels such as *TERF* "thought-terminating clichés" which are "brief, highly reductive, definitive-sounding phrases...that become the start and finish of any ideological analysis."[578]

And it is here, through the use of language, where the gender ideologues hold the most power: if they can label you as a bigot, and it sticks, then they can get you to shut up, to stop speaking your mind, and to perhaps get you to join their side in return--the side of equality and compassion nonetheless.

Soon after the backlash against Rowling's tweet, Twitter released a story examining the author's 'hurtful' and 'transphobic' statement that biological sex is real:

---

[576] Variety. (2019). JK Rowling faces backlash after defending a woman fired over transgender comments. *Twitter*.
[577] Joyce, H. (2018). The new patriarchy: How trans radicalism hurts women, children, and trans people themselves. *Quillette*.
[578] Ibid.

> "Examining why JK Rowling's recent comments are hurtful despite the author's claims of acceptance...JK Rowling was accused of transphobia after siding with a woman who was fired for her absolutist views on sex and gender. To some, Rowling's remarks did not seem worthy of backlash, so [this Twitter user] explained how generic or cliché expressions of open-mindedness can reveal regressive and judgmental attributes."[579]

The piece featured a series of tweets by a user who ascribes animus to JK Rowling's views. Rowling was not responding with true acceptance, argued the Twitter user, and was instead showing 'transphobia' through clichéd expressions of open-mindedness and acceptance. It didn't matter that JK Rowling said she wanted each individual to live their own life free of discrimination, and it didn't matter Rowling showed no animus towards trans individuals. What mattered was that she dared to claim that biological sex is 'real.'

For the social constructionist ideologues, such a statement is blasphemy, and when your ideological doctrine is blasphemed, there is no room for rational discourse. Instead, reason, science, and logic will be abandoned in favor of *gaslighting*: confusion and gambits, appeals to authority, and emotional abuse will reign supreme. While your ideological opponents shame you as a bigot for your commonly held belief, you'll be labeled as a 'transphobe' and, like Maya Forstater, fired by your employer and told by a court judge that your scientifically-grounded beliefs in the reality of biological sex are "not worthy of respect in a democratic society."[580] For Forstater, like Rowling, it did not matter that she championed diversity, inclusion, respect, and the dismantling of gender stereotypes. It did not matter she was a feminist, a fighter for women's rights, and a person who believed that "everyone should be free to live as

---

[579] Pride Toronto. (2019). Examining why JK Rowling's recent comments are hurtful despite the author's claims of acceptance. *Twitter*.
[580] Newman, D. (2019). Forstater vs CGD Europe, What the tribunal actually found. *A Range of Reasonable Responses*.

they choose without harassment or discrimination because of adopting or not adopting gender norms and stereotypes."[581]

What mattered was that she dared to criticize the holy doctrines of social constructionist ideology--the doctrines which say sex can change at whim, based on a given day or time; that there are no differences between boys and girls; and that all gender gaps are indicative of injustice.[582] Like Maya Forstater or JK Rowling, if you dare question, criticize, or debate the illiberal gate-keepers of contemporary gender theory, you will be bullied, shut down, and abused for the *scientifically illiterate bigot you are.*

Even if you are on the side of the oppressed and the marginalized; even if you fight for equality and equal rights; even if you actively champion the dismantling of constraining gender stereotypes, call out the pathologies, and let your children live their own life free of judgment; none of that matters. One comment, one criticism, or one questioning of the ideology, and you will be publicly shamed and potentially fired. All of this illiberal gate-keeping can only be reinforced and solidified if *you* acquiesce your words to the ideologues. And it is here, in the area of *language,* where the social constructionist ideologues wield the most power.

One political commentator, Chad Felix Greene (a non-binary, married gay man) describes it this way:

> "This is the power they hold," he wrote, "They label you 'anti-trans' and can use it as a weapon against other people who merely agree with your right to speak."[583]

In response to the immense backlash to Rowling's tweet that biological sex is 'real,' other feminists and LGBTQ activists spoke up against the social constructionist ideologues who believe that sex can be changed at whim. One feminist commentator rejected the idea that the gender ideologues stand for her:

---

[581] Forstater, M. (2019). What I believe. *Hiya Maya.*
[582] Questioning LGBT/CSE Education, (2019). Sexualizing Children with Sex Education. *Twitter.*
[583] Greene, C.F. (2019). In response to J.K. Rowling. *Twitter.*

> "Almost every LGBTQ group and media is demonstrating to JK Rowling and to the world at large that all they truly stand for are men's rights and their own extremist dogma. They preach inclusivity while they attack and attempt to 'correct' women's voices. They don't speak for us."[584]

As many social constructionist ideologues attempt to dismantle biological sex, feminists and LGBTQ activists alike continue to fight for the existence of male and female as biological realities while they reject the virtue signaling of social constructionists. Standing up for both sexes against a culture of censorship, these feminists continue to point out the most basic of facts: you can identify as whatever you like, but biological sex cannot be changed; you cannot change what is immutable (your internal reproductive structures produced in utero).

While you cannot change your biological sex, you *can* change your sexual characteristics through hormones and surgery, and this is what many transgender individuals choose to do so that they can live as close as possible to the opposite sex. This distinction between changing your biological sex and changing sexual characteristics is what both JK Rowling and Maya Forstater were attempting to communicate:

> "You change your appearance, your behavior, and take hormones, but you cannot ever change your capacity to produce a particular kind of gamete. This is what people mean when they say you cannot change biological sex."[585]

Social constructionists might not understand why people keep insisting that biological sex is real and immutable. Unlike what the ideologues claim, believing that biological sex is real and that there are only two sexes is not transphobic, it is not hateful, and it is not erasing the existence or struggles of either transgender or intersex individuals. There

---

[584] Joelle, M. (2019). In response to J.K. Rowling. *Twitter*.
[585] Motive, L. (2019). Cannot change biological sex. *Twitter*.

are many reasons why we need to reaffirm that biological sex is real and that there are only two sexes. Aside from the obvious medical reasons when it comes to reproduction, there are many other variables of sex and gender differences that we must understand if we are going to function as a society. The reasons for this are described by another Twitter user below:

> "Why is it important that we affirm there are only two biological sexes and that sex is immutable?" she wrote, "Why can't we just be kind, and allow that maybe the existence of transgender people and intersex people 'proves' existence of a 'third sex,' or that 'sex is a spectrum'? Because biological sex is of huge political significance for the 52% of the population who face endemic sex-based discrimination and epidemic sex-based violence. Children/young people must be protected from ethically unsound/experimental medicalization and ideological brainwashing. People with disorders of sexual development are hampered, not helped, by conflation of intersex with transgenderism, and by lack of clear factual understanding of their conditions."[586]

Understanding sex and gender differences is a critical aspect for the effective treatment of diseases and pathologies, whether these pathologies relate to the body, the brain, or both. As mentioned in previous chapters, knowing the complex sex differences inside the brains of males and females can help medical scientists develop more effective treatments for both neurological diseases and psychiatric problems which affect the sexes in different ways. But such advancements in treatment can only be undertaken if we as a society accept that male and female are biological realities and not social constructs.

And yet, knowing this, I can understand why social constructionists have the tendency to deny the reality of biological sex. In the past, categories have been used to discriminate people, and it is certainly true

---

[586] Jaspert, B. (2019). Importance of biological sex. *Twitter*.

that women have faced immense violence and injustice committed against them. Here the social constructionists think that, perhaps, if we deny biological reality and eliminate the categories, then the basis for discrimination and injustice will be dismantled. The problem, however, is that it is only through the recognition of biological categories, not in their denial, where discrimination can be fought. Injustice is not committed against women because they *identify* as women, but rather, because they *are* women. In the Middle East, for example, patriarchal nations continue to oppress women because of who they are--because they are *females*. Because of their female anatomy and reproductive functions, women who are biological females face different pressures compared to biological males; or, as described by another Twitter user:

> "Even where we are not openly oppressed," she wrote, "the risk of pregnancy and the need to find ways of combining childrearing with work make our sex something that has a huge impact on our lives, and which should have a huge impact on how society is structured. The arguments about sex as a social construct, apart from dismissing our reality, are telling us we don't have the right of self-definition or freedom of association, and we may not organize ourselves in a way that works for us and gives us dignity, privacy, and safety."[587]

This distinction between feminists who recognize the importance of biological sex and gender theorists who wish to dismantle biological sex is critical to understanding these two unique worldviews. One says that the rights of women can only be championed when biological sex is defined (feminism), and the other says that the rights of women can only be championed when biological sex is dismantled (queer theory):

> "Queer theorists see the intimate connection between biological sex and oppression, and they react by dismantling the

---

[587] Caroline, Real Feminists. (2019). In response to sex as a social construct. *Twitter*.

notion of biological sex; feminists see the intimate connection between biological sex and oppression, and they react by dismantling oppression. That's the fundamental difference between liberals and radicals; one sacrifices truth to avoid confronting power, and one confronts power to avoid sacrificing truth."[588]

Ultimately, this dismantling of the male-female category and biological sex itself is not a way to protect people from discrimination, and it's not a movement about compassion; rather, it's a way to push a certain ideology about sex and gender into the public sphere with the force of the state; or, as one commentator notes:

> "Getting rid of the categories (male and female) doesn't erase the material reality that only people with a certain type of anatomy can gestate young. This movement asks that we deny the realities of the physical world and enforce this denial through the force of law and social sanctions. They are attempting to force a psychotic (out of touch with reality) way of perceiving the world onto the public."[589]

Thus, knowing that we as a society are experiencing the proliferation of the theory which states that male and female must be dismantled, we can then better understand the backlash against statements such as "sex is real." If you view the dismantling of the male-female category as the primary method for ending discrimination, then you will react negatively to any pushback or questioning of these desired goals. In the grip of your ideological possession, you do not see that good people may believe that recognizing biological sex is the best way to end discrimination; instead, you see people reinforcing a binary which, in your eyes, is the cause of discrimination itself. And those people must be *bad*.

---

[588] Mix, J. (2017). Playing the intersex card. *Medium*.
[589] Misogynoir. (2017). Comment on 'Playing the intersex card.' *Medium*.

If we as a society wish to move past regressive and illiberal ideologies which silence political opponents, relegate moral feminists to the category of 'transphobes,' and use the force of the state to bludgeon dissidents into submission, then it's time for each of us to stand with Maya Forstater and JK Rowling: *we will not let the biological reality of male and female be taken over by radical ideologues.*

> "I spoke up because I believe we all have a responsibility to use our voice in whatever way we can," wrote Forstater, "We all pick our battles, but sometimes they pick us. I am fighting on because I don't want my story to be a cautionary tale about the high personal cost to women of having the courage to speak up, but instead to try to make it a win for freedom of thought, belief and expression, and for the heart and soul of the institutions that underpin an open society."[590]

JORDAN PETERSON AND COMPELLED SPEECH

Another gaslighting tactic from social constructionist ideologues relies on the power of federal law to make dissenters shut up. In just the past few years, freedom of speech rights have been continually under threat by ideologues who wish to have control over your words. For the postmodernists, if language is at the core of patriarchal power structures, then it is also in language where liberation can be found. If biological sex can be restructured linguistically, then the male-female category can be dismantled. Once it's been torn down, gender identity can be enshrined into law as a protected class. Thus, it is *gender identity* and not biological sex which is viewed by the gender ideologues as an innate trait which must be protected. Here someone's feelings, rather than objective reality, reign supreme.

In social constructionist language, gender identity is someone's perception of one's own sex or gender. Rather than being tied to your genetics or hormones, as we have discussed, the gender ideologues view

---

[590] Forstater, M. (2019). Sex and Gender, What I Believe. *Hiya Maya*.

gender identity as something based on an ephemeral feeling, almost spiritual in its nature. And thus any distinction between your internal identity and your external reality cannot be made. Your internal feelings must be enshrined into law and your external reality must be molded to fit within your internal feelings. Such is the logical conclusion of postmodernist deconstruction, where there is nothing outside of subjective perception, no objective reality, and no such thing as biological sex. The postmodern theories of subjective perception have now reached Canada, where gender identity and gender expression (akin to personality traits) have been written into law as protected classes.

Known as Bill C-16, the law amends the Canadian Human Rights Act and the Criminal Code to include gender identity and gender expression:

> "This enactment amends the Canadian Human Rights Act to add gender identity and gender expression to the list of prohibited grounds of discrimination. The enactment also amends the Criminal Code to extend the protection against hate propaganda set out in that Act to any section of the public that is distinguished by gender identity or expression and to clearly set out that evidence that an offence was motivated by bias, prejudice, or hate based on gender identity or expression constitutes an aggravating circumstance that a court must take into consideration when it imposes a sentence."

The stated aims and application of the bill seems innocuous enough, and yet, the definitions of what constitutes "discrimination," "prejudice," and "hate" is not entirely clear. Instead of being specific with how these critical words are used, the law is written to have the broadest possible application. The writers of Bill C-16 define *discrimination* as this:

> "Discrimination happens when a person experiences negative treatment or impact, **intentional or not**, because of their gender identity or gender expression."[591]

Think about just how broad this definition is. *Intentional or not? Negative treatment or impact?* How can you objectively define such things? What if the person *feels* like they have been negatively impacted by your words? Can they use this legislation to take legal action against you? Such a broad definition of discrimination is incredibly problematic. And yet these ill-defined descriptions continue with the definition of what constitutes *harassment*:

> "Harassment is a form of discrimination. It can include sexually explicit or other inappropriate comments, questions, jokes, name-calling, images, email and social media, **transphobic**, **homophobic**, or other bullying, sexual advances, touching and other unwelcome and ongoing behavior that **insults, demeans, harms,** or **threatens** a person in **some way**."[592]

Such broad definitions of both discrimination and harassment may lead to many legal issues. How can a person's actions be objectively determined to be *negatively impactful* and how can a person's intentions not matter? Rather than being careful with their words, the writers of Bill C-16 made sure that it had the broadest possible application:

> "The first thing you might note is that this has been written to cover the broadest possible range of possibilities," says Dr. Jordan Peterson, "That's exactly the opposite of what good legislation does. So if you combine these discrimination and harassment definitions then the policies basically allow for people to...for instance, if I *negatively impact* you with a joke, then *I'm*

---

[591] Peterson, J. (2016). 2016/09/27: Part 1: Fear and the Law. *YouTube*. 48m00s.
[592] Ibid., 51m30s.

harassing you, and that is on its way to being a violation of the Ontario Human Rights Code, and it's an offense according to the Ontario Human Rights Commission; and it's on the way to being a hate crime with the modifications of federal legislation."[593]

And yet, for Bill C-16, things get much worse than mere ill-defined descriptions of discrimination and harassment. When it comes to *hate speech*, the definition of it cannot be clearly decided upon. How do you define *hate speech*, how do you differentiate between speech which makes people uncomfortable, which is passionate and even angry, with speech that is hateful? What if the speech negatively impacts someone, even if you did not mean any harm? And this is the critical thing to understand: when it comes to the law, the definition of hate speech will not be defined by *you*, the citizen, but rather by those in *power*. It is the powerful among us who have the ability to define what speech is allowed and what speech is not. And for the postmodernists, such power over language is exactly what they desire. Bill C-16 gives postmodern social constructionists the broad sweeping legal definitions necessary for the control of language.

In September of 2016, professor of psychology at the University of Toronto Dr. Jordan Peterson released a series of videos on Bill C 16. Having studied the regimes of the Soviet Union and Nazi Germany for decades, and having produced lectures on the topics for his classes, Peterson was well aware of how dangerous it can be to utilize the force of the state to compel people's speech. And it is here, in the arena of free speech, where the social constructionist ideologues have gained power over others.

In Canada, this balance point between freedom of speech and the rights of others has not been dealt with properly; instead, gender identity and gender expression have been weaponized to compel people's speech through the force of the state. Bill C-16 has now made it illegal to create any distinction between someone's internal reality (their gender identity and expression) and their external reality (their biological sex). Any distinction

---

[593] Peterson, J. (2016). 2016/09/27: Part 1: Fear and the Law. *YouTube.* 51m30s.

made, such as refusing to call an individual by their gender identity pronouns or even discussing the reality of biological sex, now borders on hate speech, abuse, and a violation of the Criminal Code.

In the new law, gender identity is defined as "*each person's internal and individual experience of gender. It is their sense of being a woman, a man, both, neither, or anywhere along the gender spectrum,*" and gender expression is defined as "*the way in which people publicly present their gender.*"[594]

These internal and ephemeral perceptions of oneself have been written into law as protected classes, and while it is perfectly fine for someone to live their life how they wish, you should not be able to use the force of the state to make people agree with your internal belief system (the same is true with religious belief). And yet, such internal realities of gender identity and gender expression cannot be questioned; criticism and dissent regarding gender identity and gender expression can be interpreted as hate speech, and questioning them may lead to legal action against you.

No matter what you think of gender identity and gender expression, you must agree with the definitions provided in the legislation; you must agree that gender identity is an internal sense of being a woman, a man, both, neither, or anywhere along the spectrum; and you must agree that it can change based on a person's internal feelings. Anything less means you are agreeing with a "biological essentialist" view of gender, which is to say, you are discriminating against an individual's internal identity.

If you criticize the definitions of gender identity and expression now written into law; if you speak out against the idea that you can change your biological sex, that there are more than two sexes, that you can be *both* a woman and a man at the same time, and that you can also be *neither*, then you may be subject to serious legal action. Or, in the words of Dr. Jordan Peterson:

---

[594] Peterson, J. (2016). 2016/09/27: Part 1: Fear and the Law. *YouTube*. 34m00s.

> "These definitions [gender identity and gender expression] have become more than fact; they have become facts that you question at your *legal* peril."[595]

In the definitions of gender identity and gender expression written into Bill C-16, it is implicit that such categories are independent from biological sex, that there is no relationship between gender identity and sex and that you can disconnect the two. However, such a claim is untrue, as we explored in Chapter 9 on the development of gender identity. Opposite of what the social constructionists believe, we discovered that gender identity was directly tied to your genetics and hormones, which are directly linked to your biological sex. Attempts at altering a child's gender identity through socialization and genital surgery led to many suicides, as the boys rejected their "female" gender identity. And yet, the definitions of Bill C-16 do not agree with this; instead, the social constructionist view of gender identity and gender expression has been written into law:

> "[Bill C-16] is predicated on the idea that your gender is somehow independent from your biological sex, but that's a proposition, not a fact. Even if it was determined to be true, it is certainly not determined to be true to that degree now and certainly not to the degree that it should be instantiated in law; at most it's an opinion."[596]

This social constructionist view of sex and gender must be agreed to. Views critical of the idea that gender identity is both a social construct and an ephemeral feeling can be considered hate speech if the law is interpreted broadly. And when it comes to the use of gender pronouns, the social constructionist ideologues have gained power over people's use of language as well. In Ontario, Canada, if you consistently refuse to use the

---

[595] Peterson, J. (2016). 2016/09/27: Part 1: Fear and the Law. *YouTube*. 34m00s.
[596] Ibid., 36m00s.

words that the gender ideologues want you to use (even gender-neutral pronouns such as *xe/xem, ze/zer, hir/hirs,* or *zhe/zher*), such an act can be considered discrimination if the person "experiences negative treatment or impact, **intentional or not.**"[597]

If your actions and words are deemed negatively impactful to the individual, however that is defined, then you can be sent to what's called a *social justice tribunal,* an extra-judicial body which can fine you for speech crimes, and if you refuse to pay the fines, you can be sent to jail.[598] Here we can see how the postmodern social constructionists leverage marginalized groups for their own political power and influence. If you say you're on the side of the oppressed, the marginalized, and the downtrodden, then who can disagree with you? *Only bigots, that's who.*

When it comes to the use of gender pronouns, much of the issue is found not from the use of *he* or *she* but in the use of the constructed *gender-neutral* pronouns. Many of these pronouns are linguistic constructs utilized by social constructionist ideologues to dismantle the male-female category through the weaponization of language. If language can be restructured so that the male-female category can be eliminated, then perhaps discrimination can be eliminated as well. As gender identity proliferates into more and more fragmented identities, the Canadian law will require you to use such gender-neutral pronouns under threat of legal fines and jail time. Such compelled speech legislation is now the law in the Canadian Human Rights Act and the Criminal Code, and it is not an exaggeration: to refuse to use a person's preferred gender pronoun is considered a denial of their identity and is therefore an act of discrimination. Such an act may be subject to legal action in accordance with the federal legislation.

While it is perfectly reasonable to call a trans individual by their preferred opposite sex pronoun such as *he* or *she,* Dr. Peterson finds the compelled use of *gender-neutral* pronouns (such as *xe/xem* or *ze/zer*) problematic. Such pronouns are not natural linguistic creations which have

---

[597] Peterson, J. (2016). 2016/09/27: Part 1: Fear and the Law. *YouTube.* 48m00s.
[598] Dragicevic, N. (2018).

formed over time but are rather the products of social constructionist ideologues who wish to use language to shape people's perceptions, to manipulate other's actions, and to control people's speech through the force of law. Like Maya Forstater and JK Rowling, Jordan Peterson refused to give any inch of ground to the ideologues who want linguistic power over others:

> "I'm not doing it," said Dr. Peterson, "I think those gender-neutral pronouns are politically motivated; I think they are connected to an entire underground apparatus of radical left political motivations, and I think uttering those words makes me a tool of those motivations."[599]

Chad Felix Greene, a gay and non-binary political commentator, described the use of gender-neutral pronouns in a similar way to Dr. Peterson:

> "The pronoun thing is a power play," Greene writes, "It's a way for insecure people to create uncomfortable situations they can exploit for attention and personal validation. There is absolutely no reason anyone needs to be referred to as 'they' or 'zir' or anything else to feel 'seen.'"[600]

There is no problem with trying to protect marginalized groups from discrimination and injustice, and there is no problem with calling trans individuals by their preferred *he* or *she* pronoun, but we also must not give an inch of ground to ideologues who want control over other people's use of language. We need to make sure we do not infringe upon the rights of others, do not create legislation which controls people's right to speak, and do not mold our entire society's external reality to the definitions of someone's internal subjective feelings. We should allow people to live how

---

[599] Peterson, J. (2016). 2016/09/27: Part 1: Fear and the Law. *YouTube*. 42m30s.
[600] Greene, C.F. (2019). In response to gender-neutral pronouns. *Twitter*.

they want to live as long as they do not infringe upon the rights of others, and this is where things get tricky. Thus, the problem is not in trying to protect people from discrimination, but in compelling people to use the language that *you* want them to use through the force of the state. Jordan Peterson was objecting to the latter, not the former.

Though Dr. Peterson is a professor of psychology and a clinical psychologist who has made his entire career about understanding and opposing the evils of discrimination, injustice, and oppression, even he was not immune to the *fatal strategies* of the postmodernists.

Despite Dr. Peterson's continual fight against the evils of totalitarian ideologies and his continual acceptance of transgender individuals, social constructionist activists attempted to shut him down due to what they called his 'hateful,' 'bigoted,' and 'transphobic' views. Here the gender ideologues utilized their favorite gaslighting tactic: *make sure anyone who disagrees with you is portrayed as an evil bigot.*

During a free speech event at McMaster University in Ontario, Jordan Peterson was scheduled to speak with a panel of three other professors about free speech and political correctness, but due to harassment, the other three panelists dropped out and left Peterson as the only speaker. Deciding to continue regardless, the lone professor and audience members were interrupted by a set of student activists who marched into the room and began disrupting the event. A few held air horns, blowing the deafening sounds into the ears of the audience members while they circled around Dr. Peterson. Instead of dealing with his arguments against Bill C-16 with logic and reason, the social constructionist ideologues used gaslighting to get him to shut up.

"Transphobic piece of shit! Transphobic piece of shit!" they yelled at the top of their lungs as Peterson talked. "Transphobic piece of shit! Transphobic piece of shit!" they continued to yell at him, unaware that their gaslighting tactics were actually hurting their own cause.

In their ideological possession, if they could just repeat the name-calling and the bulling over and over again, then perhaps the mean old professor would finally shut up. And yet, as the shouting voices of the ideologues met his ears, Peterson spoke:

"One of the things I would ask all of you in this room to do…" Peterson said with a mic in hand as the activists surrounded him, "is to think hard before you assert that the lazy activists actually root for the people they purport to represent, because they do not; they have no legitimacy."[601]

The audience applauded as the activists continued to try and drown out his speech, and yet once again, Peterson continued. This time, he got up from his seat and began walking around the room as he looked at each audience member while the activists followed.

"I find it also quite disturbing that one of the things that people in university are taught by their activist professors is that coming out and protesting someone constitutes a moral thing to do; it's super easy," he said, as the activists continued to blow their air horns and shout at him, "One of the reasons I've been talking about individual responsibility is because this is absolutely pathetic."

An activist with an air horn walked closer to Dr. Peterson and began yelling *"NO FREEDOM FOR HATE SPEECH,"* blowing the air horn at him as he continued to talk about their ideological possession:

"I also think that when you watch people who are ideologically possessed protest, all you will hear is noise, because there is not much difference between the ways they think and *noise,"* Peterson said, calmly.

"NO FREEDOM FOR HATE SPEECH!" the activists continued to shout, "NO FREEDOM FOR HATE SPEECH!"

"There's nothing about it that is individual," Peterson said over the blaring of the air horns, "you can't talk to someone who's in a state of mind like that. People in those activist classes

---

[601] Eggplantfool. (2017). Jordan Peterson at McMaster University. *YouTube.* 14m45s.

learn they can memorize ten rather appalling statements and learn how to make noise as if that constitutes civilized discourse and a way to make a positive impact on the world."[602]

As the activists got increasingly louder, Dr. Peterson decided to take the event outside, and his audience followed him:

> "What I didn't like in Bill C-16 is the surrounding policies produced by the Ontario Human Rights Commission," he said, "One of the things that's happened that's reprehensible from a psychological perspective is that we're going to write into a law an idea of identity that is radically insufficient."[603]

While standing outside among his audience and the continual shouting of activists, Dr. Peterson lays out a critique of the social constructionist doctrine that all sex and gender characteristics vary independently from one another:

> "The way that Bill C-16 is currently formulated is that it proposes biological sex, gender identity, gender expression, and sexual orientation vary independently. The fact of the matter is that's not true, and you can demonstrate it straightforwardly. The first thing you observe is that virtually everyone whose biological sex is male or female has a gender identity of male or female. The people who do not fit into that category exist, but they are a small minority. And the reason I'm pointing out they are a small minority is because it makes the claim that those things vary independently false. And then you know that most people who have a binary biological sex and a binary gender identity also present themselves in a manner that is the same as their gender identity and their biological sex.

---

[602] Eggplantfool. (2017). Jordan Peterson at McMaster University. *YouTube.* 16m00s.
[603] Ibid., 32m40s.

You can see too that the vast majority of men who are biologically male, who identify as male, and who express themselves as male are also heterosexual. The reason I think this [idea] is a problem is because it's written into the law. The law is not the place that you instantiate a particular philosophy.

One of the arguments that people who are gay often make is that there is a biological underpinning to their homosexuality. Are we going to dispense with that? What happens when we dispense with the idea that there is a biological underpinning to homosexuality? Does that mean it can just be changed? The people fighting for homosexual rights have for the last thirty years frequently made the case that the reason their rights were necessary was because there was a reality to the condition. The law the way it's currently constituted makes arguments like that completely invalid."[604]

As Dr. Peterson continued to make an argument for LGBT rights, the activists began shouting again: "SHUT DOWN PETERSON, SHUT DOWN PETERSON! SHUT DOWN PETERSON!!" they yelled with their repetitive ideological slogans. How could they be so ignorant that they were shouting down a person who was actually for freedom of expression and individual liberty? The answer: because they were not dealing with the real person named Jordan Peterson; they were dealing with a caricature--a caricature who they viewed as a 'transphobe.'

To the social constructionist ideologues at the event, anyone who disagreed with their viewpoints that sex and gender are social constructs was not just ignorant, but *evil* and *bigoted*. And it is here, while being shouted down by activists, where Dr. Peterson argues that the people who purport to represent the marginalized, the oppressed, and the downtrodden often do not truly represent them:

---

[604] Eggplantfool. (2017). Jordan Peterson at McMaster University. *YouTube*. 33m00s.

> "When I talked to you about the letters I received from transsexual people, they're very terrified about all of this," Peterson said, as he looked back at the vehement activists and then back to his audience, "because the vast majority of them want to be as invisible as everyone else wants to be, and all this does is draw negative attention to them. They are also not happy about that fact that they're being represented *noisily* by people who have absolutely no right to speak for them."[605]

To leave the audience with something to think about, Dr. Peterson critiqued the postmodernist idea that diversity is achieved through sexual and racial characteristics and, while he was doing so, championed the idea that the individual, not the group, is what should be sovereign:

> "One thing we have to get over in a serious way is the idea that just because you're talking to someone who's black, white, transsexual, or homosexual, that you're speaking to a member of a homogenous tribe and that the opinion they have is the opinion which all those who are like them share. There's nothing that's more *racist* than that assumption!" Peterson exclaimed as the activists ironically continued to shout meaningless phrases to shut him down. Yet the courageous professor continued among all the yelling and the air horns: "It means that I don't have to know anything about you except the most boring part of your identity, and that I can predict everything that you are and everything that you think. That's a terrible thing to believe, and it's patently untrue!" he said, "There's no reason to assume that any group of people who don't fit into traditional society are any more homogenous in their general makeup and political views than any other group. All the evidence points to precisely the contrary. Identity politics is predicated on the idea that the only

---

[605] Eggplantfool. (2017). Jordan Peterson at McMaster University. *YouTube*. 37m00s.

important thing about you is what's most obvious when I first look at you. *That's an appalling proposition!*"[606]

For the student activists possessed by postmodern *fatal strategies* of disturbance, manipulation, and control, there was no room for the idea that perhaps this psychology professor had something meaningful to say. Instead, they decided to use the much easier tactics of shouting him down, manipulating language through slinging accusations of transphobia and homophobia, and enforcing their power over him so that he would shut up. However, for Dr. Jordan Peterson, he knew exactly how to respond: with calmness and courage, and because of this, the gaslighting tactics of bullying, name-calling, and disruption did not cause Jordan Peterson to fade away; on the contrary, these tactics actually helped bolster him to international fame.

Once the media got a hold of the Bill C-16 controversy and Dr. Peterson's principled stance against the compelled speech proposition, it was only a matter of time until this clinical psychologist would be launched into the mainstream. Throughout 2018, Jordan Peterson and his wife, Tammy, along with Dave Rubin, went on a lecture tour across more than 100 of the world's cities where Peterson presented his ideas about personal responsibility, personality psychology, and the importance of meaning to audiences of tens of thousands. And if it wasn't for the activists who brought large attention to his views combined with his courageous and forthright stance, it's uncertain whether he would have ever been the most influential driving force for free speech, Jungian psychology, and religious phenomenology in our postmodern era.

Thus, like Jordan Peterson, recognize when your words are being taken over by social constructionist ideologues. Such ideologues do not stand for those they purport to represent. They speak about compassion, acceptance, and tolerance while they shout down anyone who disagrees with them; and they do so with the intolerance and bigotry they supposedly despise. They are not fighters for the marginalized and oppressed; they are

---

[606] Eggplantfool. (2017). Jordan Peterson at McMaster University. *YouTube*. 38m00s.

fighters for the elites and the powerful, the gate-keepers and the manipulators, the postmodernists and the Marxists.

Instead of giving in to the demands of the ideologues who want control over your language, realize that you *can* stand up for the rights of trans individuals, treat them with dignity and respect, and call them by their preferred pronouns without giving up your inalienable freedom to be in control of how you use your own words.

Acceptance is something that each individual deserves no matter how they identity, and such acceptance must be extended to trans individuals as long as this acceptance does not infringe upon the rights of others. Because of this, compelled speech is not the way to achieve trans acceptance. Using the force of government to control someone's speech is not the road to a tolerant society; it's a road to intolerance, bigotry, and state-sponsored censorship. Those who push such regressive and illiberal laws do not need to be given any more power than they already have, because such laws will not achieve the acceptance that trans individuals desire and deserve. In the words of a self-described humanist and de-trans woman:

> "Acceptance is achieved through persuasion, not coercion," she wrote, "People who bully others who don't share their views, who support compelled speech, and who applaud when someone is kicked off a public platform or loses their job, they do not want acceptance. *They want obedience and control.*"[607]

---

[607] ImWatson91. (2019). In response to compelled speech. *Twitter*.

MEDICALIZATION OF GENDER NON-CONFORMANCE

So far, I have covered two case studies showing how social constructionist ideologues gaslight the public into believing that 1) biological sex exists on a spectrum and 2) biological sex can be changed. Obfuscating through appeals to authority by pointing to the complexity of sex determination mechanisms while using intersex infants to argue that sex exists on a spectrum, social constructionist ideologues call those who disagree with their claims *scientifically illiterate transphobes*. While doing this, the social constructionist ideology which says there is no link between biological sex and gender identity has now been written into Canadian law. One's internal gender identity, from which a proliferation of infinite identities emerge, cannot be questioned.

However, these gaslighting tactics combined with federal laws has not stopped opponents from speaking out against the idea that biological sex is separate from gender identity. Most people are not willing to bend their concept of reality, along with their language, to ideological possessed authoritarians. And yet, while the ideologues receive a severe pushback from the public, the powerful among us continue to instantiate social construction ideology into our laws and policies, and now, these beliefs are being directed at children. If you cannot convince adults that biological sex exists on a spectrum and that biological sex can be changed, then perhaps you can direct your attention to the malleable minds of children, from whom the dismantling of biological sex can be achieved.

In the early chapters of this book I discussed how social constructionist ideologues inversed the nature of sex and gender. Gender-critical feminists from the Beauvoir tradition viewed sex (male/female) as a biological reality and gender as the system of values, beliefs, and practices constructed around sex. It was the gender-critical feminists, not the strong social constructionists, who believed males and females should not be defined by gender stereotypes and societally-determined gender roles. However, beginning in the 1990s with queer theory, Judith Butler inversed this view: *biological sex* was now the system of values, beliefs, and practices *constructed* around gender. Gender identity, the internal sense of your own gender, was now your ultimate essence, not sex.

Through the inversion of the *nature* versus *nurture* dichotomy, Butler theorized that the systems around biological sex were constructed through nurture, and that one's internal gender identity was formed from some ephemeral sense of one's 'nature.' If biological sex was constructed, then perhaps it could also be *deconstructed, dismantled,* and *changed* so that discrimination based on the male-female category could be eliminated. To the horror of gender-critical feminists who believe that it is the values and beliefs around *gender*, not biological sex, which must be deconstructed, the social constructionist ideology is now being weaponized to indoctrinate gender non-conforming children into believing they were born in the wrong body. And this is where the third and final case study into the gaslighting tactics of the gender ideologues begins.

As children grow up and learn who they are, it is not uncommon for them to play with gender-atypical toys. For instance, boys may find themselves interested in people-oriented activities such as playing with dolls rather than things-oriented activities which involve playing with mechanical objects like trains and trucks. Such gender-atypical play is perfectly healthy and normal, as we should not expect children to fit within a stereotypical mold. However, in recent years, the philosophy of deconstructing gender has become rather insidious due to the conflation of gender with biological sex. Now, from the doctrines of the social constructionist ideology, gender-atypical play has quickly become both pathologized and medicalized. It has been pathologized through the idea that gender-atypical behavior means that the child has been born into the wrong body, and medicalized through the idea that such gender-atypical play requires transition to the opposite sex.

A 2019 study presented by NBC claimed that children who identify with gender-atypical toys and clothes are likely to be transgender, which is to say, to have a gender identity which is opposite of your birth sex. To justify this claim, the researchers interviewed transgender children from ages 3 to 12 and asked them how much they felt like a boy or girl based on what toys and clothes they liked:

> "The transgender kids showed strong preferences for toys and clothing typically associated with their *gender identity*, not their assigned sex, the study found. Their preferences did not appear to differ based on how long they had lived as their current gender."[608]

The kids were smart: they recognized that they enjoyed playing with toys stereotyped to be opposite of their biological sex. For instance, a boy playing with dolls may realize that mostly girls, not boys, play with dolls. Because of this, he may think that he was born in the wrong body, and if he was born in the wrong body, then perhaps he should transition to become a girl. Ironically, such thinking relies on the child to believe in rigid gender stereotypes, namely, *that only girls, not boys, play with dolls*. One might think that such a correlation between being transgender and liking opposite sex toys is strong. However, the study had one major flaw:

> "One limitation of the study is that all the transgender kids lived in families that *affirmed their current gender identity*, the study team notes."[609]

In other words, the study only interviewed kids whose parents reinforced the idea that their children were the opposite sex; it did not study actual cases of gender dysphoria, but rather, used cases where the parents reinforced gender stereotypes on their child. Without understanding that one can be gender non-conforming and not be transgender, the children actually believed themselves to be the opposite sex, and their parents, rather than dismantling gender stereotypes, *applied* stereotypes onto their children: *if you're a boy who plays with dolls, then you're actually a girl*. This is the first step in pathologizing gender non-conformance: transforming gender-atypical play into something which must be treated medically through hormone replacement therapy and genital mutilation.

---

[608] Chiu, J. (2019). Trans children sense their gender identities at young ages, study suggests. *NBC*.
[609] Ibid.

Here the study reported by NBC reaffirms the idea of Judith Butler, that biological sex is socially constructed and gender identity is innate:

> "The current study 'helps to confirm the unique and separate reality of gender and how it is distinct from biological sex and socialization. **It supports the idea that gender is inherent and separate from biological sex which would seem to then come down on the side of *nature* as opposed to nurture.**'"

The researchers claimed this study showed that gender identity comes down on the side of *nature* and that it is separate from biological sex. In other words, if you're a boy who plays with female-typical toys, then you have a gender identity of female, and therefore, *are* a female. As we have shown throughout the book, especially in Chapter 9 on the development of gender identity, such ways of thinking actually reinforce gender stereotypes.

Such logic can only result in child abuse, as it requires you to gaslight children into thinking they have been born into the wrong body. It requires that you reinforce the gender stereotypes feminists wish to dismantle, and it requires that you treat your biological boy as a *girl*. Such logic is sexist and dangerous. Just because someone plays with gender-atypical toys does not mean they are the opposite sex; or, in the words of evolutionary biologist Heather Heying:

> "Suggesting that boys or men who enjoy engaging in traditional female-typical gender norms are actually girls or women is regressive, misogynistic, reality-denying garbage."[610]

As the claims of the study reached the mainstream, others began speaking out against what they viewed as a regressive ideology of reinforcing gender stereotypes onto children:

---

[610] Heying, H. (2019). In response to NBC study. *Twitter*.

"This is backwards," said one feminist and researcher, "Playing with toys and wearing clothes that are typical for the opposite sex is the main reason a child is labeled trans. (If you don't believe me, read any interview with parents of trans kids. It's all 'he loved dolls and pink' or 'she insisted on short hair.')"[611]

One father of a young adult also responded to the NBC study with clarity in regards to stereotyping young children:

"My daughter was a typical tomboy with older brothers and thought of herself as a boy when younger. She still likes toys typical for boys. She is 24 and would be the last person to say she's a boy. *Likes and interests do not make your gender.*"[612]

Other commentators, like Chad Felix Greene (a gay, non-binary man), responded with confusion: *why are we reinforcing gender stereotypes?*

"This would require clothing and toys to be strictly male or female," he wrote, "I thought we all decided clothing and toys were gender neutral? Also what is this psychic ability children possess that enables them to determine the gender of clothing and toys so young?"[613]

"I'm confused," said another person, "If gender roles are a social construct, then why do boys and girls who may identify with traditionally opposite-gender toys/clothes need to conform to *that* gender identity, since it doesn't match their own? [Social constructionists] scream about gender roles then rigidly reinforce them."[614]

---

[611] Joyce, H. (2019). In response to the NBC study. *Twitter.*
[612] BTC. (2019). In response to the NBC study. *Twitter.*
[613] Greene, C.F. (2019). In response to the NBC study. *Twitter.*
[614] Curtis, A. (2019). In response to the NBC study. *Twitter.*

"In order to make gender identity work," wrote Greene, "the Left has had to rewrite all gender theory. Basically, if a boy likes *trucks* he *is* a boy and if he likes *dolls* he *is* a girl. To prove transgenderism, they have even resorted to supernatural 'sensing' of gender."[615]

In response to all of this, political commentator and beekeeper Matt Walsh then reinforced the idea that a feminine boy is not a transgender girl; he's simply a boy who likes feminine toys and clothes:

"If the claim is that 'transgender girls' are simply boys who identify with the things that society has arbitrarily and artificially associated with girls, then a transgender girl is by no means and in no way a girl. He would be, at most, a feminine boy."[616]

A gender-critical feminist and commentator also explored the dangers of reinforcing such gender stereotypes onto gender non-conforming children:

"Are adults seriously willing to see gender non-conforming children mistakenly set on a pathway that begins with puberty blockers leading to hormone treatments and life-changing surgeries, all of which can lead to them being sterile?"[617]

Because gender-atypical play is now associated with an 'innate' gender identity and not innocuous toy preferences, social constructionists have ironically pathologized and medicalized gender non-conformance. If a boy or girl exhibits gender-atypical behavior and preferences, then social constructionist ideology says that the child's gender identity is opposite of their birth sex; if it is opposite of their birth sex, then this gender identity must be actualized through hormone replacement therapy and genital

---

[615] Greene, C.F. (2019). In response to the NBC study. *Twitter*. 2.
[616] Walsh, M. (2019). In response to the NBC study. *Twitter*.
[617] Gender is harmful. (2019). In response to the NBC study. *Twitter*.

surgery so that they can become a full 'male' or 'female.' Ironically, nothing is more regressive, abusive, and sexist than that.

In response to the strong social constructionist ideology which relegates boys and girls to dangerous stereotypes, one feminist quickly responded with an alternative: boys and girls should be taught to accept their bodies, to accept themselves for who they are, and to realize that just because they like gender-atypical toys does not mean they are the opposite sex:

> "How about this," she wrote, "Tell your child that no one can change sex; that they were born in a perfect body and that they can do anything they like while living in that body. It defines what sex they are. It does not define what person they are."[618]

Reinforcing a strong social constructionist ideology of sex and gender means that one is also reinforcing an infantile view of what sex and gender is. Just because a boy puts on a dress does not mean he *becomes* a girl as if some religious sacrament has taken place. While it is certainly true that young children may believe they actually *become* what they act out (such as a boy becoming a girl through wearing a dress), this is because young children are not able to separate sex and gender into categories. Their understanding of sex and gender is incredibly limited:

> "Developmental studies show that young children have only a superficial understanding of sex and gender (at best). For instance, up until age 7, many children often believe that if a boy puts on a dress, he becomes a girl. This gives us reason to doubt whether a coherent concept of gender identity exists at all in young children. To such extent as any such identity may exist, the concept relies on stereotypes that encourage the conflation of gender with sex."[619]

---

[618] Woman Is Biology. (2019). In response to the NBC study. *Twitter*.
[619] Malone, W., Wright, C., & Robertson, J. (2019). No one is born in the 'wrong body.' *Quillette*.

Knowing that young children have limited concepts of sex and gender and cannot separate fantasy (such as *becoming* a girl through wearing a dress) from reality, we should not be enforcing gender stereotypes onto them and treating them as though they *are* the opposite sex simply because they enjoy gender-atypical toys; in other words, we should not be conflating biological sex with gender stereotypes. We should allow for the wide variety of behavior within the category of male and female without making children think they are stuck in the wrong body, lest we want to reinforce rigid stereotypes; or, in the words of endocrinologist William Malone and evolutionary biologist Colin Wright:

> "In the case of an adolescent female whose behavior, personality traits and preferences are more 'masculine' than most girls and most boys, she could be led to incorrectly conclude that she is really a male, born in the wrong body. That child's parents could become confused as well, noticing how 'different' their child's behavior is from their own, or from that of their peers. **In reality, that child simply exists at the end of a behavioral spectrum, and 'sex-atypical' behavior is part of the natural variation exhibited both within and between the sexes. Personality and behavior do not define one's sex.**"[620]

Thus, variation within the male and female category does not mean that a child is actually the opposite sex; it just means that they are exhibiting gender-atypical behavior, and that's okay! (Male and female behavior has plenty of overlap, but this does not mean that biological sex can be changed.) When social constructionists fail to understand this critical point about variation within male and female behavior, gendered behavior is often conflated with biological sex, and this conflation is then pushed onto children:

---

[620] Malone, W., Wright, C., & Robertson, J. (2019).

> "In most cases, the thing that is now called 'gender identity' likely is simply an individual's perception of how their own sex-related and environmentally influenced personality compares to same and opposite sexed people. Put another way, it's a self-assessment of one's stereotypical degree of 'masculinity' or 'femininity,' and it's wrongly being conflated with biological sex. This conflation stems from a cultural failure to understand the broad distribution of personalities and preferences within sexes and the overlap between sexes."[621]

If you remember how much emphasis I put on understanding *bimodal distributions,* where male and female behavior falls within two averages and significant overlap, then you can also understand how gender-atypical play can exist in boys and girls: *there is plenty of variation within males and females.* Once again, gender-atypical behavior is simply a behavior which does not fall within the average of male and female distribution; it is by definition *atypical,* and yet this does not mean that a boy or girl exhibiting such atypical behavior is actually the opposite sex. Perhaps the social constructionist ideologues can realize this critical point so we can put an end to the proliferation of the sexual, psychological, and physical abuse of children.

> "Surely the progressive position would be for children and adults with no care for their biological sex to express their gender in whatever way makes them comfortable?" writes one doctor, "So for a boy to wear a dress and still be a boy and a girl to play football and still be a girl. Why force children, before they have the capacity to choose, into a gender box? Why not just let them be, and allow them to explore gender as part of their development. Quit forcing patriarchal gender labels onto

---

[621] Malone, W., Wright, C., & Robertson, J. (2019).

children. Why are woke 'progressives' acting like the patriarchy's useful idiots?"[622]

As the social constructionists refuse to realize they are playing in to traditionalist views of gender, they continue to push transgender ideology onto young kids who exhibit gender non-conformance. Such ideological gaslighting makes children believe they are *actually* the opposite sex, which transforms harmless gender-atypical play into a pathology supposedly requiring medical treatment. Because social constructionist ideology has proliferated into the mainstream, clinics for sex reassignment have begun to experience an over-diagnosing of gender dysphoria in gender non-conforming children. For example, in the United Kingdom, the Gender Identity Development Service (GIDS) received more than 2,500 children applicants for gender reassignment in the last year alone, compared to only 77 a decade ago.[623]

Because of the proliferating medicalization of gender non-conformance, more and more children are now being treated with hormones and surgically altered to align with their perceived 'gender identity' along with their parents' encouragement and society's affirmation. In the past three years, thirty-five psychologists who refused to take part in the sexual abuse of children resigned from the GIDS; they have raised concerns about giving young children hormones and puberty blockers while altering their genitalia:

> "We are extremely concerned about the consequences for young people," said one of the psychologists, "For those of us who previously worked in the service, we fear that we have had front row seats to a medical scandal."[624]

---

[622] Thorne, A. (2019). In response to the NBC study. *Twitter*.
[623] Lockwood, S. & Lambert, H. (2019). NHS 'over-diagnosing' children having transgender treatment, former staff warn. *Sky News*.
[624] Ibid.

Staff at the GIDS are under immense legal and social pressure to treat these children as quickly as possible, with little time for a thorough evaluation into what the children are experiencing. As of now, there's only one pathway through the service, and that is through the medical alteration of the child's physiology and anatomy.[625] Half of the children visiting the clinic are put on drugs to block the mechanisms which initiate puberty; such use of puberty blockers are considered by clinicians to be an irreversible form of treatment which halts healthy sexual development. Once these drugs are administered, the child is given cross-sex hormones to further differentiate their anatomy and physiology. These kids are often as young as 12. As these children grow older, they often realize such medical intervention was a mistake and decide to detransition to their biological sex. Many detransitioners reported that the current culture was a major factor in them deciding to transition in the first place: "Mainly the thing that was fueling me was that I didn't fit in, and then I was slowly drip-fed this idea that you could change sex," said one detransitioner.[626]

As social constructionist ideologues push the idea that biological sex can be changed at whim, the pressure to transition has become so great that any questioning of the child's medical treatment from the clinicians may brand the psychologists as *transphobes:*

> "[The psychologists] described a service where they are constrained in the work they can do with a patient for fear of being called 'transphobic.' The psychologist said: 'The alarm started ringing for me…I didn't feel able to voice my concerns, or when I did I was often shut down by other affirmative clinicians.'"[627]

As the gender ideologues continue to gaslight innocent children into believing their gender-atypical play means they were born in the wrong

---

[625] Lockwood, S. & Lambert, H. (2019).
[626] Ibid.
[627] Ibid.

body, international studies have shown a strong correlation between childhood gender dysphoria and homosexuality. In fact, the correlation is so strong that 85% of children who have gender dysphoric feelings grow out of it during puberty and often come out as gay.[628]

And yet, despite this, social constructionist ideologues continue to pathologize and medicalize gender-atypical behavior to the detriment of children's later fertility:

> "Instead of offering counseling, medical professionals now are commonly telling children that they may have been 'born in the wrong body.' This new approach, called 'gender affirmation,' makes gender dysphoria less likely to resolve, pushing children down the path toward irreversible medical and surgical interventions. If aggressive transition options are pursued early in puberty, the combination of puberty-blocking drugs, followed by cross-sex hormones, will result in permanent infertility."[629]

Ironically, such 'gender affirmation' actually enflames feelings of gender dysphoria:

> "Telling a child that he or she was born in the wrong body pathologizes 'gender non-conforming' behavior and makes gender dysphoria less likely to resolve."[630]

Because it is impossible to fully change one's sex (as this would require alteration of sex-specific genes in the brain, a restructuring of sex-differentiated neurons, and a surgical replacement of all internal reproductive structures) it is nothing short of gaslighting to tell a child their gender non-conforming behavior means they are the opposite sex. Rather than dealing with the problem at hand through psychological treatment,

---

[628] Ristori, J., Steensma, T. (2015). Gender dysphoria in childhood. *International Review of Psychiatry, 28(1).*
[629] Malone, W., Wright, C., & Robertson, J. (2019).
[630] Ibid.

such gaslighting tactics of 'gender affirmation' create more pathological and maladaptive coping strategies than if the child learned to accept themselves, which is to say, to accept the sex they were born with:

> "The fact is, no child is actually born in the wrong body. Adults should expand their understanding of what normal male and female behavior and preferences look like--which would lead them to appreciate that being male or female comes with a wider range of personalities preferences, and possibilities than old stereotypes would have us believe."[631]

Ending gender stereotypes for good means 1) reinforcing the idea that biological sex cannot be changed and 2) allowing gender-atypical behavior in boys and girls without telling them they were born in the wrong body. Ending gender stereotypes also means we must put a stop to gaslighting tactics which manipulate reality to fit within an ideological agenda. Gaslighting children by telling them they are actually the opposite sex is not progressive; it's child abuse.

We must abandon the social constructionist view that sex is a malleable social construct and that it can be changed at whim. We do this not because we are *transphobic* to those who suffer from gender dysphoria, but because we wish to allow the wide diversity of human behavior to express itself without pathologizing and medicalizing gender-atypical traits.

You are not transphobic if you believe that young children should not make life-altering decisions--decisions which can create *real* gender dysphoria resulting in sexual, psychological, and physical trauma. You are not transphobic if you believe that a kid should be able to express themselves for who they are while reaffirming their biological reality of being born a male or female. And finally, you are not transphobic if you are protecting the innocence of children who need guidance and wisdom when it comes to sex and gender, not an ideology which masks its perversion

---

[631] Malone, W., Wright, C., & Robertson, J. (2019).

16 THE GASLIGHTING IDEOLOGY

through compassion and acceptance. For the sexual health of our children, it's time to put an end to the gaslighting; it's time to stop being swayed by those who wish to have power over us and our children. It's time to stop letting our words be taken over by postmodern ideologues who want nothing more than power, influence, and control.

It's time to stand with Maya Forstater, JK Rowling, and Dr. Jordan Peterson and say that biological sex is a fundamental reality of human experience, a reality which cannot be changed through medical treatment. It's time we stand for those who have been hurt by the social constructionist ideology, who have been medically and psychologically abused so its doctrines can be solidified, and it's time we stand for those who have suffered from the reality of gender dysphoria. Once we allow for the wide variation of human behavior within the categories of male and female, we will see a more tolerant, more progressive, and more enlightened society. And only when we accept that individuals differ, rather than trying to equalize and dismantle all categories through the force of law, will we finally have true liberty. Doing so will dismantle constraining gender stereotypes while reaffirming our male-female reality.

As the decades pass, let this be known: *the theory of social constructionism will see its judgment day*. Once the gaslighting ideology begins to die and those who have been sexually abused by its doctrines witness the psychological destruction laid about before them, they will turn back to our institutions and wonder why our society let them damage beyond repair the priceless qualities which nature bestowed upon them; or, in the words of Watson, a young Scottish woman who de-transitioned:

> "Referrals to Scottish Gender Services for kids aged just 4 to 10 have risen by more than 80% in a year. The coming decade will see a slew of young adults with altered bodies and no sexual function, who will turn to the National Health Service and ask, 'Why did you let us do this to ourselves?'"[632]

---

[632] ImWatson91. (2019). In response to the medicalization of gender non-conformance. *Twitter*.

17

BEYOND GENDER

What would happen if we abandoned the ideology of social constructionism and left the doctrines of the blank slate to the wolves? What would happen, if each one of us, recognized the innate qualities within each individual, qualities which shape our choices, interests, and desires? What if we recognized the critical impact that biology and psychology play in making us who we are instead of reducing ourselves down to mere cogs in a machine of social forces? What if our biology and our psychology has much more to do with our choices and our actions than the transient forces of a cultural zeitgeist? What would happen if we recognized biology for what it was, rather than dismissing it?

In the eyes of the social constructionists, the postmodernists, the gender theorists, and the blank slate philosophers, such an embrace of biology would lead to more discrimination, more injustice, and more oppression. But what if such an understanding of biology and its interaction with social forces is the critical step towards less discrimination, less injustice, and less oppression? If we truly understood the complexity of how biology, psychology, and society shapes each individual, would we instead become more tolerant and accepting of others rather than more intolerant, divisive, abusive, and controlling? Would understanding the primordial forces of our biology help us become more empathic, more caring, and more willing to hear the opinions of others different from our own? What would happen, for instance, if we deeply understood the complexity of behavior within and between men and women? What if we recognized that, while there are important differences which make men and women different, there are many more important similarities? What would happen if we understood how men and women often differ in important traits yet share considerable overlap in others? What if we could recognize the biological reality of male and female while allowing for a wide variation of behavior within these categories? And what if this variation has more to do with biology than culture? After all, if nature has bestowed upon each individual priceless qualities, then perhaps understanding these qualities will help us better understand ourselves and others. Such a reality is within our grasp, closer than we realize.

What if we could recognize the differences between men and women, and the similarities, while rejecting harmful stereotypes? What if we could allow both men and women to explore their interests without social engineering from either side? What would happen if we allowed each boy and girl to express themselves for who they were without confusing them into gender dysphoria? What if we recognized that having interests atypical of your sex does not make you the opposite sex, but rather, like any trait, makes you a unique male or female? What if we could break forth from gender stereotypes and simply live our lives while still recognizing our fundamental biology of being a man or a woman? What if we could embrace the idea that just because you are a boy who likes to play with female-typical toys, does not mean you are actually a girl trapped in the wrong body? In a sense, what if we could move *beyond gender*?

What if we could embrace the idea that boys and girls will often have an interest in toys played by the opposite sex? What if we could move past rigid stereotypes of what it means to be a man and what it means to be a woman and embrace a view that goes beyond gender, a view which allows for the wide diversity of behavior among males and females while still recognizing the biological substrates?

Can we at the same time embrace biology and reject gender stereotypes? I believe the answer is *yes*. Can we embrace biology and reject discrimination? Yes. Can we understand that the differences between men and women, the healthy ones, are often important to our psychological and sexual health as we, at the same time, reject the pathologies of the masculine and the feminine? Yes. And finally, can we embrace biology and move beyond rigid definitions of gender? Yes. In fact, we can and should do all these things.

And yet, for the social constructionist ideologue, such integration of biology cannot take place. The sociocultural landscape is the only relevant force in human action and choice. Put aside any evidence presented throughout the book, and instead, consider this: what if such a social constructionist view continued to proliferate? Can social engineering really achieve the utopian dreams of the ideologues? Are individuals really as malleable as the social constructionists claim? What if the construction of a biological framework for understanding individual differences is a much more effective weapon against tyranny? What if understanding the biological substrates of our being allows us to form stronger arguments for the rights of all individuals? What if, at the end of the day, the social constructionist ideology of the blank slate is morally bankrupt? What if the idea that humans are born without a nature is built upon an unstable foundation of enticing lies? Instead, what if we could embrace a model grounded in biology, psychology, and society, integrating a variety of fields?

Our first explanation of the Gender Equality Paradox relied on the doctrines of social constructionism: *gender differences are merely the result of social conditioning.* And yet, as we explored further into the book, we began to find that the Gender Equality Paradox was not merely the result of social conditioning; instead, such a paradox was actually the result of men's and women's own choices and inclinations--choices which often led them to different career paths and inclinations which often gave them different personality traits.

Yet for the social constructionists, such differences, no matter how free or gender-equal the society, must be the result of social forces. And as we explored the depths of the social constructionist philosophy, we began to understand its core tenets: that subjective feelings reigned supreme; that biology played a small to non-existent force on human action; and that ultimately, there was no truth to our external reality except for the more ephemeral internal reality from which the only objective truth could be ascertained. *The internal truth, rather than external reality, was the ultimate shaper of the objective world.*

PARADOXES OF SOCIAL CONSTRUCTIONISM

Thus, throughout the social constructionist ideology of the blank slate, we discovered numerous paradoxes which could not be solved--paradoxes about reality, about gender identity, and about sex and gender:

> "On the one hand, [social constructionists] claim that the real self is something other than the physical body, in a new form of Gnostic dualism, yet at the same time they embrace a materialist philosophy in which only the material world exists. They say that gender is purely a social construct, while asserting that a person can be 'trapped' in the wrong gender.
>
> They say there are no meaningful differences between man and woman, yet they rely on rigid sex stereotypes to argue that 'gender identity' is real, while human embodiment is not. They claim that truth is whatever a person says it is, yet they believe there is a real self to be discovered inside that person. They promote a radical expressive individualism in which people are free to do whatever they want and define the truth however they wish, yet they try ruthlessly to enforce acceptance of transgender ideology.
>
> If gender is a social construct, how can gender identity be innate and immutable? How can one's identity with respect to a social construct be determined by biology in the womb? How can one's identity be unchangeable (immutable) with respect to an ever-changing social construct? And if gender identity is innate, how can it be 'fluid'?
>
> If the categories of 'man' and 'woman' are objective enough that people can identify as, and be, men and women, how can gender also be a spectrum, where people can identify as, and be,

both, or neither, or somewhere in between? What does it even mean to have an internal sense of gender? What does gender feel like? What meaning can we give to the concept of sex or gender—and thus what internal 'sense' can we have of gender—apart from having a body of a particular sex?

Apart from having a male body, what does it 'feel like' to be a man? Apart from having a female body, what does it 'feel like' to be a woman? What does it feel like to be both a man and a woman, or to be neither? Why should feeling like a man—whatever that means—make someone a man? Why do our feelings determine reality on the question of sex, but on little else? Our feelings don't determine our age or our height.

Of course, a transgender activist could reply that an 'identity' is, by definition, just an inner sense of self. But if that's the case, gender identity is merely a disclosure of how one feels. Saying that someone is transgender, then, says only that the person has feelings that he or she is the opposite sex. Gender identity, so understood, has no bearing at all on the meaning of 'sex' or anything else. But transgender activists claim that a person's self-professed 'gender identity' is that person's 'sex.'

Gender identity can sound a lot like religious identity, which is determined by beliefs. But those beliefs don't determine reality. Someone who identifies as a Christian believes that Jesus is the Christ. Someone who identifies as a Muslim believes that Muhammad is the final prophet. But Jesus either is or is not the Christ, and Muhammad either is or is not the final prophet, regardless of what anyone happens to believe. So, too, a person either is or is not a man, regardless of what anyone—including that person—happens to believe. The challenge for transgender activists is to present an argument for why transgender beliefs determine reality.

Determining reality is the heart of the matter, and here too we find contradictions. On the one hand, transgender activists want the authority of science as they make metaphysical claims, saying that science reveals gender identity to be innate and unchanging. On the other hand, they deny that biology is destiny, insisting that people are free to be who they want to be. Which is it? Is our gender identity biologically determined and immutable, or self-created and changeable? If the former, how do we account for people whose gender identity changes over time? Do these people have the wrong sense of gender at some time or other? And if gender identity is self-created, why must other people accept it as reality? If we should be free to choose our own gender reality, why can some people impose their idea of reality on others just because they identify as transgender?"[633]

No matter how we analyze it, these contradictions and paradoxes inside the social constructionist ideology cannot be solved through simple theorizing. What we need is scientific evidence, logic, and reason, not rhetorical and aesthetic arguments. Yet, here among the doctrines of social constructionism we find the core tenets of postmodernism: there is no objective truth, no way to ascertain objective knowledge, reason is a tool to oppress marginalized groups, and there is no such thing as *progress*.[634] Thus, the doctrines of postmodern social constructionism rely not on reason or logic, but on *authenticity, feeling,* and *indeterminacy*.[635] Reality is nothing more than one's subjective feelings, and if reality is nothing more than one's *internal* reality, then this reality must be accepted, not questioned:

> "At the core of the ideology is the radical claim that feelings determine reality. From this idea come extreme demands for society to play along with subjective reality claims. [Gender] ideologues ignore contrary evidence and competing interests, they disparage alternative practices, and they aim to muffle

---

[633] Anderson, R. (2018). Transgender ideology is riddled with contradictions. Here are the big ones. *Heritage Foundation.*
[634] See chapter, "The Postmodern Tyranny."
[635] Bonevac, D. (2013).

skeptical voices and shut down any disagreement. The movement has to keep patching and shoring up its beliefs, policing the faithful, coercing the heretics, and punishing apostates, because as soon as its furious efforts flag for a moment or someone successfully stands up to it, the whole charade is exposed. That's what happens when your dogmas are so contrary to obvious, basic, everyday truths."[636]

In response to any question of their religious-like doctrines, we showed how the social constructionist ideologues used gaslighting tactics to manipulate, control, and silence voices of dissent.[637] Outside of this ideology, however, there is a more enlightening world at our disposal.

BEYOND THE BLANK SLATE

If we wish to reject the dogma of the blank slate and the social constructionist ideology, then we should be willing to accept that individuals differ. They differ in their traits, desires, interests, preferences, and ultimately, their choices; each of which are influenced by a mix of biological, psychological, and social forces. If we wish to end discrimination, we cannot embrace a blank slate view of the world. Embracing the blank slate view means we can mold individuals in ways we see fit. Such a theory was tried throughout the 20th century, as utopian dreams of an equal state proliferated throughout the world, and yet, in its wake, hundreds of millions of people died.

We can no longer believe that humans are without a nature, for if we do, then such pathological beliefs will continue to damage us and destroy our societies from within. We cannot mold individuals for our own aims without serious consequences to the individual and to society, for our human nature will always return with a fierce vengeance. Thus, if we wish to end discrimination, we must accept that there are innate differences in people which cannot be altered through social forces. Such an acceptance

---

[636] Anderson, R. (2018).
[637] See chapter, "The Gaslighting Ideology."

of individual differences is often difficult for those who believe that acknowledging differences among people will lead to discrimination, but a critical scientific examination of individual differences will lead to the exact opposite; it will lead to more tolerance and more acceptance of others:

> "It's a staggering thing to understand how different we are innately," said Dr. Jordan Peterson, "and there's tremendous resistance to that idea even on the political front. That attitude is often described now by the people who are obsessed, ideologically, with the notion that we're only a consequence of our socialization. They call that 'biological essentialism' and think about it as politically inappropriate, but it's not at all."[638]

It is not biological essentialism to understand how individuals, including men and women, differ. In fact, our innate differences give us not just an intrinsic value which cannot be infringed upon but also a greater respect for others who do not share our perspectives:

> "The fact that you are a certain way gives you an intrinsic value and also, a certain intrinsic dignity," said Peterson, "You cannot even force yourself in some ways; you have to negotiate with yourself. And to know that people are the way they are, with some parameters for change, gives you respect for the diversity of opinion and perception that manifests itself in the world. It's really incumbent on you, if you are a conservative for example, to understand that there are unalienable advantages to a liberal perspective. Now, that doesn't mean that the liberal perspective is correct 100% of the time, but it's correct *some* of the time, and so is the conservative perspective. Wildly creative people tend to have contempt for conventional people. Often, the reverse is true too."

---

[638] Peterson, J. (2019). Introduction to Personality Psychology. *Discovering Personality.* 15m00s.

"But that's not helpful," said Peterson, "because we need conventional people to keep the conventions in place, but we also need innovative people to make change when it's necessary. That means that those two different types of people have to learn not only to put up with each other, but to appreciate the fact that those differences that can cause conflict actually exist. And so I would say that that genuine understanding of the range of diversity and the depth of that diversity also brings about something that approximates genuine tolerance and maybe the willingness to listen too. It's actually interesting once you understand how much people differ to talk to someone who is temperamentally different than you are because they don't see the world the same way, and if you listen, you might get some clues about what advantages their idiosyncratic or particular perceptual frame brings to that particular situation."[639]

And so, if individuals differ in their biology, psychology, perceptions, and interests, then equality of outcome is impossible to achieve. Because of this, we cannot end discrimination through equalizing individual outcomes; instead, we can put an end to discrimination through providing each individual with equal opportunities to aim for their desired goals. Such a doctrine of *equality of opportunity*, not equality of outcome, presupposes that biology and society interact in complex ways to produce differences in human choice. Equality of opportunity allows these unique aspects of our selves to interact in a free, diverse framework. Liberty is at the core, but not equality. True equality is when individuals share the same traits, desires, interests, and outcomes. Such a thing can never happen without the force of the state. Thus, what we should be striving for is *equality under the law* and equality of opportunity when it comes to people's choices.

---

[639] Peterson, J. (2019). Introduction to Personality Psychology. *Discovering Personality*. 16m00s.

Knowing this, solutions to gender inequality can be found in producing more equality of opportunity for men and women while understanding the reality of sex and gender differences: that in some cases, the outcomes between the sexes will not be the same; such outcomes in occupational disparities, for example, may not entirely equalize because men and women, on average, have different interests when it comes to their career choices. Such disparities between men and women are healthy as long as these choices are the result of liberty and not discrimination or injustice.

And so, first, we must recognize that male and female exist as biological realities if we wish to fight against discrimination. Denying sex differences, gaslighting the public into believing that sex exists on a spectrum, and manipulating science to say that biological sex can be changed is not *progressive*; such practices are illiberal and regressive, further complicating the already complex topic of sex and gender through lies, fabrications, and rhetoric. It makes fighting for men's and women's liberty much more difficult. After all, how can we champion what we cannot define, especially if the male-female category is seen as unreliable, defunct, and oppressive? Evolutionary psychologist Geoffrey Miller elucidates this paradox:

> "On one hand you get a wholesale denial of sex differences by a lot of the Left," said Miller, "On the other hand, you get a demonization of men, as everything that's wrong in history is due to men and patriarchy. On one hand there is no essential sex differences, and on the other hand the core essential sex difference is *men are evil*. And I do not know how we're going to reconcile those views. It makes a very difficult intellectual landscape to navigate."[640]

---

[640] Rebel Wisdom. (2019). Sex and evolutionary psychology, Geoffrey Miller & Diana Fleischman. *YouTube.* 10m00s.

Second, to fight discrimination and injustice correctly, we must recognize that males and females differ, on average, when it comes to their personalities, occupational interests, life choices, and even sexuality. Any disparity between males and females must therefore be analyzed within the context of many variables, including female choice. Understanding these sex differences in psychology and interest is important for our society's sexual health. Such an understanding of one's sexuality is also critically important for adolescent males who are unsure what to make of their own desires:

> "I think to understand your sexuality when you're an adolescent male..." said Miller, "it's virtually impossible if you don't know some animal behavior and you don't understand how sexual selection shapes males and females differently and how it's done that for at least four thousand species of mammals and three hundred species of primates. This is really deep stuff. It's programmed into us to do things like seek sexual variety and seek sexual novelty."[641]

Third, we must understand that femininity and masculinity are not mere cultural constructs but are, in large measure, patterns of behavior which have evolved over billions of years across sexually reproducing animals. While we should be willing to tear down the cultural caricatures of the masculine and the feminine, we should not try and eliminate the masculine and feminine archetypes themselves, the patterns of behavior we have evolved to embody. Masculinity and femininity must be embraced by every individual: we must learn to integrate our feminine side if we are a man, and also learn to integrate our masculine side if we are a woman. Such integration will produce a healthy and interactive syzygy, where neither the feminine nor the masculine will be pathologized, which is to say, to become harmful. Thus, we should not reject healthy masculine and feminine traits as cultural constructs; instead, we should embrace them as biological

---

[641] Rebel Wisdom. (2019). 17m45s.

realities. The evolution of masculine and feminine behaviors cannot be dismissed through sociocultural forces, for these archetypal patterns are much deeper than we realize:

> "It stretches far back enough into the past that we can find [the feminine and the masculine] across the animal kingdom," said evolutionary biologist Bret Weinstein, "If we look into plants, it is separately evolved but nonetheless, the same pattern emerges there. So we should alter the dynamics of sexuality for modern times, and we should do so with fairness in mind. I am married to a woman who views the world in masculine terms; she lives in a way that makes sense in masculine terms: she goes to the Amazon, she enjoys adventure. I am all in favor of women doing things that are not traditionally female, but I am not in favor of pretending that femininity was invented by men to keep women chaste or something like that. Masculinity and femininity were not invented by humans at all."[642]

By accepting the idea that masculine and feminine behavioral patterns are the result of biology, and not society, we can work to better integrate the masculine and feminine archetypes into our psyches. Doing so will produce a healthier and more well-rounded individual who can effectively embody both masculine and feminine traits.[643]

Fourth, to fight against discrimination and stereotyping, we must also recognize the variation within males and females, which is to say, the plenty of overlap between male and female traits. There is an incredible diversity within the male-female category, and through recognizing such diversity, we can put an end to rigid gender stereotypes, for not all males embody the average male and not all females embody the average female. Through this we must break down stereotypes and barriers to individual self-expression and not judge male and female behavior according to

---

[642] Rebel Wisdom. (2018). 'Jordan Peterson, sex, and ideology' with Bret Weinstein. *YouTube.* 10m50s.

[643] See chapter, "Gender Stereotype Pathology."

stereotypical thinking. We should allow individuals to express themselves and flourish, integrating both the feminine and the masculine when appropriate.

Fifth and finally, we must recognize that we can never achieve equality of outcome across all hierarchies. Engineering desired outcomes through socialization will not necessarily result in equal outcomes. Instead, we must embrace the idea that men and women, boys and girls, often differ in interests, and these differences in interest often lead them to different career paths. And that's okay! Through giving men and women, boys and girls, equality of opportunity, and not judging them based on gender stereotypes, we can finally achieve true liberty. But it is only through liberty (allowing people to choose their life-paths without sociocultural hindrance), not social engineering, where discrimination and injustice can be put to an end.

Perhaps both men and women can achieve true liberty once we finally allow them to express their true selves without social engineering. In fact, what if trying to eliminate all gender differences between men and women by treating them *more equally* only results in continued differences, as we have explored in the Gender Equality Paradox? And what if the social constructionist ideology does more harm than good? To move beyond the blank slate once and for all, consider the following video essay by psychologist Dr. Jordan Peterson which explores similar themes we have covered throughout the book:

> "In 2010, Sweden was voted 'most gender-equal society' by the World Economic Forum," notes Peterson, "A group of Swedish activists nonetheless concluded that this was not good enough: instead, they want to erase gender differences altogether and bring about *gender neutrality*. A new genderless pronoun, *hen*, has recently entered the Swedish language. A new children's book employs this pronoun; some clothing stores have done away with separate girls' and boys' sections. Social critics have stated that natural free play among children is wrong, because it is where gender stereotypes are 'born' and 'cemented,' while the

Swedish Green Party has suggested placing gender pedagogues in preschools to act as 'gender behavior watchdogs.' Who could possibly think that this would be anything but horrible?

Here's a counter idea to consider: what if gender roles themselves are virtuous? What if they represent ideals to aspire to? Consider these words: *ambitious, self-reliant,* and *tough*; and some others: *loving, unselfish,* and *kind.* The former are masculine virtues, traditionally and statistically. The latter, feminine. In the world of ideological theorizing, these qualities exist as arbitrarily imposed cultural prejudices; in the natural world: shaped by billions of years of evolution, gender differences are real and deep.

What if it is too much to ask each child to be a good girl, and a good boy, at the same time? What if such virtues cannot be initially developed without specialization? After all, real expertise at something takes 10,000 hours of practice. Consider this too: what if truly satisfying sex requires a certain male boldness and female acceptance? Is it really culture that drives the common romance fantasy of the dominant, aggressive male (billionaire, doctor, pirate, or vampire) who can only be tamed by the love of the right woman? What if the interaction between well-integrated men and women, in long-term monogamous relationships, allows male and female virtues to flourish equally, while limiting their exaggeration, so that men do not become pushy, aggressive, and arrogant, and women do not become indecisive, dependent, and resentful? What if this promotes the broadening of already established gender identity, not early, but later in life, when it's appropriate?

Finally and most ominously, what if the hypothetically well-meaning rejection of sexual differences produces a kick-back? *The reactionary over-development of masculine and feminine features instead of mere uniformity.* We already know that boys in a fatherless family tend to develop hyper-masculine traits; the girls in fatherless families hit puberty earlier and preteen girls are dressing in an increasingly provocative fashion. Here's another idea: no Clint Eastwood, more Freddy Kruger; no Marilyn Monroe, more Abercrombie & Fitch '*Feeling Lucky Thong Underpants for 10 year olds.*'

Only fools think that planned radical social change necessarily produce the desired result. Who would have ever imagined an explosion of teenage pregnancies thirty years after the birth control pill? As Tom Waites says, '*You can drive out nature with a pitchfork, but it always comes roaring back again.*'"[644]

BEYOND IDEOLOGY

And so, as we move beyond the blank slate and understand the reality of sex and gender differences and similarities, we must also move beyond the comfortable confines of ideology. We must instead embrace the pursuit of truth and knowledge, wisdom and courage. We must abandon rigid ways of thinking and pathological belief systems if we wish to have a functioning society. And yet to move beyond ideology is no easy task: it requires a critical eye, a courageous heart, and a willingness to integrate, *not reject*, useful criticism; such is the task of a free spirit.

The ideologue, in contrast, is no such free spirit but is rather a chained spirit, locked within their unerring convictions. There is no self-examination, no critical eye, and no intellectual courage and fortitude. Instead, their mind seeks out safety and comfort; theory explains everything: they are not merely convinced but *convicted* that their theory is correct. In many cases they are not liars or manipulators but are rather innocent minds gripped by an enticing ideology. They are, in a sense, spiritually convicted that their ideas are not just right, but moral and virtuous. Such a phenomenon is described by the video essayist from *Academy of Ideas*:

---

[644] The Agenda with Steve Paikin. (2012). Agenda insight: Beyond Gender? *YouTube*.

"A lie is an outward expression of a falsehood one inwardly knows to be false, meaning the liar can still know the truth. A conviction, on the other hand, is an inward certainty one has attained the truth, and thus in many cases, gives way to an arrogance that enmeshes one in a web of delusion and falsehood, and cuts one off from the possibility of moving towards knowledge.

'The claim that truth is found and that ignorance and error are at an end is one of the most potent seductions there is,' wrote Nietzsche in *The Will to Power*, 'Supposing it is believed, then the will to examination, investigation, caution, experiment is paralyzed... 'Truth' is therefore more fateful than error and ignorance, because it cuts off the forces that work toward enlightenment and knowledge.'"[645]

Moving beyond ideology requires that we are wary of proclaiming that we have the ultimate 'Truth,' for such a proclamation can be deceiving to ourselves and others. As Nietzsche states, 'Truth' can be more fateful than ignorance, for it can block the pursuit of enlightenment and knowledge. And here, we find that the social constructionist ideologues often champion the unbreakable doctrine of 'Truth' in exchange for ideological rigidity and intellectual atrophy. Unlike the ideologues, we must embrace skepticism and experimentalism if we wish to uncover the complexities of sex and gender. We must be willing to question our own beliefs, become free spirits, and transform into our own greatest skeptic:

"Employing the experimentalism advocated by Nietzsche involves becoming your own greatest critic, subjecting your cherished convictions to constant assessment, and attacking every so called 'Truth' you believe, in order to determine how strong its foundations really are. It involves actively seeking out and experimenting with new ideas, trying them on for size, so to

---

[645] Academy of Ideas. (2017). Nietzsche and truth: Skepticism and the free spirit. *YouTube*.

speak, and continually updating and improving your judgments about the world. After all, the ability to change our beliefs is a unique and precious capacity of the human mind--one of its defining features--but it must be continually exercised to prevent atrophy. Nietzsche's experimentalism trains this capacity, and is thus a counterforce against the attraction people feel towards conforming to narrow faiths and dogmatic visions of the world. As Nietzsche stated in *Dawn of Day*, 'The snake which cannot cast its skin has to die. As well the minds which are prevented from changing their opinions; they cease to be mind.'"[646]

To break free from ideological rigidity and intellectual atrophy, we must expose our ideas to contrarian perspectives. We must be willing to update and improve our knowledge, our beliefs, and our values through exposure to opposing ideas and philosophies. Such a practice is terrifying, as it can reveal entire gaps in our knowledge, in our beliefs, and even in our value systems. But without such a practice of exposure, we can never hope to grow and mature--we can never hope to become a *skeptic*:

> "Make it a rule never to withhold or conceal from yourself anything that may be thought against your own thoughts. Vow it! This is the essential requirement of honest thinking. You must undertake such a campaign against yourself every day,' wrote Nietzsche in *Dawn of Day*. Most people are unable to abide by this daily practice as their personal identity becomes tied up with certain beliefs they hold on faith. Such people become fearful of new and challenging ideas, seeing them as a threat to their character and worldview. The skeptic, in contrast, adopts a more profitable approach by maintaining a proper distance from his beliefs. He is therefore able to play with ideas, move in and out of them with grace and suppleness, and use them as tools in the service of a heroic goal."

---

[646] Academy of Ideas. (2017). Nietzsche and truth: Skepticism and the free spirit. *YouTube*.

Such an ability to play with ideas and step back from your deeply held beliefs is perhaps the greatest intellectual skill of them all. It allows you to consider different ideas objectively, analyze their pros and cons, and even explore the strengths and weaknesses of your own. When doing so, you are truly a free spirit, as no one idea can take hold of you. For the social constructionist ideologues, however, an idea has taken hold of their psyches; they are bound spirits, prisoners of their own beliefs. If they are willing to step back and analyze their ideas with a skeptical mind, then they too can break free from ideological possession and become free spirits:

> "While the vast majority of people are 'bound spirits,' prisoners of beliefs that have been inculcated into them by their parents, governments, and religions, the free spirit is one who has liberated himself from these chains: 'The term 'free spirit' here is not to be understood in any other sense; it means a spirit that has become free, that has taken possession of itself,' wrote Nietzsche in *Ecce Homo*.
>
> In contrast to bound spirits, whose weakness motivates them to censor and label as *dangerous* ideas which challenge their worldview, the free spirit, as 'a monster of courage and curiosity...a born adventurer and discoverer' (Nietzsche), is driven to grasp even the treacherous truths which would destroy the weak."[647]

Thus, once we break free from our ideological chains, we will become, in Nietzsche's words, a "monster of courage and curiosity, a born adventurer and discoverer." Such courage can only manifest itself if we make the search for enlightenment and knowledge a continual priority in our lives--if we put our ideas to the test and develop the most thorough critical eye. Unfortunately, it seems, most people do not want to expend the energy required for such a courageous task, for they often do not want to engage with the possibility that their ideas might be wrong. But it is only

---

[647] Academy of Ideas. (2017). Nietzsche and truth: Skepticism and the free spirit. *YouTube*.

through making such an effort to examine one's beliefs where true maturity and wisdom can flourish. Those who search for enlightenment and knowledge will be free from the chains of ideology; they will have the free spirit on their side, unable to be enticed by well-sounding but bankrupt ideas, and they will develop a depth of intellectual understanding and humility unusual for the common man. Such a free spirit also becomes the ultimate enemy of the ideologues who see intellectual depth and freedom as a danger to their continual cultivation of ignorance, power, and control:

> "But with the spirit of the search for truth on their side, the free spirit also understands that in the search for truth it is not only terribleness that one will find, but truths that are liberating both to the individual and society, and which, for various reasons, are often hidden away or deemed blasphemous by mainstream opinion. In seeking these truths, the free spirit--the 'genuine and solitary philosopher'--becomes an enemy of all those who attempt to promote ignorance for the sake of gaining power over others. As master of his mind, the free spirit forms an internal vault untouchable by those who wish to deceive, and thus, perhaps unknowingly, keeps alive the flame of truth even in darker ages of ignorance, censorship, and tyranny."[648]

It is thus the free spirit, not the bound spirit, who is the master of his mind, who understands his own beliefs, and who critically examines them; quite the opposite from the bound spirit, who is the slave to his own mind, who does not understand his own beliefs, and who keeps them contained in a locked vault free from critical examination. Therefore, it is the free spirit, not the bound spirit, who threatens the ideologue's wishes of power, influence, and control:

> "As Nietzsche penned: 'Where there have been powerful societies, governments, religions, public opinions, in short

---

[648] Academy of Ideas. (2017). Nietzsche and truth: Skepticism and the free spirit. *YouTube*.

wherever there has been tyranny, there the solitary philosopher has been hated; for philosophy offers an asylum to a man into which no tyranny can force its way, the inward cave, the labyrinth of the heart: and that annoys the tyrants."[649]

To annoy the tyrants, the great illiberal gatekeepers of enlightenment and knowledge, is to question the ideological zeitgeist of the time, not for the sake of disagreement, but for the critical pursuit of truth and wisdom. The social constructionist ideology is the gatekeeper of our time. Silencing dissent, shutting down voices of disagreement, banning speakers from college campuses, and slinging offensive accusations to those who dare question the ideology are all tactics of illiberal gatekeeping. Such tactics are used, ironically, not to annoy the tyrants, but to serve them. For in serving the tyrants, the ideologue also reinforces and solidifies the present system. So too is the case with social constructionists, who have inculcated their doctrines throughout our institutions, universities, businesses, and governments. Any critical examination of the doctrines of social constructionist ideology are akin to blasphemy, as questioning the status quo through free thought and open inquiry also means to question the legitimacy of the tyrants and their grip on the collective unconscious.

To maintain their ideology, strong social constructionists reject scientific evidence in exchange for the comfort of utopian daydreams. They reject the biological reality of male and female, that males and females are not entirely the same, that biology and psychology have roles to play in creating the wide diversity of human behavior, and that some gender differences may be the result of free choice and human prosperity. To maintain such hegemony over our institutions, the ideologues must eliminate any and all gender differences regardless of whether these disparities are the result of liberty and freedom of choice. If such gender differences in personality, interests, and occupational preferences can be eliminated, then the ideology of social constructionism can be proven true. And for the ideologues who wish nothing more than to see their theory

---

[649] Academy of Ideas. (2017). Nietzsche and truth: Skepticism and the free spirit. *YouTube*.

solidified as the ultimate 'Truth,' there can be no critical examination of their own beliefs, no questioning of their values, and no integration of new insights.

Ideology is comfortable. It requires no intellectual work, no challenge of your deeply held beliefs, and no courageous questioning of the status quo. Thus, as the postmodern ideologue rejects reason, logic, and skepticism--the very keys which can free them from their ideological chains--the vitality of their beliefs grow cold. Because they are not willing to adapt their convictions to the scientific evidence, the chains will remain, and the postmodern social constructionist ideologue will face the ultimate death of intellectual atrophy. As the light fades and the sun finally sets on their decaying ideology, the social constructionists will choose to hold onto the doctrines of the blank slate until the end of time. For the social constructionist who is unwilling to free themselves from their ideological chains and accept a more nuanced model for sex and gender differences and similarities, the Gender Equality Paradox will remain forever unsolved.

Even though understanding biology is crucial for the sexual, psychological, and physical health of our civilization, and even though understanding sex and gender differences can help us dismantle constraining stereotypes and allow people to express their true selves, the social constructionist ideologue will continue to gaslight, manipulate, and reject any evidence which challenges their religious dogma of the blank slate. In their intellectual atrophy, they will find themselves in the chains of their own making--the true bound spirits--unwilling to experience the wonders of knowledge and enlightenment. While they protect their theory from critical examination through gaslighting, manipulation, and abuse, they will attempt to mold their external reality to fit within their internal subjective feelings.

But, like the postmodernists, all the ideologues have is *words*. It is only through language where the ideologues can hijack the free spirit of the courageous philosopher and construct a false image of skepticism and curiosity; they will attempt to convince you with their persuasive words, their knowledgeable dispositions, and their effective gambits, but do not be fooled: *the ideologues are mere apparitions, for they hold no substance.*

When the host of a Norwegian documentary visited social constructionists at the Nordic Gender Institute, he found, not free spirits in search of enlightenment and knowledge, but bound spirits closed off from the beauty of intellectual exploration. Unlike the biologists and psychologists who explained gender differences through biology, psychology, and society, the social constructionists could only allow for one variable: *socialization*.

When one gender sociologist was pressed to consider the biological and psychological research for the causes of gender differences, he pushed back: "So far science has not been able to prove a genetic origin to gender differences," he said.[650]

The host's eyes grew wide as he leaned towards the sociologist. "That's because you do not recognize the studies which show this," he said, calmly.

"But they have a missing link!" the man responded with eyes wide and frustration in his face.

"How do you know it is not innate?" the host responded, "How do you know there are no gender differences?"

"My hypothesis is that there are none," said the sociologist, "Science has not shown any."

After this discussion, the host visited another gender sociologist at the institute. Here he showed her that gender differences in toy preferences have been replicated in nonhuman mammals, and that prenatal androgens seem to have a large effect on the differentiation of male and female psychology, often leading men and women to different career paths. But, despite this evidence, the sociologist doubled down: *all gender differences are the result of social forces,* she said.

"What is your scientific basis to say that biology plays no part in the two genders' choice of work?" the host asked as he leaned forward.

"My scientific basis?" the sociologist asked.

The host nodded.

"I have what you would call a *theoretical* basis," she said, as she looked back at the host, *"There's no room for biology in there for me."*

---

[650] Eia, H. (2010). Hjernevask (Brainwash), "The Gender Equality Paradox." *YouTube*.

18

BIBLIOGRAPHY

Every scientific paper has an extensive bibliography which references the research utilized throughout the paper. In *The Gender Paradox*, I used hundreds of sources from the fields of biology, personality psychology, evolutionary psychology, neuroendocrinology, neurology, and sociology, sociology of gender, anthropology, and philosophy. These widely diverse sources are referenced here, ordered by the major categories within sex and gender research for more effective reference. The references within the categories are organized by date published, rather than alphabetical order. I've divided the categories into two major sections: Biopsychosocial Literature and Sociology of Gender Literature.

Biopsychosocial Literature

INTRO TO SEX DETERMINATION, SEX AND GENDER DIFFERENCES

Brown, D. (2002). Human Universals, New York: McGraw-Hill.

Moore, D. (2003). The Dependent Gene: The Fallacy of "Nature vs. Nurture." Holt Paperbacks; M. Ridley, Nature via Nurture: Genes, Experience, and what makes us Human. Harper Collins.

Dewing, P., Shi, T., et al. (2003). Sexually dimorphic gene expression in mouse brain precedes gonadal differentiation. *Molecular Brain Research, 118*, 82.

Williams, T., Carroll, S. (2009). Genetic and molecular insights into the development and evolution of sexual dimorphism.

Sekido, R., Lovell-Badge, R. (2009). Sex determination and SRY: Down to a wink and a nudge? *Trends in Genetics, 25(1)*, 19-29.

Zhao, D., et al. (2010). Sometic sex identity is cell autonomous in the chicken. *Nature, 464*, 1-8.

Matson, C., et al. (2011). DMRT1 prevents female reprogramming in the postnatal mammalian testis. *Nature, 476*, 101-107.

Matson, C., Zarkower, D. (2012). Sex and the singular DM domain: Insights into sexual regulation, evolution, and plasticity. *Nature Reviews in Genetics, 13*, 163-174.

Gamble, T., Zarkower, D. (2012). Sex determination. *Current Biology, 22(8)*.

Wood, W. & Eagly, A. (2013). Biology or culture alone cannot account for human sex differences and similarities. *Psychological Inquiry, 24(3)*, 241-247.

Zarkower, D. (2013). DMRT genes in vertebrate gametogenesis. *Current Topics in Developmental Biology, 102*, 327-356.

Wood, W. & Eagly, A. (2013). The nature-nurture debates: 25 years of challenges in understanding the psychology of gender. *Perspectives on Psychological Science, 8(3)*, 340-357.

Hyde, J. (2014). Gender similarities and differences. *Annual Review of Psychology, 65*, 373-398.

Schmitt, D. (2016). Sex and gender are dials (not switches). *Psychology Today.*

F.P.F.W. (2017). Biological sex differences: bones & muscles. *Fair Play for Women.*

Damore, J. (2017). Google's Ideological Echo Chamber. *Damore-Google Manifesto.*

Schmitt, D. (2017). The truth about sex differences. *Psychology Today.*

Schmitt, D. (2017). On that Google memo about sex differences. *Psychology Today.*

Wootson, C. (2017). A Google engineer wrote that women may be unsuited for tech jobs. Women wrote back. *Washington Post.*

CONCEPTION TO CHILDHOOD

Berenbaum, S., Hines, M. (1992). Early androgens are related to childhood sex-typed toy preferences. *Psychological Science, 3(3)*, 203-207.

Pasterki, V., et al. (2005). Prenatal hormones and postnatal socialization by parents as determinants of male-typical toy play in girls with congenital adrenal hyperplasia. *Child Development, 76(1)*.

Cohen-Bendahan, C., Beek, C., Berenbaum, S. (2005). Prenatal sex hormone effects on child and adult sex-typed behavior. *Neuroscience and Biobehavioral Reviews, 29.*

Hassett, J. Siebert, E., Wallen, K. (2008). Sex differences in rhesus monkey toy preferences parallel those of children. *Hormones and Behavior, 54,* 359.

Alexander, G. M., et al. (2009). Sex differences in infants' visual interest in toys. *Archives of Sexual Behavior, 38.*

Jerome, B. (2010). Reproduction and Development. *Visual Learning Systems, In Amazing Human Body, Narrated by Nina Keck.*

Beltz, A., Swanson, J., Berenbaum, S. (2011). Gendered occupational interests: Prenatal androgen effects on psychological orientation to Things versus People. *Hormones and Behavior, 60.*

Bao, A., Swaab, D. (2011). Sexual differentiation of the human brain: Relation to gender identity, sexual orientation and neuropsychiatric disorders. *Frontiers in Neuroendocrinology, 32.*

Zimmer, R., Riffell, J. (2011). Sperm chemotaxis, fluid shear, and the evolution of sexual reproduction. *PNAS, 108(32).*

Hines, M., Constantinescu, M. (2012). Relating prenatal testosterone exposure to postnatal behavior in typically developing children. *Child Development Perspectives, 6(4).*

Sperling, M. (2014). *Pediatric Endocrinology 4th Edition.* Elsevier, Inc. 107-156.

Sharma, R., Seth, A. (2014). Congenital Adrenal Hyperplasia: Issues in diagnosis and treatment in children. *Indian Journal of Pediatrics, 81(2),* 178-185.

Escudero, P., et al. (2014). Sex-related preferences for real and doll faces versus real and toy objects in young infants and adults. *Journal of Experimental Child Psychology, 116,* 367-379.

Hines, M., et al. (2015). Early androgen exposure and human gender development. *Biology of Sex Differences, 6(3),* 1-10.

Hines, M., et al. (2016). Prenatal androgen exposure alters girls' responses to information indicating gender-appropriate behavior. *Philosophical Transactions, 371.*

Lonsdorf, E. (2017). Sex differences in nonhuman primate behavioral development. *Journal of Neuroscience Research, 95,* 231-221.

Smith, L. (2018). *Your pregnancy at week 3.* Medical News Today.

Todd, N. (2018). Conception & Pregnancy, Ovulation, and Fertilization. *WebMD.*

Genetics Home Reference. (2019). SRY gene, sex development. *U.S. National Library of Medicine.*

BRAIN AND COGNITION

Bachevalier J., Hagger, C. (1991). Sex differences in the development of learning abilities in primates. *Psychoneuroendocrinology, 16,* 177-188.

Kimura, D. (1992). Sex differences in the brain. *Scientific American.*

Pinker, S. (2004). Why nature and nurture won't go away. *Daedalus 133(4).*

Falter, C.M., et al. (2006). Testosterone: Activation or organization of spatial cognition. *Biological Psychology, 73,* 132-140.

Pinker, S. (2006). The blank slate. *General Psychologist.*

Cahill, L. (2006). Why sex matters for neuroscience. *Nature Reviews Neuroscience.*

Pfannkuche, K.A., Bouma, A., Groothuis, T.G. (2009). Does testosterone affect lateralization of brain and behaviour? A meta-analysis in humans and other animal species. *Philosophical Transactions Royal Society London B, Biological Sciences, 364 (1519),* 929–942.

Shepard, K., et al. (2009). Genetic, epigenetic, and environmental impact on sex differences in social behavior. *Physiology & Behavior, 97,* 157-170.

Peper, J.S., et al. (2009a). Sex steroids and brain structure in pubertal boys and girls. *Psychoneuroendocrinology, 34,* 332-342.

Lippa, R., Collaer, M., Peters, M. (2010). Sex differences in mental rotation and line angle judgments are positively associated with gender equality and economic development across 53 nations. *Archives of Sexual Behavior, 39,* 990-997.

Ngun, T., et al. (2010). The genetics of sex differences in brain and behavior. *Frontiers in Neuroendocrinology, 32(2011).*

Blakemore, S.J., Burnett, S., & Dahl, R. (2010). The role of puberty in the developing adolescent brain. *Human Brain Mapping, 31.*

Hines, M. (2011). Gender development and the human brain. *Annual Review of Neuroscience, 34.*

Ellis, L. (2011). Identifying and explaining apparent universal sex differences in cognition and behavior. *Personality and Individual Differences, 51.*

Berenbaum, S., Beltz, A. (2011). Sexual differentiation of human behavior: Effects of prenatal and pubertal organizational hormones. *Frontiers in Neuroendocrinology, 32.*

Bao, A., Swaab, D. (2011). Sexual differentiation of the human brain: Relation to gender identity, sexual orientation and neuropsychiatric disorders. *Frontiers in Neuroendocrinology, 32.*

Tomasi, D., Volkow, N. (2011). Laterality patterns of brain functional connectivity: Gender effects. *Cerebral Cortex, 22,* 1461.

Uematsu, A., et al. (2012). Developmental trajectories of amygdala and hippocampus from infancy to early adulthood in healthy individuals. *PLoS ONE, 7(10).*

Khazan, O. (2013). Male and Female Brains Really Are Built Differently. *The Atlantic.*

Little, A. (2013). The influence of steroid sex hormones on the cognitive and emotional processing of visual stimuli in humans. *Frontiers in Neuroendocrinology, 34.*

McHenry, J., et al. (2013). Sex differences in anxiety and depression: Role of testosterone. *Frontiers in Neuroendocrinology, 35.*

Jantz, G. (2014). Brain differences between genders. *Psychology Today.*

Yang, C., Shah, N. (2014). Representing Sex in the Brain, One Module at a Time. *Neuron, 82,* 263.

Celec, P., et al. (2015). On the effects of testosterone on brain behavioral functions. *Frontiers in Neuroscience, 2.*

Davey, R., Grossmann, M. (2016). Androgen receptor structure, function, and biology: from bench to bedside. *Clinical Biochemist Reviews, 37(1),* 3-15.

Bayless, D., Shah, N. (2016). Genetic dissection of neural circuits underlying sexually dimorphic social behaviors. *Philosophical Transactions B, 371.*

Chekroud, A., et al. (2016). Patterns in the human brain mosaic discriminate males from females. *PNAS, 113(14).*

Marrocco, J., McEwen, B. (2016). Sex in the brain: hormones and sex differences. *Dialogues in Clinical Neuroscience, 18,* 373-383.

Goldman, B. (2017). Two minds: the cognitive differences between men and women. *Stanford Medicine.*

Zhang, C., et al. (2017). Functional connectivity predicts gender: evidence for gender differences in resting brain connectivity. *Human Brain Mapping, 39,* 1765.

Brooks, C., et al. (2017). Sex/gender influences on the nervous system: Basic steps towards clinical progress. *Journal of Neuroscience Research, 95(14-16).*

Herting, M., Sowell, E. (2017). Puberty and structural brain development in humans. *Frontiers in Neuroendocrinology, 44,* 122-137.

Amen, D. (2017). Gender-based cerebral perfusion differences in 46,034 functional neuroimaging scans. *Journal of Alzheimer's Disease.*

Williams, J. (2018). The amygdala: definition, role & function. *Study.com*.

Anderson, N., et al. (2018). Machine learning of brain gray matter differentiates sex in a large forensic sample. *Human Brain Mapping, 40,* 1496-1506.

Sepehrband, F., et al. (2018). Neuroanatomical morphometric characterization of sex differences in youth using statistical learning. *NeuroImage, 172,* 217-227.

Van Putten, M., et al. (2018). Predicting sex from brain rhythms with deep learning. *Scientific Reports, 8,* 1.

Cortes, L, Cisternas, C., Forger, N. (2019). Does gender leave an epigenetic imprint on the brain? *Frontiers in Neuroscience, 13(173),* 3.

Luo, Z., et al. (2019). Gender identification of human cortical 3-D morphology using hierarchical sparsity. *Frontiers in Human Neuroscience, 13(29).*

Cherry, K. (2019). Hippocampus role in the limbic system. *Verywellmind.com*.

BEHAVIOR

Goy, R., et al. (1988). Behavioral masculinization is independent of genital masculinization in prenatally androgenized female rhesus macaques. *Hormones and Behavior, 22,* 552-571.

Wallen, K. (1996). Nature needs nurture: The interaction of hormonal and social influences on the development of behavioral sex differences in rhesus monkeys. *Hormones and Behavior, 30,* 364-378.

Eagly, A. & Wood, W. (1999). The origins of sex differences in human behavior: Evolved dispositions versus social roles. *American Psychologist, 54(6),* 408-423.

Berenbaum, S., Bailey, J. (2003). Effects on gender identity of prenatal androgens and genital appearance: evidence from girls with congenital adrenal hyperplasia. *Journal of Clinical Endocrinology and Metabolism, 88,* 1102-1106.

Romeo, R. (2003). Puberty: A period of both organizational and activational effects of steroid hormones on neurobehavioral development. *Journal of Neuroendocrinology, 15,* 1185-1192.

Archer, J. (2004). Sex differences in aggression in real-world settings: A meta-analytic review. *Review of General Psychology, 8(4),* 291-322.

Wallen, K. (2005). Hormonal influences on sexually differentiated behavior in nonhuman primates. *Frontiers in Neuroendocrinology, 26(1),* 7-26.

Else-Quest, N., Hyde, J., et al. (2006). Gender differences in temperament: A meta-analysis. *Psychological Bulletin, 132(1),* 33-72.

Silverman, I., Choi, J., and Peters, M. (2007). The hunter-gatherer theory of sex differences in spatial abilities: Data from 40 countries. *Archives of Sexual Behavior, 36,* 261-268.

Mortenson, T. (2011). For every 100 girls. *Pell Institute for the Study of Opportunity in Higher Education.*

Fisher, A. (2013). Boys vs. girls: What's behind the college grad gender gap? *Forbes.*

Bailey, J., Vasey, P., Diamond, L., et al. (2016). Sexual Orientation, Controversy, and Science. *Psychological Science in the Public Interest, 17(2),* 56.

Scientific American Editors. (2017). The new science of sex and gender. *Scientific American.*

Horwitz, S. (2017). Truth and myth on the gender pay gap. *Foundation for Economic Education.*

Kristiansen, J., Sandnes, T. (2018). Women and men in Norway. *Statistics Norway.*

(2019). Gender Economic Inequality. *Inequality.org.*

PERSONALITY PSYCHOLOGY

C.G. Jung. (1990). *The Archetypes and the Collective Unconscious (Collected Works of C.G. Jung Volume 9 Part 1).* Bollingen Series XX: Princeton University Press, 82.

McCrae, R., Oliver, J. (1992). An introduction to the five-factor model and its applications. *Journal of Personality, 60(2),* 184.

Edinger, E. (1992). *Ego and Archetype.* C.G. Jung Foundation Book Series (Book 4), Shambhala.

Feingold, A. (1994). Gender differences in personality: a meta-analysis. *Psychological Bulletin, 116(3),* 429-456.

Lippa, R. (1998). Gender-related individual differences and the structure of vocational interests: the importance of the people–things dimension. Journal of *Psychology and Social Psychology, 74(4),* 996-1009.

Costa, P., et al. (2001). Gender differences in personality traits across cultures: Robust and surprising findings. *Journal of Personality and Social Psychology, 81(2).*

Boeree, C. G. (2002). A bio-social theory of neurosis. *Shippensburg University.*

Chapman, B., et al. (2007). Gender differences in Five Factor Model personality traits in an elderly cohort. *Personality and Individual Differences, 43,* 1594-1603.

Schmitt, D., et al. (2008). Sex differences in Big Five personality traits across 55 cultures. *Journal of Personality and Social Psychology, 94(1),* 179.

Su, R., & Rounds, J., et al. (2009). Men and things, women and people: a meta-analysis of sex differences in in interests. *Psychological bulletin, 135(6),* 859-884.

Mathews, G., et al. (2009). Personality and congenital adrenal hyperplasia: Possible effects of prenatal androgen exposure. *Hormones and Behavior, 55.*

Lippa, R. (2010). Sex differences in personality traits and gender-related occupational preferences across 53 nations: Testing evolutionary and socio-environmental theories. *Archives of Sexual Behavior, 39,* 619-636.

Weisberg, Y., et al. (2011). Gender Differences in Personality across the Ten Aspects of the Big Five. *Frontiers in Psychology, 2(178).*

Solc, V. (2013). Father Archetype. *Therapy Vlado.*

Elliott, Z. (2017). *Sex Differences: A Land of Confusion.* Lulu Press. 39.

Peterson, J. (2017). Jordan Peterson on Diversity. *Bite-sized Philosophy, YouTube.*

Peterson, J. (2018). Jordan Peterson explains Jung's animus and anima. *Bite-sized Philosophy, YouTube.*

Peterson, J. (2018). Identity Politics & The Marxist Lie of White Privilege. *Sovereign Nations, YouTube.* 48:00.

Unbelievable? (2018). Jordan Peterson vs Susan Blackmore: Do we need God to make sense of life? *YouTube.*

Peterson, J. (2019). Introduction to Personality Psychology. *Discovering Personality.*

Peterson, J. (2019). Extraversion: Enthusiasm and Assertiveness. *Discovering Personality.*

Peterson, J. (2019). Agreeableness: Compassion and Politeness. *Discovering Personality.*

Peterson, J. (2019). Conscientiousness: Industriousness and Orderliness. *Discovering Personality.*

Peterson, J. (2019). Openness to Experience: Intellect and Openness. *Discovering Personality.*

Peterson, J. (2019). Men & Women: Personality Differences, Lecture Notes 6. *Discovering Personality.*

PERFORMANCE

Halpern, D., et al. (2007). The science of sex differences in science and mathematics. An introduction to the five-factor model and its applications. *Journal of Personality, 60(2),* 184.

Lindberg, S., Hyde, J. (2010). New trends in gender and mathematics performance: A meta-analysis. Psychological Bulletin, 136(6).

Stoet, G., Geary, D. (2010). The gender equality paradox in STEM education. *Psychological Science.*

Eia, H. (2010). Hjernevask (Brainwash), "The Gender Equality Paradox." *YouTube.*

Wang, Eccles, and Kenny (2013). Not lack of ability but more choice: Individual and gender differences in choice of careers in science, technology, engineering, and mathematics. *Psychological Science, 20(10),* 1-6.

Stoet, G., and Geary, D. C. (2015). Sex differences in academic achievement are not related to political, economic, or social equality. *Intelligence, 48,* 137-151.

Jussim, L. (2017). Why brilliant girls tend to favor non-STEM careers. *Psychology Today.*

Sociology of Gender Literature

INTERSEX INFANTS AND THE JOHN/JOAN CASE

Money, J., Ehrhardt, A. (1972). *Man and Woman, Boy and Girl.* Baltimore: John Hopkins University Press.

Money, J., Tucker, P. (1975). *Sexual Signatures: On Being a Man or Woman.* Boston: Little, Brown, 98.

Money, J. (1985). The conceptual neutering of gender and the criminalization of sex. *Archives of Sexual Behavior, 14(3),* 280.

Kessler, S. (1990). The Medical Construction of Gender: Case Management in Intersexed Infants. *Journal of Women in Culture and Society,* 16(11), 25.

Diamond, M., Keith, H. (1997). Sex reassignment at birth: a long term review and clinical implications. *Archives of Pediatrics and Adolescent Medicine, 151.*

Thorne, B. (1993). *Gender play: Girls and boys in school.* New Brunswick, NJ: Rutgers University Press. 2.

Colapinto, John (2001). *As Nature Made Him: The Boy Who Was Raised as a Girl.* New York: HarperCollins. 33-34.

Bullough, V. (2003). The Contributions of John Money: A Personal View. *The Journal of Sex Research, 40(3)*, 232.

Ehrhardt, A. (2007). John Money, Ph.D. *The Journal of Sex Research, 44(3)*, 223-224.

Gaetano, P. (2017). David Reimer and John Money gender reassignment controversy: The John/Joan Case. *The Embryo Project Encyclopedia*.

CONSTRUCTION OF SEX AND GENDER

Connell, R. W. (1987). *Gender and power: Society, the person and sexual politics.* Cambridge: Polity, 188.

West and Zimmerman, (1987). Doing gender. *Gender and Society, 1(2)*.

Strate, L. (1992). Beer commercials: A manual on masculinity. *Men, Masculinity, and the Media*, Sage Publications, 534-535.

Johnson, A. (1997). Patriarchy, the System. *The Gender Knot: Unraveling our Patriarchal Legacy*, 96.

Karin, M. (1998). Becoming a Gendered Body: Practices of Preschools. *American Sociological Review, 63(4)*, 494.

Fausto-Sterling, A. (2000). *Sexing the body: Gender politics and the construction of sexuality.* Basic Books.

Kimmel, M. (2000). Masculinity as homophobia: Fear, shame, and silence in the construction of gender identity.

Lindsey, E., Mize, J. (2001). Contextual differences in parent-child play: Implications for children's gender role development. *Sex Roles, 44(3/4)*, 1-22.

Kane, E. (2006). 'No way my boys are going to be like that!' Parents' responses to children's gender nonconformity. *Gender & Society, 20(2)*.

Blakemore, J., Hill, C. (2007). The child gender socialization scale, a measure to compare traditional and feminist parents. *Sex Roles, 58*, 192-207.

Moss-Racusin, C. A., et al. (2010). When men break the gender rules: status incongruity and backlash against modest men. *Psychology of Men & Masculinity, 11(2)*, 141.

The Agenda with Steve Paikin. (2012). Agenda insight: Beyond Gender? *YouTube*.

McLaughlin, H., et al. (2012). Sexual harassment, workplace authority, and the paradox of power. *American Sociological Review, 77(4)*, 635.

Howard, L., Rose, J. (2013). Raising African-American boys: An exploration of gender and racial socialization practices. *American Journal of Orthopsychiatry, 83(2,3)*, 218-230.

Sherwani, A. et al. (2014). Hysterectomy in a male? A rare case report. *International Journal of Surgery Case Reports, 5(2014)*, 1285-1287.

Tenebaum, H., Leaper, C. (2014). Parent-child conversations about science: The socialization of gender inequities? *Developmental Psychology, 39(1)*, 34-47.

Dreger, A. (2014). The social construction of sex. *Pacific Standard*.

Lawson, K., et al. (2015). Links between family gender socialization experiences in childhood and gendered occupational attainment in young adulthood. *Journal of Vocational Behavior, 90*, 26-35.

Ristori, J., Steensma, T. (2015). Gender dysphoria in childhood. *International Review of Psychiatry, 28(1)*.

Brannon, L. (2016). *Gender: Psychological Perspectives*. Routledge Press, 162.

Carlson, A. (2016). Sex, biological functions and social norms: A simple constructivist theory of sex. *Nordic Journal of Feminist and Gender Research, 24(1)*, 18-29.

Escalante, A. (2016). Gender nihilism: An anti-manifesto. *Libertarian Communist*.

Halpern, H., Jenkins-Perry, M. (2016). Parents' gender ideology and gendered behavior as predictors of children's gender-role attitudes: A longitudinal exploration.

Ludden, D. (2016). When Sex and Gender Don't Match. Psychology Today.

Boe, J., Woods, R. (2017). Parents' Influence on Infants' Gender-Typed Toy Preferences. *Sex Roles*.

McNamara, B. (2017). Period Activist Cass Clemmer Responds To Hate After Posting Period Photo. *Teen Vogue*.

Nichols, J. (2017). Women Aren't The Only Ones Who Get Periods. HuffPost.

Coleman, N. (2017). Trans man gives birth to baby boy. CNN.

Scientific American Editors. (2017). The new science of sex and gender. *Scientific American*.

(2017). Questions and Answers about Gender Identity and Pronouns. *Ontario Human Rights Commission*.

(2018). Feminist Perspectives on Sex and Gender. *Stanford Encyclopedia of Philosophy*.

Joyce, H. (2018). The new patriarchy: How trans radicalism hurts women, children, and trans people themselves. *Quillette*.

Kollmayer, M., et al. (2018). Parents' Judgments about the Desirability of Toys for Their Children. *Sex Roles*.

Ainsworth, C. (2018). Sex redefined: The idea of 2 sexes is overly simplistic. *Scientific American*.

(2018). Simone de Beauvoir. *Stanford Encyclopedia of Philosophy*.

McLaughlin, H. (2018). Gendered institutions. *Sociology of Gender, Oklahoma State University*.

McLaughlin, H. (2018). What is sex? *Sociology of Gender, Oklahoma State University*.

McLaughlin, H. (2018). Gender Socialization: How and When do We Learn the Rules? *Sociology of Gender, Oklahoma State University*.

McLaughlin, H. (2018). Creating Social Change. *Sociology of Gender, Oklahoma State University*.

McLaughlin, H. (2018). Doing gender. *Sociology of Gender, Lecture at Oklahoma State University*.

Chiu, J. (2019). Trans children sense their gender identities at young ages, study suggests. *NBC*.

Malone, W., Wright, C., & Robertson, J. (2019). No one is born in the 'wrong body.' *Quillette*.

POSTMODERNISM

Nietzsche, F. (1882). *The Gay Science*.

Baudrillard, J. (1976). *Symbolic Exchange and Death*, Ian Hamilton Grant (trans.), London: Sage Publications, 1993.

Derrida, J. (1994). *Specters of Marx*. Routledge.

Hicks, S. (2004). *Explaining Postmodernism: Skepticism and Socialism from Rousseau to Foucault*. Scholargy Publishing, Inc.

Bonevac, D. (2013). Postmodernism. *University of Texas, Austin*.

(2015). Postmodernism. *Stanford Encyclopedia of Philosophy*.

Sarkis, S. (2017). 11 warning signs of gaslighting. *Psychology Today*.

Academy of Ideas. (2017). Nietzsche and truth: Skepticism and the free spirit. *YouTube*.

Anderson, R. (2018). Transgender ideology is riddled with contradictions. Here are the big ones. *Heritage Foundation*.

Peterson, J. (2018). Equity: When the left goes too far. *Jordan Peterson*.

Bowles, N. (2018). Jordan Peterson: Custodian of the patriarchy. *New York Times*.

19

GLOSSARY

After writing this book, it became apparent that it needed a thorough glossary to briefly explain each of the main concepts and terminology used throughout. Such a glossary can help those wanting to learn the often complex terms regarding sex and gender while providing a quick and easy reference point. From social constructionism and biological essentialism, to gender identity, gender expression, and intersex, this glossary will give you a brief run-down of each major concept in *The Gender Paradox*. For definitions and references to additional concepts with their page numbers, see the Index.

Activational effects
- Sex hormones during puberty such as testosterone and estrogen activate dormant neural circuits which were organized in the womb. These activational effects during puberty produce changes in sexual behavior, psychology, and the brain.

Androgen
- A group of hormones that play a role in male-typical psychological traits and reproductive activity. Androgens affect the fetal neuroanatomy and neurocircuitry, producing structural changes in the brain such as *suboptimal arousal* and *lateralization*. Girls exposed to high levels of prenatal androgens often have a condition called *Congenital Adrenal Hyperplasia* and exhibit male-typical psychological traits such as an interest in things over people and an affinity for manipulating objects in three-dimensions.

Anima
- The set of feminine traits in a male, often repressed in the unconscious due to societal conditioning.

Animus
- The set of masculine traits in a female, often repressed in the unconscious due to societal conditioning.

Archetypes
- A collectively-inherited unconscious idea, pattern of thought, or image that is universally present in individual psyches (Jungian psychology).

Average Differences
- Used to describe differences within and between groups which are normally distributed. On a given trait, many people will fall near the average, while a handful will fall on either extreme. Average differences in sex and gender are useful for understanding generalized patterns of behavior.

Big Five
- The five-factor model in personality psychology developed from linguistic factor analysis. The Big Five is used to describe essential traits of human psychology, which are Extraversion, Agreeableness, Conscientiousness, Neuroticism, and Openness to Experience.

Bimodal distribution
- A distribution with two modes, or averages, which are normally distributed. Similar to average differences, bimodal distributions explain patterns of behavior across two populations, such as males and females. Sex and gender differences arrange themselves in bimodal distributions, not spectrums or binary switches. Males and females, therefore, share considerable overlap in a given trait.

Biological essentialism
- The belief that an individual's traits (personality, psychology, and behavior) are fixed and innate, rather than a product of society and culture. The exact inverse of social constructionism.

Biological sex
- Two distinct reproductive capacities which produce two functionally different gametes. Small, mobile gametes are sperm. Large, generally immobile gametes are ova (eggs). Males produce sperm. Females

produce ova. Disorders of sexual development produce abnormalities in chromsomes, hormones, and genitals.

Biopsychosocial model
- An interdisciplinary model that looks at the interconnection between biological, psychological, and sociocultural factors and how they influence individual behavior. Biopsychosocial model views sex and gender differences as products of biology, psychology, and society, each of which interact together.

Congenital Adrenal Hyperplasia
- A condition where a fetus, often an XX zygote, is exposed to abnormally high levels of prenatal androgens, which in turn produce abnormal genitals such as an enlarged clitoris and fused labia in girls. Girls with CAH exhibit male-typical psychological traits from infancy to adulthood despite socialization.

Cognitive Social Learning Theory
- A theory which argues that sex and gender differences are produced through external behavioral reinforcements and punishments which become internalized as 'innate' traits.

Deconstruction
- Postmodernist philosopher Jacques Derrida's method of textual analysis, used to trace the linguistic connections between texts. Used to expose arbitrary and subjective language patterns.

Disorders of Sexual Development
- Medical conditions involving the reproductive system which produce abnormal sexual characteristics such as extra or missing sex chromosomes, gonadal deficiencies, and anatomical abnormalities. DSDs are incredibly rare, at around 0.05% of births, and do not constitute new sexes. *Congenital Adrenal Hyperplasia* is an example of a DSD.

Egalitarianism
- The principle that all people are equal and deserve equal rights and opportunities.

Enlightenment
- A European intellectual movement of the late 17th and 18th centuries emphasizing reason and individualism rather than tradition. It was heavily influenced by 17th-century philosophers such as Descartes,

Locke, and Newton, and its prominent exponents include Kant, Goethe, Voltaire, Rousseau, and Adam Smith.

Epigenetics
- The study of changes in organisms caused by modification of gene expression rather than alteration of the genetic code itself. Embodiment of gender roles may have epigenetic effects on DNA.

Equality of opportunity
- A political philosophy, or ideal, that believes people ought to be able to compete on equal terms, or on a "level playing field," for social, economic, and political positions. Similar to Egalitarianism.

Equality of outcome
- Opposite of Equality of Opportunity. A political philosophy, or ideal, that believes individual outcomes in social, economic, and political positions should be equal, regardless of skill, merit, or individual choice. Similar to Equity.

Equity
- In the non-financial sense, the quality of being fair and impartial. In the sociopolitical sense, the idea that representations of individuals across social, economic, and political positions must be equalized. For example, people who believe in equity will push for equal representations of men and women in a given hierarchy.

Evolutionary Theory
- For sex and gender, a subset of natural selection which produces sexually differentiated adaptations between males and females used for specific reproductive capacities and sex roles.

Feminine archetype
- Jungian term for the universal ideal feminine behavioral patterns:
- *"Maternal [care] and sympathy; the magic authority of the female; the wisdom and spiritual exaltation that transcend reason; any helpful instinct or impulse; all that is benign, all that cherishes and sustains, that fosters growth and fertility. The place of magic transformation and rebirth, together with the underworld and its inhabitants, are presided over by the mother."*

Femininity
- Qualities or attributes regarded as characteristic of women. Feminine qualities can be developed in a man through *anima* integration.

Feminized
- A neuroscience term relating to a brain structure which has been made female-typical due to exposure to estrogen and progesterone. Specific regions of the brain can experience feminization through estrogen specific receptors, such as the hippocampus, which is related to learning and memory.

Gender binary
- The classification of gender into two distinct, opposite, and disconnected forms of the masculine and the feminine.

Gender Equality Paradox
- The social science phenomenon that occurs when gender differences in personality and interests grow larger as a society becomes more gender-equal and gender egalitarian.

Gender Expression
- A person's behavior, mannerisms, interests, and appearance that are associated with gender within the categories of femininity and masculinity.

Gender Identity
- A person's sense of one's own gender, which can align with one's biological sex or can differ from it. Most individuals' gender identities align with their biological sex.

Gender Identity Disorder (GID)
- A psychological disorder where an individual exhibits marked and persistent identification with the opposite sex and persistent discomfort (dysphoria) with his or her own sex or sense of inappropriateness in the gender role of that sex.

Gender Inequality Index
- A United Nations Development Programmed (UNDP) index which measures the human development costs of gender inequality. The higher the GII value, the more disparities between men and women. Scandinavian countries have the lowest GII (most gender equality), whereas parts of the Middle East and Africa have the highest GII (least gender equality).

Gender Roles
- A set of behaviors, values, and attitudes deemed appropriate for both males and females. Gender roles are often seen as separate and binary rather than integrative and interactionist.

Gender similarities hypothesis
- A statistical analysis of sex and gender differences by psychologist Janet Hyde which states that men and women are more alike than different across all traits. 78% of gender differences are zero to small, whereas 22% are moderate to large.

Gender stereotypes
- An often harmful phenomenon where gendered behavioral patterns are generalized and viewed as fixed and binary. Male and female behaviors are then judged according to these patterns. Extending to individuals an observation of behavioral differences between groups is stereotyping. Individual behavior, being more variable *within* populations than *between* them, makes stereotyping a slippery slope.

Genotypic
- The genetic makeup of an organism or group of organisms with reference to a single trait, set of traits, or an entire complex of traits. The sum total of genes transmitted from parent to offspring. Males and females have distinct genotypic traits from sex chromosomes.

Greater male variability hypothesis
- A theory which argues that the greater variability in traits seen in men is a consequence of the fact that men are more disposable as they are less likely to reproduce successfully. This is exemplified in IQ statistics. Though males and females have the same average IQ levels, there tends to be both more male geniuses and more male idiots. Females tend to have a more balanced distribution of IQ.

Hermaphroditism
- An organism that has complete or partial reproductive organs and produces gametes normally associated with both male and female sexes. Functional hermaphroditism (where the two reproductive capacities function simultaneously) does not exist in humans. But vestigial structures, which do not function, can exist internally.

Hormone treatment
- Administration of hormones, sometimes sex hormones in the case of some transgender individuals, to alter one's physiology and morphology to more closely resemble the opposite sex.

Indefinite variation
- An argument used by social constructionists and gender theorists which states that biological sex exists on a spectrum and that individuals cannot be sorted into male and female categories.

Individual differences
- Differences in neuroanatomy, neurocircuitry, personality, behaviors, and interests among individuals. There is more variation *within* groups of people than there is *between* groups. Diversity is achieved through individual differences, not group differences.

Inequality
- Difference in size, degree, circumstances, etc.; the existence of an imbalance regardless of whether the difference resulted from injustice or free choice. Low representation of women in STEM fields is an example of gender inequality.

Injustice
- A lack of fairness or justice; in the field of sex and gender research, injustice would be discrimination of someone due to their sex or gender.

Interactionist approach
- The idea (from Eagly and Wood) that the best approach to studying sex and gender differences and similarities is to analyze how biology and society interact to mold men and women.

Intersex
- A generic term for disorders of sexual development (DSDs) which produce chromosomal, hormonal, and/or gonadal abnormalities. Intersex does not mean someone is a new sex (which would require the formation of a third gamete type), but rather that the default female or male developmental pathways were altered. Most intersex infants grow up to identify as male or female and usually do not experience gender identity disorder. (See Congenital Adrenal Hyperplasia).

Jacques Derrida
- French postmodernist intellectual from the Marxist tradition. Believed that the West could be conceptualized as Logocentric, or logic-centered,

and that this Logocentrism could be deconstructed through textual analysis.

Janet Hyde
- Psychologist and Women's Studies Professor who developed the *Gender Similarities Hypothesis*.

Jean-Francois Lyotard
- French postmodernist intellectual who described postmodernism as "incredulity towards meta-narratives."

John Money
- Famous sexologist of the 1950s and 1960s who performed surgical and psychological experiments on intersex infants and infants with damaged genitalia. He believed infants were born psychosexually neutral and that the appearance of the genitals was critical to normal psychosexual development. Genetics and hormones were theorized to be independent from gender identity. His famous experiment, the John/Joan case, led to thousands of male infants being castrated and raised as girls. The John/Joan case led to numerous suicides, namely that of David Reimer and his twin brother.

Jordan Peterson
- Professor of Psychology at the University of Toronto, Dr. Peterson is an expert in individual differences, religious mythologies, and personality psychology. He gained notoriety for his outspoken opposition to Bill C-16 in Canada, a compelled-speech proposition which wrote gender identity and gender expression into law as innate/protected classes. Author of *12 Rules for Life: An Antidote to Chaos* and *Maps of Meaning*. His lectures on personality psychology and Maps of Meaning can be found on YouTube.

Judith Butler
- Feminist and queer theorist of the 1990s who is a main proponent of the strong social constructionist view that sex and gender are both indistinct social constructs. Differences between men and women are the result of power structures created through language.

Kane
- Professor of Sociology and Gender and Sexuality Studies at Bates College. Teaching interests encompass inequalities of race, class, gender and sexuality; gender and family; sociology of childhood; research methods;

public opinion; social psychology; community-based research; and community-engaged learning.

Kimmel
- A sociologist specializing in gender studies who believes that men's lives are structured around relationships of power.
- *"Men's lives are structured around relationships of power and men's differential access to power, as well as the differential access to that power of men as a group. Our imperfect analysis of our own situation leads us to believe that we men need more power, rather than leading us to support feminists' efforts to rearrange power relationships along more equitable lines."*

Lateralization
- A neuroscience term used to describe alterations in neurocircuitry which emphasize certain sides of the brain. Lateralization can occur through prenatal androgens, which causes structural systems to be rewired so that they favor visuospatial cognition. Occurs in more males than females. Females tend to be less lateralized and more interconnected between the left and right hemispheres.

Male-typical
- Traits, behaviors, and interests which are usually observed in males but can also be seen in females.

Masculine archetype
- Jungian term for the universal ideal masculine behavioral patterns:
- *"The father archetype, in its logos function, exerts his influence on the human mind in order to transform undifferentiated--concrete--emotionality bind to body, to less material and abstract form of images and mental representations, but also, in its specific complex-bound function, serves as regulator of boundaries, restrictions, and social values that are imposed as rules and laws."*

Masculinity
- Qualities or attributes regarded as characteristic of men. Masculine qualities can be developed in a woman through *animus* integration.

Masculinized
- A neuroscience term relating to a brain structure which has been made male-typical due to exposure to androgens. Specific regions of the brain can experience masculinization through androgen-specific receptors,

such as the amygdala, which is related to fear, emotion, and sex drive. Masculinized brain structures can also relate to *lateralization*.

Michel Foucault
- French postmodernist intellectual of the Marxist tradition who believed reason was a rhetorical device used to marginalize oppressed groups.
- *"Reason is a power that defines itself against an other, a faculty seen operating outside of reason is not allowed to speak for itself and is at the disposal of a power that dictates the terms of their relationship."*

Neuroanatomy
- The study of the structure and organization of the nervous system, including the brain.

Neurocircuitry
- The study of the system of neural connections within the nervous system, including the brain, and how they interact.

Neuroendocrinology
- The study of hormones' effects on the nervous system, especially the brain and its functions.

Organizational effects
- Effects of sex hormones on the fetal brain structure (neuroanatomy and neurocircuitry) which further differentiates male and female psychology before birth.

Origins of sex and gender differences
- Scientific research field which studies how sex and gender differences and similarities have arisen through biological, psychological, and social processes.

Ovaries
- A female reproductive organ in which ova or eggs are produced, present in humans and other vertebrates as a pair.

Parental encouragement
- A process of social conditioning through behavioral reinforcement which produces learned behaviors, specifically gender-appropriate behaviors.

Patriarchy
- A system of society or government in which men hold the power and women are largely excluded from it. Iran represents the epitome of a *patriarchy*. Feminists often call the West a *patriarchy*.

Performance
- A subset of sex and gender research, performance deals with sex differences in cognitive tasks and abilities such as mathematics, spatial ability, and verbal learning. Males and females share considerable overlap in these tasks. Some abilities, however, show differential specialization.

Personality psychology
- A branch of psychology that studies personality and its variation among individuals. It is a scientific study which aims to show how people are individually different due to psychological forces.

Phenotypic
- Relating to the observable characteristics of an individual resulting from the interaction of its genotype with the environment.

Postmodernism
- A philosophy of relativism, skepticism, and cynicism which rejects the concepts of truth, knowledge, reason, and historical progress. Epistemic certainty and objective meaning are thrown out the window in exchange for an air of uncertainty and subjectivity in regards to every aspect of reality.

Psychopathology
- The scientific study of mental disorders. Men and women tend to experience different types of psychopathologies.

Puberty
- The activational period during which adolescents reach sexual maturity and become capable of reproduction.

Rhesus macaques
- Small non-human primates which show similar toy preferences to those seen in children.

Sex determination
- A biological system that determines the development of sexual characteristics in an organism. Sex in humans is determined through two distinct developmental pathways initiated by the presence or absence of the SRY gene. (For a diagram of this, see Chapter 7.)

Sex difference research
- An overarching scientific field which includes a variety of professions interested in understanding the origins, complexity, and development of sex differences between males and females.

Sex hormones
- Male-specific and female-specific hormones which produce structural differences in the fetal brain, as well as activational effects during puberty. Sex hormones mediate sexual behaviors and maintain sexually dimorphic characteristics and interests. Both sexes have testosterone and estrogen, just in differing degrees.

Sex prediction
- A neuroscience term relating to methods used to determine sex through measurements of an individual's brain. For example, 3D cortical morphology mapping has made predicting sex in the brain 97% accurate.

Sex roles
- Different from gender roles; relates to sexually dimorphic reproductive and sexual behaviors in males and females, such as breastfeeding.

Sex-differentiated
- A broad term for any trait which can be differentiated by sex. For example, neuronal structures in the brain can develop sex-differentiated properties through exposure to sex hormones.

Sex-specific gene
- A gene which activates through action of sex hormones or sex chromosomes. Researchers have discovered more than 50 sex-specific genes in the brain.

Simone de Beauvoir
- A French feminist philosopher and social theorist who provided the foundational methods and framework for the critiques of patriarchy laid out by contemporary gender theory. She believed that men and women were different yet should be treated equally.
- *"One is not born but becomes a woman."*

Social constructionism
- A theory of knowledge which states that differences in behavior, personality, and interests among individuals are the result of learned behaviors. Strong social constructionists tend to ignore biological explanations. Exact inverse of *biological essentialism*.

Sociocultural Theory
- A theory which proposes that a society's division of labor by gender drives all other psychological gender differences which are not the result of sex-category. This theory acknowledges the existence of biological differences between men and women and says that gender differences are mapped onto sex-category through the culture's division of labor.

Sociology of Gender
- A prominent subfield of sociology. Social interaction directly correlated with sociology regarding social structure. One of the most important social structures is status. This is determined based on position that an individual possesses which effects how they will be treated by society.

Spatial ability
- The ability to manipulate and visualize objects in three-dimensions. It's a common sex-differentiated cognitive ability which has a greater tendency to be more specialized in males than females. However, females exposed to higher androgens often exhibit this specialization as well. Tends to relate to lateralization.

Spatial location memory
- The ability to remember one's location based on objects and landmarks. It's a common sex-differentiated cognitive ability which has a greater tendency to be more specialized in females than males. Can be affected by estrogen exposure in humans and rats.

Spectrum
- A common social constructionist term which describes both sex and gender as *spectrums,* where every individual represents a unique category. However, this is statistically incorrect. Biological sex (defined through gametes) exists on a mostly binary distribution, whereas sexual characteristics and gendered behaviors tend to exist on *bimodal distributions.*

SRY gene
- The presence of a gene known as the *sex-determining region Y protein* (SRY) on the Y chromosome causes the development of testes in males, which then release sex-specific hormones.

STEM
- The fields of science, technology, engineering, and mathematics. Disparities between men and women in these fields in the West tends to

be high, yet these disparities tend to not exist in more traditional and patriarchal nations such as those found in the Middle East and Africa. Iran's science graduates are 67% female.

Suboptimal arousal
- The neurological tendency to get bored with one's environment relatively quickly and seek out new stimuli. More common in males than in females due to high prenatal androgen exposure. Relates to lateralization.

Systematizing
- A male-typical cognitive trait relating to the arrangement of data and objects into organized systems. Males tend to show greater interests in such tasks compared to females.

Syzygy
- An interactive and corresponding pair which cannot be separated, often occurring in nature. Examples are yin and yang, chaos and order, moon and sun, feminine and masculine.

Testes
- A male reproductive organ in which spermatozoa are produced, present in humans and other vertebrates as a pair.

The Genderbread Person
- A social constructionist diagram which claims anatomical sex, sex assigned at birth, gender, gender identity, gender expression, and sexual orientation vary independently from one another.

Things-People Dimension
- A psychological dimension which measures a person's interest in things-*oriented* activities versus people-oriented activities. The Things-People dimension is one of the largest sex differences between males and females, with an effect size of $d = 0.93$.

Toy preferences
- Studies on sex and gender differences often focus on toy preferences in infants and toddlers. Toy preferences are affected by prenatal androgen exposure, sex-specific genes, and socialization. Girls with Congenital Adrenal Hyperplasia show male-typical toy preferences.

Transgender
- Denoting or relating to a person whose sense of personal identity and gender does not correspond with their birth sex.

Visual interest
- A term in sex difference literature which relates to how long an infant or toddler stares at a given object, particularly things vs people-based objects.

Visuospatial cognition
- Relates to spatial ability; more specifically, a neuroscience term describing the link between the visual and spatial systems in the brain and how they're wired. Visuospatial cognition is often wired differently in males than in females.

Wage gap
- An economic statistic showing that women earn about 80 cents to every dollar a man earns, calculated by taking all of the average incomes of full-time workers and separating it by one variable: sex. The wage gap reduces to a few percentage points once other variables, such as differences in job types, life choices, and hours worked, are considered.

West and Zimmerman
- Sociologists and gender theorists who developed the term *doing gender*; theorized that gender is not just a set of traits but a performative act.

Women's equality
- Could mean multiple things. 1) Women's legal equality and equal treatment. 2) Women's equality across all hierarchies, where women are equally represented in every social, economic, and political position.

20

INDEX

A

activational effects, 110, 115, 224, 226, 245, 251, 252
adrenarche, 225, 226
aggression, 33, 34, 43, 100, 120, 136, 138, 149, 183, 187, 202, 241, 243, 246, 285
Agreeableness, 87, 112, 264, 265, 270, 272, 273, 275, 276, 283, 285, 287, 289, 323, 329, 357, 358, 442, 452
amygdala, 147, 235, 237, 238, 241, 243, 244, 328, 356, 439, 440, 460
androgen receptors, 232, 236, 237, 239, 240, 243, 244
anima, 208, 209, 210, 212, 220
animus, 208, 209, 210, 212, 219, 220
anxiety, 245, 246, 247, 261, 262, 287
archetypes, 203, 204, 206, 211, 212, 213, 214, 215, 216, 217, 219, 220, 252, 256, 328, 354, 421, 422
autism, 153, 205, 246
average differences, xiii, 18, 31, 228, 278

B

Baron-Cohen, 153, 189, 190, 278
Big Five, xi, 255, 261, 264, 265, 267, 268, 269, 270, 273, 276, 277, 289, 291
bimodal distribution, 106, 107, 109, 113, 118, 119, 125, 135, 248, 273, 290
binary, x, 2, 3, 4, 5, 6, 8, 9, 10, 11, 12, 41, 44, 50, 67, 107, 109, 110, 111, 114, 115, 117, 129, 131, 165, 166, 167, 181, 182, 195, 197, 198, 203, 211, 215, 323, 343, 344, 345, 346, 347, 350, 353, 366, 370, 371, 373, 376, 380, 388, 391, 400, 452, 455, 456, 463

biological essentialism, 3, 97, 126, 323, 418, 451, 462
biological sex, 3, 4, 5, 6, 7, 9, 10, 11, 12, 16, 19, 24, 32, 49, 120, 129, 139, 141, 160, 167, 168, 169, 170, 193, 248, 250, 252, 256, 284, 304, 325, 326, 327, 340, 343, 344, 345, 346, 347, 348, 352, 360, 366, 367, 368, 369, 370, 371, 372, 373, 374, 375, 376, 377, 378,379, 380, 381, 382, 384, 385, 386, 391, 396, 397, 398, 399, 403, 404, 406, 408, 409, 420, 455, 457
biopsychosocial model, 11, 33, 35, 49, 80, 96, 98, 102, 105, 115, 120, 123, 125, 126, 129, 135, 141, 143, 194, 198, 289, 299, 324, 325, 330, 331
bone density, 230
bone structure, 13, 134, 140, 228
brain structure, 13, 39, 134, 135, 140, 151, 155, 157, 177, 182, 186, 199, 216, 236, 238, 244, 250, 252, 253, 257, 438

C

CAH, 11, 135, 153, 154, 155, 156, 158, 159, 161, 170, 186, 187, 188, 189, 191, 238, 239, 243, 251, 285, 286
career, ix, 45, 76, 83, 168, 200, 201, 259, 290
Carl Jung, 204
chromosomes, ix, 9, 12, 38, 41, 105, 124, 131, 133, 135, 140, 144, 145, 171, 172, 173, 239, 248, 325
cognitive social learning, 200, 250, 251
Cognitive Social Learning Theory, 120, 122, 124, 143, 158, 159, 167, 289
conception, 40, 102, 131, 143, 145, 174, 251, 252, 253
Congenital Adrenal Hyperplasia, 11, 135, 152, 153, 155, 186, 285, 302, 327, 345, 351, 437, 451, 453, 457, 464
Conscientiousness, 264, 265, 266, 270, 276
cortex, 147, 150, 235, 236, 249

D

David Reimer, 40, 176, 192, 193, 216, 256
David Schmitt, 114, 270, 280
Deconstruction, 62, 63
degrees, xiv, 83, 88, 102, 109, 121, 166
depression, 245, 246, 271, 288
dials, 107, 109, 110, 114, 115, 119, 125
discrimination, xii, 3, 4, 17, 19, 25, 32, 34, 54, 64, 67, 76, 78, 79, 80, 82, 84, 87, 88, 89, 90, 100, 102, 117, 165, 166, 290, 295, 296, 297, 298, 299, 306, 310, 312, 316,

330, 339, 340, 360, 368, 375, 378, 379, 380, 382, 383, 384, 387, 388, 389, 397, 411, 412, 417, 419, 420, 421, 422, 423, 457

disorders of sexual development, 11, 15, 344, 345, 346, 371, 373, 378, 457

disparity, 50, 69, 80, 81, 82, 88, 109, 278, 279, 290, 299, 306, 308, 309, 310, 314, 316, 319, 330, 358, 360, 421

diversity, xiii, 10, 24, 90, 91, 114, 140, 143, 258, 259, 260, 261, 270, 291, 292

division of labor, 123, 124, 278, 463

DNA, 140, 144, 151, 223, 224, 231, 251

dolls, 42, 44, 178, 181, 183, 184, 186, 191, 197, 215, 285

E

Eagly and Wood, 125, 126

effect size, 118, 119, 270, 272, 275, 276

egalitarian, x, xi, xii, 71, 77, 81, 102, 161, 215, 260, 277, 278, 280, 281, 283, 284, 288, 289

electroencephalograms, 249

embodiment, 27, 49, 50, 251

engineering, 83, 101, 198, 199, 200, 201, 278

Enlightenment, 28, 30, 54, 55, 56, 61, 63, 69, 70, 71, 72, 76, 77

epigenetics, 71, 96, 123, 139, 140, 151, 231, 283

equality of opportunity, 32, 101, 260, 291, 319, 419, 420, 423

equality of outcome, 32, 67, 161, 291, 312, 364, 419, 423

equity, 24, 37, 67, 69, 72, 78, 88, 89, 90, 93, 100, 102, 161, 213, 290, 295, 297, 310, 312, 319, 320, 340, 454

estrogen, 134, 135, 150, 169, 178, 179, 226, 227, 231, 234, 236, 238, 240, 244, 246

Evolutionary Theory, 120, 121, 122, 124, 143

Extraversion, 264, 265, 270, 276, 289

F

fatal strategies, 65, 66, 67, 72

female-typical, 111, 113, 135, 149, 150, 151, 158, 159, 160, 161, 177, 187, 188, 189, 191, 200, 233, 237, 238, 239, 243, 245, 246, 247, 248, 250, 251, 285, 286, 312, 313, 326, 327, 328, 332, 346, 355, 360, 399, 412, 455

feminine, xii, 33, 34, 35, 36, 40, 43, 44, 45, 48, 111, 134, 141, 154, 174, 178, 183, 191, 199, 201, 202, 203, 204, 205, 206, 207, 208, 209, 210, 211, 212, 213, 214, 215, 216, 217, 218, 219, 220, 247, 251, 252, 256, 355

feminine archetype, 206, 207, 211, 212, 214, 219

femininity, 25, 36, 45, 110, 202, 203, 214, 348, 349, 353, 354, 355, 404, 421, 422, 455

fMRI, 138

follicle-stimulating hormone (FSH), 226

G

gaslighting, 340, 343, 361, 365, 366, 370, 373, 374, 375, 381, 389, 394, 396, 397, 405, 407, 408, 409, 417, 420, 431, 447

gender binary, 41, 182

gender differences in personality, 93, 257, 265, 268, 269, 270, 271, 272, 275, 276, 278, 279, 280, 281, 282, 284, 285, 288, 289, 290

gender dysphoria, 177, 179, 367, 368, 398, 405, 407, 408, 409, 412

Gender Equality Paradox, xi, xii, xiii, xiv, 77, 223, 253, 277, 278, 298, 315, 316, 322, 323, 324, 326, 331, 340, 432, 443, 455

gender expression, 44, 107, 108, 159, 194, 203, 256

gender gaps, xii, 2, 84, 87, 340, 343, 358, 360, 361, 363, 364, 376

gender identity, 12, 13, 38, 39, 40, 41, 107, 108, 134, 147, 155, 156, 160, 162, 165, 166, 167, 168, 169, 170, 171, 172, 173, 174, 175, 176, 177, 180, 181, 182, 184, 190, 191, 192, 193, 194, 195, 210, 223, 252, 256, 327, 346, 348, 350, 351, 364, 368, 369, 371, 381, 382, 383, 384, 385, 386, 387, 391, 396, 397, 398, 399, 400, 401, 402, 403, 404, 405, 414, 415, 424, 437, 439, 440, 444, 451, 457, 458, 464

gender identity disorder, 169, 170, 192, 193, 256

gender inequality, 53, 77, 78, 84, 297, 310, 420, 455, 457

Gender Inequality Index, x, 277, 319

gender similarities hypothesis, 118

gender stereotypes, 42, 43, 44, 46, 102, 117, 118, 182, 194, 195, 197, 198, 200, 201, 202, 203, 206, 210, 211, 214, 216, 219, 220, 252, 256, 268, 295, 310, 328, 353, 355, 368, 375, 376, 396, 398, 399, 400, 401, 403, 408, 409, 412, 422, 423

gender-appropriate, 115, 135, 159, 167, 187, 188, 200, 219

gender-atypical, 214, 278, 295, 300, 330, 346, 397, 398, 399, 401, 402, 403, 404, 405, 406, 407, 408

gender-equal, x, xi, xii, xiv, xv, 1, 2, 51, 77, 81, 231, 253, 260, 275, 276, 280, 281, 284, 287, 289, 298, 303, 312, 315, 316, 319, 322, 323, 329, 330, 331, 350, 358, 359, 360, 364, 413, 423, 455

gender-typical, 2, 139, 158, 200, 215, 299, 322, 350

gender-unequal, xiii, 278, 279, 284, 322

genetics, ix, 11, 40, 41, 71, 110, 123, 124, 138, 139, 141, 162, 172, 173, 194, 229, 252, 253, 256, 257, 260, 290, 291
genotypic, 145, 200
gestation, 38, 99, 114, 121, 133, 146, 156, 158
Glass Ceiling, v, 310, 311, 312, 314
gonadarche, 225, 226, 227, 234, 235, 236
greater male variability hypothesis, 274
grip strength, 115, 229

H

height, 106, 113, 144, 252
hermaphroditism, 14, 38, 130
hierarchy, 20, 26, 50, 51, 53, 54, 57, 58, 59, 60, 62, 63, 64, 65, 67, 69, 70, 71, 72, 78, 92, 243
hippocampus, 150, 235, 237, 238, 240, 243, 328, 356, 439, 455
hormone treatment, 170, 193, 285
hypothalamic-pituitary-adrenal, 225
hypothalamic-pituitary-gonadal, 225, 227, 232
hypothalamus, 147, 225, 234

I

ideology, 6, 25, 37, 56, 68, 89, 100, 126, 141, 161, 297, 304, 316, 323, 331, 339, 340, 353, 361, 363, 365, 366, 369, 370, 374, 376, 380, 396, 397, 399, 401, 402, 405, 408, 409, 411, 413, 414, 416, 417, 422, 423, 425, 426, 429, 430, 431, 445, 447
in utero, 110, 131, 198, 256, 261, 325, 326, 327, 328, 356, 377
income, 71, 78, 79, 81
indefinite variation, 10, 16, 19, 20, 23
individual differences, iii, 90, 198, 250, 255, 258, 260, 261, 264, 269, 289
inequality, x, xiv, 25, 32, 46, 50, 51, 53, 67, 68, 69, 71, 76, 77, 78, 80, 82, 84, 90, 93, 96, 100, 102, 105, 113, 114, 118, 119, 122, 123, 124, 125, 161, 298, 299, 312
infancy, 153, 159, 167, 184, 186, 189, 231, 237, 251, 260, 261, 283, 285
injustice, xii, 4, 17, 18, 19, 32, 53, 54, 67, 69, 78, 82, 84, 88, 117, 209, 290
interactionist approach, 125
intersex, 5, 8, 10, 11, 12, 14, 38, 39, 40, 41, 130, 131, 169, 170, 171, 172, 327, 343, 344, 345, 346, 347, 348, 351, 366, 367, 371, 372, 373, 377, 378, 380, 396, 451, 457, 458
IQ, 18, 97, 266, 267, 273, 274, 456

J

Jacques Derrida, 62, 129
Janet Hyde, 117
Jean-Francois Lyotard, 57
JK Rowling, 370, 373, 374, 375, 376, 377, 381, 388, 409
John Money, 7, 39, 40, 41, 171, 172, 173, 176, 180, 181, 182, 192, 195, 216, 291
Jordan Peterson, vi, 90, 92, 125, 203, 209, 210, 258, 265, 272, 381, 383, 384, 385, 388, 389, 390, 391, 392, 393, 394, 409, 418, 422, 423, 442, 447, 458
Judith Butler, 19, 23, 24, 26, 31, 32, 367, 396, 399, 458

K

Kane, 44, 45, 167
Kimmel, 213

L

language, xiv, 7, 8, 9, 17, 20, 25, 27, 51, 54, 60, 62, 63, 67, 72, 73, 93, 97, 129, 148, 155, 166, 183, 189, 204, 205, 219, 240, 241, 262, 264, 265, 291
lateralization, 148, 199, 205, 226, 233, 239, 240, 241, 243, 244, 248
liberal democracies, 28, 78, 81, 88, 89, 92, 96, 102, 125, 141, 290, 291
liberty, iii, xv, 101, 290, 364, 392, 409, 420, 423, 430
linguistics, 308, 313, 329
Logocentrism, 64
logos, 64, 207, 209, 459
low testosterone, 237, 246
luteinizing hormone (LH), 226

M

male-typical, 111, 113, 114, 124, 134, 136, 147, 149, 150, 151, 154, 155, 158, 159, 160, 161, 162, 178, 180, 181, 182, 183, 186, 187, 188, 190, 191, 199, 200, 226, 231, 233, 237, 238, 239, 243, 245, 246, 247, 248, 250, 251, 285, 286, 287, 296, 302, 313, 326, 327, 328, 332, 345, 346, 351, 355, 360, 436, 451, 453, 459, 464
Marxist, 26, 27, 89, 91, 92, 93
masculine archetype, 207, 208, 210, 213, 216, 217, 218, 219, 220
masculinity, 45, 51, 53, 64, 110, 115, 202, 203, 213, 214, 217, 218, 348, 349, 353, 354, 355, 404, 421, 444, 455
masculinized, 152, 170, 177, 189, 210, 232, 239, 240, 241, 243, 248, 286, 371

math ability, 122, 216, 307, 308, 309, 330
Maya Forstater, 367, 370, 375, 376, 377, 381, 388, 409
McLaughlin, xi, xii, 6, 44, 45, 215, 219
meta-analysis, 83, 118, 119, 135, 140, 183, 185, 199, 269
meta-narrative, iii, 57, 60, 61, 62, 64, 68, 95, 117
method artifacts, 280
mice, 150, 151, 232, 287
Michel Foucault, 64
Mullerian ducts, 145, 152
muscle mass, 25, 229, 230, 232

N

neural, 13, 32, 38, 97, 134, 136, 138, 150, 160, 162, 182, 224, 233, 234, 239, 248, 252
neuroanatomy, 105, 124, 223, 224, 225, 232, 233, 235, 236, 237, 238, 239, 243, 244, 247, 248, 249, 250, 251, 252, 283, 284, 299, 300, 324, 325, 326, 451, 457, 460
neurocircuitry, 223, 224, 225, 232, 233, 234, 235, 237, 238, 239, 241, 243, 244, 247, 248, 249, 250, 251, 252, 283, 284, 299, 300, 324, 325, 451, 457, 459, 460
neuroendocrinology, xiii, 151, 162, 184, 194, 238, 243, 247, 251, 253
neurological, 34, 37, 135, 184, 237, 245, 246, 247, 255, 261, 300, 312, 313, 324, 325, 329, 330, 378, 464
neurons, 55, 138, 147, 153
Neuroticism, 264, 266, 270, 276, 283, 287, 289
Nietzsche, 56, 69, 426, 427, 428, 429, 430, 446, 447
Norway, x, xi, xii, 93, 277, 303, 319, 320, 321, 322, 360, 441

O

occupational preferences, xi, 93, 158, 185, 278, 299, 303, 304, 314, 320, 323, 324, 329, 330, 351, 352, 358, 359, 360, 430, 442
Openness to Experience, 267, 273, 274
organizational, xii, 13, 110, 134, 149, 151, 153, 155, 159, 160, 161, 162, 185, 186, 224, 234, 236, 251, 266
origins of sex and gender differences, xv, 115, 117, 120, 124, 141, 143, 159
ovaries, 3, 9, 10, 15, 130, 135, 137, 138, 225, 227, 228, 231, 252, 326, 345
ovulation, 13, 134

P

parental encouragement, 167, 187, 188, 189, 200, 216

patriarchy, 19, 49, 51, 59, 63, 64, 67, 72, 88, 90, 295, 296, 297, 307, 363, 374, 405, 420, 446, 447, 460, 462

people-oriented, 36, 81, 300, 302, 303, 305, 306, 307, 309, 312, 313, 316, 324, 329, 330, 358, 359, 360, 397, 464

perceptual structures, 255, 256, 258, 259, 260, 281, 282, 283, 289, 324, 325, 329, 330

performance, 33, 36, 37, 45, 50, 51, 122, 138, 141, 155, 157, 216, 230, 243

personality, iii, xi, 1, 2, 36, 44, 71, 90, 96, 97, 99, 100, 102, 105, 139, 149, 166, 210, 224, 243, 247, 253, 255, 256, 257, 258, 259, 260, 261, 262, 263, 264, 267, 268, 269, 270, 271, 272, 274, 275, 276, 277, 278, 279, 280, 281, 282, 283, 284, 285, 287, 288, 289, 291, 300, 303, 304, 314, 316, 320, 325, 329, 335, 350, 357, 358, 360, 382, 394, 403, 404, 413, 430, 435, 441, 442, 452, 455, 457, 458, 461, 462

personality psychology, iii, 96, 253, 258, 269, 291

phenotypic, 38, 131, 132, 133, 145, 200

physiology, 13, 49, 98, 193, 203, 204, 224, 225, 231, 256, 258

postmodernism, 51, 54, 56, 57, 58, 62, 67, 68, 69, 95, 364, 365, 416, 458

pregnancy, 76, 144, 152, 158, 186, 188, 228, 244, 287

prenatal androgen, 114, 115, 147, 148, 155, 156, 157, 158, 159, 160, 161, 171, 187, 189, 227, 240, 271, 286

prenatal androgens, 12, 109, 110, 114, 115, 124, 134, 148, 153, 155, 156, 157, 158, 159, 160, 161, 162, 170, 176, 180, 182, 184, 185, 186, 187, 188, 189, 190, 191, 192, 194, 198, 199, 223, 226, 231, 233, 251, 253, 283, 285, 287, 299, 302, 314, 328, 432, 440, 451, 453, 459

preschool, 45, 46, 48, 49, 50, 152

progesterone,, 135, 246

proteins, 132, 140, 224, 229, 232, 237, 239, 240

psychopathology, 13, 134, 245, 250

psychosexually neutral, 39, 170, 172, 173, 177, 182, 194

puberty, 14, 15, 102, 110, 115, 135, 147, 150, 171, 181, 182, 193, 224, 225, 226, 227, 229, 230, 231, 232, 233, 234, 235, 236, 237, 239, 240, 243, 244, 245, 246, 250, 252, 256

puberty blockers, 181, 182, 193

R

rats, 190, 243, 287

religion, 24, 27, 45, 56
rhesus macaques, 184, 186, 244
rhetoric, 54, 57, 59, 60, 66, 67, 68, 91, 140, 161, 363, 420, 431
rightward shift, 148, 149, 191, 199, 223
risk-taking, 148, 243, 245, 282

S

Scandinavia, 277, 278, 281
sex determination, 133, 137, 366, 372, 396
sex difference research, 18, 19, 27, 37, 107, 110, 141
sex hormones, 3, 13, 114, 131, 133, 134, 136, 138, 139, 140, 144, 147, 149, 150, 151, 152, 158, 159, 223, 224, 226, 230, 231, 232, 233, 234, 235, 238, 244, 245, 246, 250, 251, 252, 257, 284, 299, 300, 325, 327, 328, 356, 406, 407, 439, 457, 460, 462
sex prediction, 248
sex roles, 279, 283, 284
sex-differentiated, 134, 135, 138, 139, 140, 147, 153, 162, 182, 183, 185, 194, 223, 224, 226, 229, 232, 233, 234, 239, 240, 244, 245, 247, 250, 252
sexism, 34, 68, 81, 89, 91, 181
sex-specific gene, 138, 147, 151, 153, 223, 257
sex-typed, 13, 133, 134, 140, 146, 153, 154, 155, 157, 158, 167, 183, 184, 185, 188, 194, 251
sexual behavior, 147, 149, 235, 244
Simone de Beauvoir, 24, 25, 26, 27, 28, 31, 32, 34, 58, 64, 171, 446, 462
social constructionism, x, 1, 3, 6, 7, 10, 20, 24, 38, 51, 54, 57, 67, 69, 89, 93, 102, 126, 172, 176, 182, 198, 204, 214, 223, 248, 253, 271, 276, 281, 297, 303, 316, 339, 340, 369, 409, 411, 413, 416, 430, 451, 452
social constructionists, 3, 4, 6, 7, 10, 11, 12, 16, 18, 30, 31, 37, 38, 40, 41, 43, 51, 53, 54, 58, 61, 67, 71, 77, 78, 84, 88, 89, 90, 91, 92, 93, 96, 97, 100, 101, 102, 105, 110, 114, 122, 141, 143, 158, 160, 161, 162, 166, 167, 172, 182, 186, 194, 195, 198, 204, 213, 214, 219, 223, 231, 244, 245, 250, 252, 253, 255, 260, 274, 275, 276, 287, 289, 290, 354
socially constructed, xi, 19, 23, 32, 34, 38, 40, 41, 42, 50, 51, 53, 54, 114, 168, 170, 172, 175, 182, 213, 256
Sociocultural Theory, 120, 123, 124, 143
Sociology of Gender, xi, xii, xiii, 6, 45, 219
spatial ability, 154, 157, 190, 199, 243, 288

spatial location, 155, 190, 241

spectrum, 3, 4, 6, 13, 16, 106, 107, 109, 130, 168, 247, 265, 269, 343, 344, 345, 346, 347, 348, 366, 371, 372, 373, 378, 385, 396, 403, 414, 420, 457

sperm, 10, 13, 75, 121, 134, 144, 157, 226

SRY gene, 132, 140, 252

STEM, v, 50, 84, 88, 201, 278, 296, 298, 299, 301, 303, 305, 306, 307, 308, 309, 310, 313, 314, 315, 316, 322, 330, 340, 358, 359, 360, 443, 457, 463

stereotype, 197, 198, 200, 201, 203, 211, 214, 217, 218

suboptimal arousal, 147, 149, 191, 223, 233

systematizing, 111, 149, 199, 241

syzygy, 205, 207, 211, 220

T

TERF, 373, 374

testes, 3, 10, 14, 15, 130, 132, 134, 135, 137, 145, 146, 150, 152, 171, 173, 225, 227, 238, 252, 299, 326, 463

The Genderbread Person, 107, 171

things-oriented, 300, 301, 302, 303, 305, 306, 307, 309, 324, 329, 330, 351, 356, 358, 359, 397, 464

things-people dimension, 185, 300, 301, 303, 304, 305, 306, 309, 464

toy preferences, 43, 44, 158, 159, 183, 184, 185, 186, 187, 188, 189, 190, 191, 194, 199, 215, 287, 302, 324, 327, 351, 352, 401, 432, 436, 437, 461, 464

transgender, 7, 10, 11, 17, 165, 169, 193

transphobes, 373, 381, 396, 406

trucks, 42, 44, 184, 215

U

utopia, 37, 53, 56, 72, 89, 101, 168, 319, 320, 340

V

variability, 3, 4, 99, 115, 171, 194, 198, 203, 243, 247, 250, 259, 260, 274, 275, 326, 456

verbal skills, 112, 122, 148, 241, 307, 308, 309, 330

visual interest, 184, 185

visual stimuli, 244, 251

visuospatial cognition, 199, 240, 241

vocal pitch, 114, 115, 229

W

wage gap, 79, 80
West and Zimmerman, 33, 35, 36, 37, 44, 46, 349
Wolffian duct, 145
women's equality, 26, 277

X

XX, 9, 11, 135, 144, 206
XY, 15, 133, 135, 171, 173, 176

Z

zygote, 131, 144

www.ingramcontent.com/pod-product-compliance
Lightning Source LLC
Chambersburg PA
CBHW021347210526
45463CB00001B/7